King of the World

King of the World

THE PADSHAHNAMA

AN IMPERIAL MUGHAL MANUSCRIPT FROM THE ROYAL LIBRARY, WINDSOR CASTLE

MILO CLEVELAND BEACH & EBBA KOCH

WITH NEW TRANSLATIONS BY WHEELER THACKSTON

AZIMUTH EDITIONS · SACKLER GALLERY

PUBLISHED ON THE OCCASION OF THE EXHIBITION
King of the World:
a Mughal manuscript from the Royal Library, Windsor Castle
ORGANIZED BY THE ARTHUR M. SACKLER GALLERY
IN CONJUNCTION WITH THE ROYAL LIBRARY, WINDSOR CASTLE

NATIONAL MUSEUM OF INDIA, NEW DELHI
27 JANUARY 1997–28 FEBRUARY 1997

THE QUEEN'S GALLERY, BUCKINGHAM PALACE, LONDON
14 MARCH 1997–27 APRIL 1997

ARTHUR M. SACKLER GALLERY, SMITHSONIAN INSTITUTION, WASHINGTON, DC
18 MAY 1997–13 OCTOBER 1997

METROPOLITAN MUSEUM OF ART, NEW YORK
18 NOVEMBER 1997–8 FEBRUARY 1998

LOS ANGELES COUNTY MUSEUM OF ART, CALIFORNIA
26 FEBRUARY 1998–17 MAY 1998

KIMBELL ART MUSEUM, FORT WORTH, TEXAS
31 MAY 1998–23 AUGUST 1998

INDIANAPOLIS MUSEUM OF ART, INDIANA
6 SEPTEMBER 1998–29 NOVEMBER 1998

ISBN 1 898592 11 X

PRODUCED BY AZIMUTH EDITIONS LIMITED
33 LADBROKE GROVE, LONDON W11 3AY
DESIGNED BY ANIKST ASSOCIATES
COLOUR PHOTOGRAPHY BY EVA ZIELINSKA-MILLAR
ADDITIONAL PICTURE RESEARCH BY CARRIE HAINES
REPRODUCED AND PRINTED BY PJ REPRODUCTIONS, LONDON
BOUND BY HARTNOLLS, BODMIN, CORNWALL

THE COVER SHOWS A DETAIL FROM
The delivery of presents for Prince Dara-Shikoh's wedding (see **21–22**)
THE FRONTISPIECE SHOWS A DETAIL FROM
Jahangir receives Prince Khurram on his return from the Mewar campaign (see **5**)
MAP OF INDIA ON PAGE 10 DRAWN BY DIANE DIXSON-O'CARROLL
JACKET DESIGNED BY LORNA RABY

Contents

FOREWORD

by Milo Cleveland Beach
Director, Arthur M. Sackler Gallery

During the sixteenth century, reports by missionaries and traders brought word to Europe of the power and wealth of the Grand Mogul, as they termed the Muslim monarchs who then controlled much of northern India. From the early seventeenth century, these often fanciful verbal accounts were supplemented with drawn and painted images. Sir Thomas Roe, the first English Ambassador to the Mughal (Mogul) court, and his party returned to London in 1619 with portraits that were then engraved and published, while the East India Company trader Peter Mundy was among those amateur artists who made sketches, as early as in the 1630s, of what they observed first-hand. In 1638, following established diplomatic custom, the Emperor Shah-Jahan inscribed an illustrated manuscript 'as a gift to the glorious and exalted King of England' and sent it (probably through a visiting merchant named Bornford) to Charles I. And as contacts with India increased, so did the flow of imagery. By the late eighteenth century, the Nawab of Oudh repeated Shah-Jahan's precedent by asking Lord Teignmouth, Governor-General of India, to carry as a gift to George III a history of Shah-Jahan's reign with text and illustrations made for that emperor. This was the *Padshahnama*, or 'Chronicle of the King of the World,' and since that time it has remained one of the greatest works in the Royal Collection.

With the gracious permission of Her Majesty Queen Elizabeth II, all the illustrations from the *Padshahnama* manuscript are being fully published and exhibited for the first time on the occasion of the 50th Anniversaries of the Independence of India and Pakistan. The Arthur M. Sackler Gallery of the Smithsonian Institution was honored by an invitation to oversee the publication and to co-organize the American presentation; the Gallery is especially grateful to Oliver Everett, Librarian, and Theresa-Mary Morton, Curator of Exhibitions, The Royal Library, for providing continual access to the manuscript and generous hospitality and support over the last three years. The conservator Alan Donnithorne also made important observations.

At the Arthur M. Sackler Gallery, Patrick Sears, Associate Director, oversaw the design of the exhibition and its didactic materials, working closely with Richard Franklin, Head, Department of Design and Installation, and Lucia Pearce, Head, Department of Education, and their staff. The didactic texts were edited with great skill by Susannah Gardiner. Becky Gregson handled registrarial concerns with her customary efficiency, while Sarah Newmeyer oversaw all contractual negotiations. At Azimuth Editions, Julian Raby enriched the catalogue through his scholarship and publishing expertise, assisted by Alison Effeny, who edited the catalogue text. PJ Reproductions, supervised by Lorna Raby, produced superlative colour plates, and the volume was designed with great sensitivity by Anikst Associates.

The cooperation of the Directors and staff of the other exhibition venues in the United States – Philippe de Montebello, Metropolitan Museum of Art; Graham W.J. Beal, Los Angeles County Museum of Art; Edmund Pillsbury, Kimbell Art Museum; and Bret Waller, Indianapolis Museum of Art – is also gratefully acknowledged here, as is the important funding provided by the Smithsonian Institution Special Exhibitions Program, and the grant of an indemnity by the Federal Council on the Arts and Humanities.

Casual visitors to India encounter the monuments of Shah-Jahan more frequently than those of any other period; the Taj Mahal and the forts in Agra and Delhi are on every tourist itinerary. The *Padshahnama* provides an unrivaled opportunity to understand better the peoples and ceremonies that enlivened those spaces, and it is particularly gratifying that its publication and exhibition can be presented in honor of the Golden Jubilee of India and Pakistan.

FOREWORD

by Oliver Everett
Librarian, Windsor Castle

The Royal Library at Windsor Castle contains, as one nineteenth-century visitor put it, 'a feast of treasures and curiosities.' British Kings and Queens have collected or been given a huge variety of objects over the centuries and these have come to form the present Royal Collection. The Library at Windsor now houses a wide selection of these objects, amongst them a group of some 40 Islamic manuscripts. These were all acquired by gift and are not, therefore, a deliberately formed group. Nevertheless, many are of substantial importance and several are exceptional by any standards. Their dates of creation range from 1492 to the nineteenth century: notable strengths include miniature paintings and calligraphy of the Mughal period in India and the Safavid period in Persia. Queen Victoria received many of them but it was that avid collector, King George III, who was given some of the finest specimens, including perhaps the outstanding item in the collection, the *Padshahnama*, the 44 illustrations from which are the subject of this catalogue and the exhibition it accompanies.

The *Padshahnama* was given in 1799 by the Nawab of Oudh to George III, who placed it in his extensive library in Buckingham House (the future Buckingham Palace). When King George IV gave his father's library to the British Museum in 1823, the *Padshahnama*, like a number of other treasures, was not included. It was placed in the Royal Library at Windsor Castle when the library was established here by King William IV in the early 1830s.

Because the manuscript was bound in a volume, it has not hitherto been possible to exhibit the illustrations except as a single opening of the book. In 1994, however, conservation work necessitated unbinding the volume and it became possible to show all the illustrations together for the first time. The Royal Collection is most grateful to Dr Milo Beach, Director of the Arthur M. Sackler Gallery and Freer Gallery of Art, Smithsonian Institution, Washington, DC, and renowned scholar of Mughal painting, for his lead in helping to organize the exhibitions and for co-authoring with Dr Ebba Koch and Professor Wheeler Thackston this superb catalogue.

It is particularly apt that the timing of this unique exhibition coincides with the 50th Anniversary of the Independence of India and Pakistan, and that it can be shown in the National Museum in New Delhi as one of the events to mark the Anniversary. The exhibition will later be shown in The Queen's Gallery, Buckingham Palace, and subsequently at the Arthur M. Sackler Gallery, Washington, the Metropolitan Museum of Art, New York, the Los Angeles County Museum of Art, the Kimbell Art Museum, Fort Worth, and the Indianapolis Museum of Art.

The Royal Collection is most grateful to the late Fred Marsh and to John Holmes and John Mumford of the India Office Library Conservation Department for their work on the *Padshahnama* manuscript; to colleagues in the institutions in India and the United States where the exhibition will be held; to the British Council in New Delhi; and to the sponsors worldwide who have supported the exhibition so generously.

AUTHORS' ACKNOWLEDGEMENTS

When Milo Beach invited me in Spring 1993 to join him and Wheeler Thackston in the publication of the Windsor *Padshahnama*, I much appreciated this occasion to draw together my ideas on a body of miniatures which had fascinated me since I was first introduced to them by Robert Skelton in 1977, soon after the beginning of a ten-year stay in India, devoted in particular to the study of Shah-Jahani art and architecture. I thank Robert Skelton for generously allowing me to use his slide archive and his library, and for sharing his vast knowledge of Indian art. I also gratefully acknowledge the assistance of Dr Muhammad Yunus Jaffery in the reading and translation of Mughal sources, and of Richard A. Barraud in the preparation of measured drawings. I also wish to thank the Archaeological Survey of India, as well as the Department of Archaeology and Museums, Government of Pakistan, for the permission to carry out a survey of Mughal architecture, and, in Vienna, the Fonds zur Förderung der wissenschaftlichen Forschung, the Bundesministerium für Wissenschaft und Forschung, and the Jubiläumsfonds der österreichischen Nationalbank for supporting my research. For access to their collections and libraries I thank in particular, in London, the British Library, the School of Oriental and African Studies, and the Victoria and Albert Museum; and in New York, the Metropolitan Museum of Art and the Institute of Fine Arts. My husband Benno patiently read the manuscript and I thank him for his support. EK

I am especially grateful to Dr Thomas Lentz, Deputy Director of the Arthur M. Sackler Gallery and the Freer Gallery of Art, who assumed increased administrative burdens at the galleries during the period that this publication was in preparation. Together with Dr Vidya Dehejia, Curator of South and Southeast Asian Art, and Dr Massumeh Farhad, Associate Curator of Islamic Near Eastern Art, he also generously contributed scholarly advice and counsel. Tobias Beach gave skilled editorial help in the early stages, while John Henry Rice provided continuing editorial comment, and assembled, checked and cross-checked the Bibliography, Glossary, Biographical listing, and the many references. He also helped to organize the exhibition and adapt the catalogue text for the exhibition's didactic materials. Elaine Gill served as the efficient and very accurate coordinator for all matters relating to the text, while Lily Kecskes, Head Librarian, and John Tsantes, Head of the Department of Photography at the Arthur M. Sackler Gallery, assured the ready availability of all research materials.

The involvement of Dr Ebba Koch and Dr Wheeler Thackston goes far beyond the text areas defined with their names, and I am grateful for their generosity with information. The following people have also over the years provided both information and help: Kazim Ali Khan, Nawab of Rampur; Eberhard Fischer, Museum Rietburg, Zurich; Sven Gahlin, London; Dr B.N. Goswamy, Chandigarh University; Howard Hodgkin, London; Dr Naval Krishna, Mehrangarh Fort Museum, Jodhpur; Terence McInerney, New York; Elizabeth Moynihan, Washington, DC; Dr Ellen Smart, San Diego Museum of Art; Stuart Cary Welch, Harvard University Art Museums; and Dr James L. Westcoat, University of Colorado at Boulder.

All the authors are grateful to J.P. Morgan & Co. for funds given to support aspects of research for the publication.

Without the support of my wife Robin, and our children, Olga, Toby, and Sophie Beach, my involvement with this publication would not have been possible. MCB

Over the years, Robert Skelton, former Keeper of the Indian Section at the Victoria and Albert Museum, has generously provided essential information and challenging ideas about the *Padshahnama*, and much of what has been written in this publication is constructed on the foundations provided by his research. With the greatest pleasure, the authors dedicate their work on this project to him.

NOTE ON TRANSLITERATION

In spelling personal names and technical terms for which there are no real English equivalents, we have striven for simplicity, omitting macrons and other diacritical markings. We have, therefore, Abid for 'Ābid and *Diwan-i Khass* for *Dīwān-i Khāṣṣ*. Geographical names are spelled in accordance with normal English usage, for example, Delhi and Lahore not Dihlīand Lāhaur. (WT)

PUBLISHER'S NOTE

Each illustrated page in the *Padshahnama* manuscript has been assigned a number, given in bold type, which relates to entries in both the section on *The Manuscript* and the *Catalogue*. All comparative figures and detail illustrations are numbered consecutively. Dimensions are in metric measurements throughout, height before width; details of media and provenance not given in the figure captions may be found in the *List of figures* on page 237. Christian Era dates are preceded by the abbreviation AD and are given according to the Gregorian calendar: any Muslim dates are preceded by the abbreviation AH.

The Mughal Empire during the reign of Shah-Jahan

MOVEMENTS OF THE IMPERIAL CAPITAL DURING THE REIGN OF SHAH-JAHAN[1]

CAPITAL	DATES OF OCCUPANCY (AD)	DATES OF OCCUPANCY (AH)[2]	SJN[3]
AGRA	(accession) 14 February 1628 – 18 November 1629	8 Jumada II 1037 – beginning of Rabi‘ II 1039	17; 35
BURHANPUR	mid-March 1630 – late March 1632 (?)	late Rajab 1039 – early Ramadan 1041 (?)	37; 81–82
AGRA	19 June 1632 – early February 1634	1 Dhu'l-Hijja 1041 – early Sha‘ban 1043	83; 121
LAHORE	7 April 1634 – mid-May 1634	beginning of Shawwal 1043–late Dhu'l-Qa‘da 1043	122; 122
KASHMIR	15 June 1634 – 1 October 1634	18 Dhu'l-Hijja 1043 – 8 Rabi‘ II 1044	124; 138
LAHORE	26 November 1634 – 1 October 1635	5 Jumada II 1044 – 18 Rabi‘ II 1045	143; 155
DAULATABAD	28 February 1636 – 21 July 1636	21 Ramadan 1045 – 17 Safar 1045	169; 189
MANDU	31 August 1636 – 16 October 1636	29 Rabi‘ I 1046 – 16 Jumada I 1046	191; 192
AGRA	16 January 1637 – 27 August 1638	18 Sha‘ban 1046 – 16 Rabi‘ II 1048	202; 247
LAHORE	22 November 1638 – 6 March 1639	middle of Rajab 1048 – 1 Dhu'l-Qa‘da 1048	251; 256
KABUL	late May 1639 – late August 1639	late Muharram 1049 – late Rabi‘ II 1049	257; 258
LAHORE	mid-October 1639 – 19 February 1640	late Jumada II 1049 – 26 Shawwal 1049	261; 266
KASHMIR	30 April 1640 – late September 1640	9 Dhu'l-Hijja 1049 – early Jumada II 1050	268; 273
LAHORE	16 November 1640 – 11 November 1642	1 Sha‘ban 1050 – 18 Sha‘ban 1052	274; 298
AGRA	mid-January 1643 – 24 January 1645	late Shawwal 1052 – 26 Dhu'l-Qa‘da 1054	299; 321
KASHMIR	27 April 1645 – 25 September 1645	1 Rabi‘ I 1055 – 4 Sha‘ban 1055	324; 330
LAHORE	4 November 1645 – 5 April 1646	15 Ramadan 1055 – 18 Safar 1056	331; 338
KABUL	6 June 1646 – 20 September 1646	22 Rabi‘ II 1056 – 9 Sha‘ban 1056	340; 363
LAHORE	17 November 1646 – 25 March 1647	8 Shawwal 1056 – 18 Safar 1057	365; 372
KABUL	5 May 1647 – 31 August 1647	last day of Rabi‘ I 1057 – last day of Rajab 1057	372; 394
LAHORE	3 November 1647 – 17 November 1647	5 Shawwal 1057 – 19 Shawwal 1057	395; 396
AGRA	11 January 1648 – 6 April 1648	15 Dhu'l-Hijja 1057 – 12 Rabi‘ I 1058	404; 406
DELHI	18 April 1648 – 19 November 1648	24 Rabi‘ I 1058 – 3 Dhu'l-Qa‘da 1058	408; 413
LAHORE	28 December 1648 – 15 March 1649	12 Dhu'l-Hijja 1058 – 1 Rabi‘ I 1059	414; 416
KABUL	16 May 1649 – 17 September 1649	4 Jumada II 1059 – last day of Sha‘ban 1059	423; 433
LAHORE	25 October 1649 – 17 December 1649	18 Shawwal 1059 – 12 Dhu'l-Hijja 1059	443; 445
DELHI	18 January 1650 – 22 February 1651	15 Muharram 1060 – 1 Rabi‘ I 1061	445; 453
LAHORE	27 March 1651 – 20 May 1651	4 Rabi‘ II 1061 – 29 Jumada I 1061	454; 456
KASHMIR	13 June 1651 – 18 August 1651	23 Jumada II 1061 – 1 Ramadan 1061	458; 459
LAHORE	26 September 1651 – 26 February 1652	10 Shawwal 1061 – 16 Rabi‘ I 1062	459; 463
KABUL	13 April 1652 – 16 August 1652	4 Jumada I 1062 – 11 Ramadan 1062	466; 473
DELHI	12 December 1652 – 20 November 1653	11 Muharram 1063 – 29 Dhu'l-Hijja 1063	476; 490
AGRA	7 December 1653 – 21 December 1653	16 Muharram 1064 – last day of Muharram 1064	490; 490
DELHI	4 January 1654 – 4 October 1654	14 Safar 1064 – 22 Dhu'l-Qa‘da 1064	494; 501
AJMER	6 November 1654 – 24 November 1654	25 Dhu'l-Hijja 1064 – 14 Muharram 1065	502; 502
DELHI	8 January 1655 – 28 October 1657	last day of Safar 1065 – 20 Muharram 1068	505; 544
AGRA	5 December 1657 – 18 June 1658 (Shah-Jahan imprisoned)	27 Safar 1068 – 17 Ramadan 1068	544; 553

1. The following chart represents a list of the major movements of Shah-Jahan's capital. It does not include many minor excursions which the Emperor – and with him the seat of imperial rule – made.

2. Muslim dates are expressed according to Shah-Jahan's lunar calendar.

3. All dates are taken from *Shah Jahan Nama*, references to the page numbers of which are listed below.

The Manuscript

WHEN LORD TEIGNMOUTH was at Luknow, a Book was produced to him out of the Nabob's Library, as a most splendid specimen of oriental manuscripts and its acceptance was pushed upon him. Lord Teignmouth declined receiving it with an observation that it was fit for a royal Library. The observation however suggested the Idea, that as a literary curiosity it might be acceptable to the King of Great Britain, & he mentioned afterwards to the Minister at Luknow, that he would not hesitate to accept it, in the Idea of depositing it in the royal Library if his Majesty would think proper to allow it. The Book, with five others selected by the Minister, as elegant specimens of Persian writing, were sent to Calcutta, after Lord Teignmouth's Departure, & forwarded by his Attorneys to Europe. They have been lately received by Lord Teignmouth, and are now in his possession; he will be happy to be honored with his Majesty's orders respecting them.[1]

In a note dated June 1799 and entitled 'Description of Persian Manuscripts designed by Lord Teignmouth as a present to His Majesty,' the five manuscripts are listed, with an entry on the *Padshahnama* that reads:

SHAHJEHAN-NAMA.
The History of the reign of the Emperor Shahjehan – Son of Jihanghuir, & Father of Aurangzebe.

This is the most splendid Persian Manuscript I ever saw. Many of the faces are very well painted & some of them are portraits. The first is the portrait of Timur or Tamerlane, & the second that of Shahjehan. This was the Book which was shewn to me at Luknow, & I was there informed that the deceased Nabob Asophuddoulah purchased it for 12000 Rupees, or about £1500.[2]

1. Letter from Lord Teignmouth, dated June 1799, Royal Library, Windsor Castle, MS.1368.
2. Royal Library, Windsor Castle, MS.1367.

Fig. 1 Detail of **16**.

Fig.2 Folios 1b–2a of the Windsor *Padshahnama* manuscript, showing the *shamsas* and 18th-century margins (see 1–2).

Fig.3 Folios 2b–3a of the manuscript, showing the portraits of Timur and Shah-Jahan with illuminated frames and 18th-century margins (see 3–4).

INTRODUCTION

by Milo Cleveland Beach

The *Padshahnama* manuscript in The Royal Library, Windsor Castle, is both an extraordinary artistic achievement and a major historical chronicle. The text, written by Abdul-Hamid Lahawri, documents the reign of the Mughal emperor Shah-Jahan (1628–1658). The Windsor copy is incomplete, for only the first third of the narrative (covering the first decade of the Emperor's rule) has survived, but it is the only known version that includes illustrations of the period. Painted by the greatest of the imperial artists, these 44 illustrations provide an enthralling view of a court unrivaled in its time for magnificence, wealth and power.

The volume consists of 239 folios, written on gold-flecked tan paper of a thinness and evenness of texture that proclaims its high quality. Each of the text folios, trimmed of its original outer margins, was set at a later date into windows cut from paper of creamy-white color, and these new borders were decorated with a repeated design in gold (Fig.65).[1] A colophon on the last page of the volume states that the text was copied by the scribe Muhammad-Amin of Mashhad, and that he completed this assignment in AH 1067 (AD 1657–8). No date is given for the remargination, but a suggested chronology for the assembly of the manuscript is proposed below (pages 126–9).

The text is written as a continuous narrative on both sides of a single thickness of paper, although the scribe left space at those locations where illustrations are now to be found. These spaces allowed for either single- or double-page paintings. The location of the pictures in immediate proximity to the related narrative passages was important, but in several cases illustrations were placed in such rapid succession that there was insufficient text to fill the adjacent pages. In those instances, rather than the usual twenty-one lines of text the scribe wrote only eleven lines, or even as few as three, on the intermediate pages (Figs 4–5). It is worth noting that this spacing of text lines, together with the fact that the text is complete, provides firm evidence that the calligrapher was working within a predetermined scheme and that assignment of the location of the illustrations was made at the same time (and therefore during the reign of Shah-Jahan), rather than as an arbitrary result of some later refurbishment of the book. And there are two other important conclusions. First, the choice of scenes to be shown clearly outweighed in importance the visual awkwardness produced by an occasional density of paintings and thinness of text. In other words, consistency in the design of the book was subservient to the intended program of illustrations. And second, while scholarly references to 'dispersed pages' from the Windsor Castle *Padshahnama* have persisted despite mounting contrary evidence, the text is complete, and an illustration is found at every point where an illustration was intended. In other words, there are no folios missing from the Windsor Castle volume.

The recent examination of the relationship of the illustrations to the text gives us substantial information about the assembly of the manuscript. While many illustrations specifically describe the scenes recorded in the adjacent text by Abdul-Hamid Lahawri, others are only generically appropriate; descriptions of *darbar*s (receptions or audiences) are always accompanied by *darbar* scenes, for example, but careful examination shows that the relevant ceremony is not always shown. A double page depicting Shah-Jahan being weighed during celebrations that include musicians and dancers (**12–13**) is a clear instance. The scene is placed to correspond with text for an event held at Fatehpur-Sikri on 27 November 1628, but the man dressed in orange at

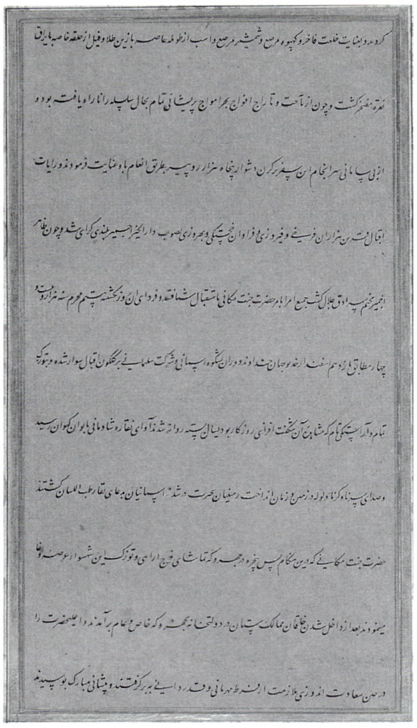

Figs 4–5. Folios 96b and 49b from the Windsor *Padshahnama* manuscript, with eleven and three lines of script respectively.

The Mughal Emperors

Babur
1487–1530;
r. 1526–1530

Humayun
1508–1556;
r. 1530–40; 1555–56

Akbar
1542–1605;
r. 1556–1605

Jahangir
1569–1627;
r. 1605–1627

Shah-Jahan
1592–1666;
r. 1628–1658

Awrangzeb
1618–1707;
r. 1658–1707

Bahadur Shah
1643–1712;
r. 1707–1712

Farrukhsiyar
1683–1719;
r. 1713–1719

Muhammad Shah
1702–1748;
r. 1719–1748

the right edge of the left-hand half is easily recognizable as Muhammad-Ali Beg (*cf.* Fig.87), a Persian ambassador who only arrived at court on 22 March 1631. In fact, his reception on that occasion is illustrated at the appropriate place, twenty-seven folios later (**17**). Confirmation that the double-page must actually depict a different ceremony than the weighing at Fatehpur-Sikri is given by the ages of the imperial princes, who are shown considerably older than in their portrayal at the accession ceremony held only eight months earlier (**10–11**), and also by the prominence of Afzal Khan Allami, the courtier standing just to the Emperor's right. It was only his appointment as *Diwan* (Prime Minister) in March 1629 that gave him the status to stand next to the Emperor at court ceremonies. Muhammad-Ali Beg was actually present at a weighing ceremony on 23 October 1632, and the various closely observed details indicate that this is the scene shown, despite the content of the adjacent text. Since historical accuracy has been considered a standard trait of Mughal painters, one has to ask how this mistake could have happened and whether it is the sole instance of such misplacement.

The first historical illustration (**5**) provides one of several additional examples. According to the related narrative, it depicts Prince Khurram (the future Shah-Jahan) taking leave of his father, the Emperor Jahangir, before the Prince's campaign against the Rajput Rana Amar Singh of Mewar, and the subsequent defeat of the Rana is clearly described in the following double-page illustration (**6–7**) and its accompanying text. Plate **5**, however, contains a highly unusual detail, for the three courtiers in the center have focused their attention not on the enthroned Emperor and his son but on a dark-skinned nobleman in a yellow robe. Most of the men in this scene are identified by minute inscriptions, and this particular figure is labeled 'Rana Karan.' Karan Singh Sisodia, to whom the inscription refers, was the son of Rana Amar Singh, and the memoirs of Jahangir describe his arrival at court accompanying Khurram on his return from Mewar: 'When he [Khurram] had finished the rites of service, he had his vow and alms displayed and said that if it was ordered, Karan should be allowed to prostrate himself and perform *körünüsh*. I ordered him brought in. The *bakhshis* brought him in with the usual ceremonials.'[2]

This is clearly the episode depicted,[3] and there can be no doubt that this scene, too, is out of place within the Windsor volume. A similar situation exists with another Windsor double page (**8–9**) placed alongside the description of Khurram's return to his father's court in Ajmer following the defeat of Rana Amar Singh. The one time when Jahangir rose from his throne and stood to embrace his son – as shown here – was, however, on Khurram's return from a later campaign in the Deccan. Confirmation of the correct identification of this scene is given by the presence at the bottom left of tribute bearers wearing turbans of a type associated with the Deccani kingdoms.

A different kind of problem exists with *A royal procession* (**34**), for it is the only scene in which a member of the imperial family or an important noble does not appear. Furthermore, its compo-

Fig.6 Detail of the left-hand edge of 13, showing figures extending over the marginal ruling.

sition suggests strongly that it is only the left half of what was a double-page composition. The absence of the more important part of the scene, which would certainly have included members of the imperial entourage, has removed the specific features that would allow the scene to be historically identified. While *A royal procession* is half of a double-page illustration placed in the space where only a single-page painting was intended, *Jahangir receives Prince Khurram on his return from the Mewar campaign* (**38**) and *Jahangir presents Prince Khurram with a turban ornament* (**39**) provide an example of the opposite procedure, for they are separate single-page illustrations placed together in spaces left open for the double-page presentation of one episode.

These inconsistencies can be explained in part by a careful physical examination of the book. It has been mentioned that the scribe Muhammad-Amin left blank pages within the body of the text for the inclusion of illustrations, and the painters would have been expected to execute their work on the actual reverse of the text page, whether or not the folio was surrounded by borders of different paper. This was standard procedure, but it is not the way illustrations were placed into the Windsor *Padshahnama*. Instead, the illustrations – which, as we shall see shortly (page 18), were culled from an available stock of paintings – are on separate sheets of a thicker paper which have been glued onto the blank reverse surfaces of the text panels. This collage construction may, in fact, have provided the incentive for the remargination of the volume, for the new, broad outer borders help to disguise the edges of the inserted illustrations, which were then framed by marginal rulings executed on the new paper. In some cases, the edges of the applied illustrations were trimmed to allow original details to extend over the new marginal rulings on the present pages. *The weighing of Shah-Jahan on his forty-second lunar birthday* (**12–13**) is one example; the turbans of two figures at the far left extend beyond and over the present rulings (Fig.6). Moreover, one can see on these tiny extensions evidence of different marginal rulings that can only be explained as the remains of an earlier folio on which the image was originally placed. More conspicuously, along the top and right-hand edges of the painting showing *The submission of Rana Amar Singh of Mewar to Prince Khurram* (**6**), there remains the gold marginal ruling which originally surrounded the illustration; along the right-hand edge of the right half of *The wedding procession of Prince Dara-Shikoh* (**23**), but inside the new marginal rulings, there is an untrimmed strip of plain paper that would have held the gold margin for that illustration, but which seems not to have been colored, suggesting that this painting may never have been put into a manuscript; and *The surrender of the fort at Udgir to Khan Dawran* (**40**) shows an original gold ruling around the entire illustration, inside the rulings later made specifically for the *Padshahnama* folio. In each case, the remaining fragments indicate that the original marginal rulings were painted directly onto the same sheet as the illustration.

We have so far noted three problems. First, the uneven fit of text and related images; second, the use of images that relate to the type of episode described but are not always a record of the specific event; and third, an unusual technique of pasting in illustrations, 'collage construction.'

At this point, we should consider again the entire manuscript. While none of the illustrations in the Windsor volume is inscribed with a date, several are signed or bear contemporary attributions, and the majority can be placed within the careers of several major imperial painters. They can also be independently dated, beginning as early as *circa* 1630, even though Abdul-Hamid Lahawri, the author of the *Padshahnama*, did not receive the commission to write his text until about 1639. It is certain, therefore, that the majority of paintings were not originally intended to adorn this particular volume, which – according to its colophon – would not have been finished and ready to accept illustrations much before 1657, nor could they have been intended for any copy of Lahawri's chronicle at the time they were painted. What we can conclude, however, is that at the time this manuscript was assembled, the decision had been made that new paintings would not be commissioned but that illustrations would be taken from earlier projects – perhaps from historical texts made obsolete by Abdul-Hamid Lahawri's new chronicle, or possibly even from works on hand but made for no specific project. That duplicate versions of the same event are included here (for example, **5** and **38**) implies that a stock of paintings was available from which illustrations could be chosen as required.

It is known that other scenes, of identical size and character to those included in the *Padshahnama* and depicting events described in the text of the Windsor volume, were used for distinctly different projects. *Shah-Jahan enthroned with Mahabat Khan and a shaykh* (Fig.142), for example, illustrates a ceremony within the program of events for Shah-Jahan's coronation; it is signed by Abid, one of the most important artists of the Windsor book, and specifically dated to the second regnal year, that is, AD 1629–30. In about 1650, however, it was placed in an imperial album, effectively denying it any future role as a manuscript illustration. The double-pages showing *A night celebration of the Prophet's birthday* (Fig.144), another ceremony described in the Windsor manuscript, were available to be taken to Iran in 1739, the year Nadir Shah sacked the Mughal capital at Delhi. There, along with other imperial illustrations including further historical scenes from the period of Shah-Jahan's rule (for example, Figs 143, 148, 155, 159 and 166), they were placed in an album known now as the St Petersburg Album.[4] In other words, from the mid-seventeenth to at least the mid-eighteenth century, a group of individual paintings of historical events of Shah-Jahan's reign – all of virtually identical size and character – was available at the imperial court for recycling into manuscripts or albums as needed.

The opening illuminations also need to be considered. Folios 1b and 2a (**1–2**, and Fig.2) consist of an elaborately beautiful illuminated double *shamsa* (sunburst design), the usual frontispiece for imperial manuscripts and albums. The following double page on folios 2b and 3a (**3–4** and Fig.3) reveals portraits of Timur and Shah-Jahan, also within illuminated frames. (Here it must be remembered that one reads a Mughal manuscript from right to left, and therefore back-to-front according to the European system.) The illustration of Timur (**3**) is painted directly onto the reverse of (that is, on the same sheet of paper as) the left half of the opening double-*shamsa* (**2**); but since the painted design of the *shamsa* is larger in size than the illuminated portrait panel on its reverse, the portrait could not be neatly trimmed to fit into its new margins.

The original tan paper thus extends over the marginal rulings, although the artist of the borders attempted to disguise this discrepancy by painting the new gold designs over and onto this original paper. This is visual evidence that folios 2a and 2b are on a single thickness of paper. Moreover, to maintain perfect symmetry for the opening double-page portraits, the facing depiction of Shah-Jahan was identically trimmed, although this was not structurally necessary. In the normal course of events, we would also expect the opening text on the reverse of folio 3a to be written on the same single thickness of paper. This is not the case, however. The text on folio 3b has been glued to the reverse of folio 3a; thus there is no proof that the opening illuminations and portraits were intended originally for this copy of the *Padshahnama* text.

The book as it exists today was therefore constructed of four separate parts. The opening illuminations and the portraits of Timur and Shah-Jahan, which are physically inseparable; the pages of text, which resulted from one project of copying out the historical narrative; the illustrations, a group of earlier paintings most of which were made before the text was either composed or copied out, and inserted into areas of the text left blank by the scribe; and the outer borders, which provided new margins for the text pages, served to cover the edges of the inserted illustrations, and neatly created a visually unified volume.

The intention of this preliminary discussion has been to provide a brief explanation for the historically illogical sequence of illustrations that now exists in the Windsor *Padshahnama*. The table below summarizes the situation. It gives the episode within the manuscript text that corresponds to the place where each illustrated folio is located, together with the identification (when this has been possible) of the actual subjects depicted by the artists. Throughout this publication, the titles for illustrations will accord with the actual subjects rather than with the associated text passages. Further discussion of the paintings is included in both the essays and the individual catalogue entries.[5]

1. The placement of the text panels and paintings within borders of a different paper is not, by itself, evidence of a later date for the borders. Folios of the great *Akbarnama* manuscript divided primarily between the British Library, London, and the Chester Beatty Library, Dublin, were also inserted in this way into borders of a different paper, although the scribal notations and attributions on those borders suggest that they belonged to the original volume; see Arnold & Wilkinson 1936, I, pp.4–12 and II, pls 6–37 and frontispiece. Folios in other manuscripts, such as the earlier *Akbarnama* in the Victoria & Albert Museum, London, on the other hand, are single, unbroken sheets of paper. That the original *Padshahnama* text folios did have different margins, however, is indicated by the presence of original catchwords which were trimmed from the earlier margins and then glued within the marginal rulings of the text panels.
2. *Jahangirnama*, fol.108a; translated by Wheeler Thackston. See also *Tuzuk-i-Jahangiri*, I, pp.276–77.
3. See also the text on page 28.
4. This album has recently been fully published; see Akimushkin *et al.* 1996.
5. Material published here was the basis for The Benjamin Zucker Lecture on Mughal Art given by the author at the Victoria & Albert Museum, London, on 24 October 1995. See also Beach 1995b.

EVENTS ILLUSTRATED IN THE WINDSOR PADSHAHNAMA

Pl. no.	Folio no.	Actual subject of illustration	Painter	Text reference for actual subject	Episode described in *Padshahnama* text
1–2	1b–2a	*Shamsas*	Unknown artist		
3–4	2b–3a	Timur and Shah-Jahan	Unknown artist		
5	43b	Jahangir receives Prince Khurram on his return from the Mewar campaign (19 February 1615)[1]	Balchand	*Padshahnama*, I/1, pp.174ff *Shah Jahan Nama*, p.6 *Tuzuk*, I, pp.276–78 *Jahangirnama*, fols 107b–108a	The departure of Prince Khurram for the Mewar campaign (26 December 1613)
6–7	46b–47a	The submission of Rana Amar Singh of Mewar to Prince Khurram (5 February 1615)	La'lchand (**6**) 'Kashmiri Painter' (attribution) (**7**)	*Padshahnama*, I/1, pp.171–72 *Shah Jahan Nama*, p.6 *Tuzuk*, I, pp.273–274	The submission of Rana Amar Singh of Mewar to Prince Khurram (5 February 1615)
8–9	48b–49a	Jahangir receives Prince Khurram on his return from the Deccan (10 October 1617)	Ramdas (**8**) Murar (**9**)	*Shah Jahan Nama*, p.7 *Tuzuk*, I, pp.393–95 *Jahangirnama*, fols 155b–156a	Jahangir receives Prince Khurram on his return from the Mewar campaign (19 February 1615)
10–11	50b–51a	Shah-Jahan receives his three eldest sons and Asaf Khan during his accession ceremonies (8 March 1628)	Bichitr (**10**) Ramdas (attribution) (**11**)	*Padshahnama*, I/1, pp.177–82 *Shah Jahan Nama*, pp.20–21	Shah-Jahan receives his three eldest sons and Asaf Khan during his accession ceremonies (8 March 1628)
12–13	70b–71a	The weighing of Shah-Jahan on his 42nd lunar birthday (23 October 1632)	Bhola	*Padshahnama*, I/1, pp.440–41 *Shah Jahan Nama*, p.87	The weighing of Shah-Jahan at the *Nawroz* ceremonies (19 March 1629)
14	72b	The presentation of Prince Dara-Shikoh's wedding gifts (4 February 1633)	Balchand	*Padshahnama*, I/1, pp.454–56 *Shah Jahan Nama*, pp.90–91	Uncertain
15	92b	A'zam Khan captures Fort Dharur (January 1631)	Balchand	*Padshahnama*, I/1, pp.339–43 *Shah Jahan Nama*, pp.52–53	A'zam Khan captures Fort Dharur (January 1631)
16	94b	The death of Khan Jahan Lodi (3 February 1631)	Abid	*Padshahnama*, I/1, pp.350–51 *Shah Jahan Nama*, p.56	The death of Khan Jahan Lodi (3 February 1631)
17	98b	Shah-Jahan receives the Persian ambassador, Muhammad-Ali Beg (26 March 1631)	'Kashmiri Painter' (attribution)	*Padshahnama*, I/1, pp.365ff *Shah Jahan Nama*, pp.62–63	Shah-Jahan receives the Persian ambassador, Muhammad-Ali Beg (26 March 1631)
18	102b	The siege of Qandahar (May 1631)	Payag	*Padshahnama*, I/1, pp.374–77 *Shah Jahan Nama*, pp.65–67	The siege of Qandahar (May 1631)
19	116b	Europeans bring gifts to Shah-Jahan (July 1633)	Unknown artist	*Padshahnama*, I/1, pp.534–35 *Shah Jahan Nama*, p.117	The capture of Port Hoogly (June–October 1632)
20	117a	The capture of Port Hoogly (June–October 1632)	Unknown artist	*Padshahnama*, I/1, pp.433–39 *Shah Jahan Nama*, pp.86–87	The capture of Port Hoogly (June–October 1632)
21–22	120b–121a	The delivery of presents for Prince Dara-Shikoh's wedding (November–December 1632)	Bishandas (attribution)	*Padshahnama*, I/1, pp.452–53 *Shah Jahan Nama*, p.90	The delivery of presents for Prince Dara-Shikoh's wedding (November–December 1632)
23–24	122b–123a	The wedding procession of Prince Dara-Shikoh (12 February 1633)	Murar (**23**) Unknown artist (**24**)	*Padshahnama*, I/1, pp.457ff *Shah Jahan Nama*, p.91	The wedding procession of Prince Dara-Shikoh (12 February 1633)
25–26	124b–125a	Shah-Jahan honoring Prince Dara-Shikoh at his wedding (12 February 1633)	Bulaqi, son of Hoshang	*Padshahnama*, I/1, pp.458ff *Shah Jahan Nama*, p.91	Shah-Jahan honoring Prince Dara-Shikoh at his wedding (12 February 1633)
27–28	126b–127a	The wedding procession of Prince Shah-Shuja' (4 March 1633)	Bhola (attribution)	*Padshahnama*, I/1, pp.460ff *Shah Jahan Nama*, p.93	The wedding procession of Prince Shah-Shuja' (4 March 1633)
29	134a	Prince Awrangzeb facing a maddened elephant named Sudhakar (7 June 1633)	Unknown artist	*Padshahnama*, I/1, pp.489–92 *Shah Jahan Nama*, pp.95–96	Prince Awrangzeb facing a maddened elephant named Sudhakar (7 June 1633)

30	135b	Prince Khurram attacking a lion (late 1610)	Balchand	*Padshahnama*, I/I, pp.493–95 *Shah Jahan Nama*, p.5 *Tuzuk*, I, p.186 *Jahangirnama*, fols 74b–76a	Prince Khurram attacking a lion (late 1610)
31	144a	The siege of Daulatabad (April–June 1633)	Murar	*Padshahnama*, I/I, pp.484ff *Shah Jahan Nama*, pp.105–106	The siege of Daulatabad (April–June 1633)
32	147b	The departure of Prince Shah-Shuja' for Kabul (16 March 1638)	Murar	*Padshahnama*, II, pp.35–37 *Shah Jahan Nama*, p.224	The departure of Prince Shah-Shuja' for the Deccan (August 1633)
33	165a	Shah-Jahan hunting	Unknown artist	Uncertain	Uncertain
34	166b	A royal procession	'Kashmiri Painter' (attribution)	Uncertain	The royal entourage traveling from Agra to Lahore (March 1634)
35	174a	The capture of Orchha by imperial forces (October 1635)	Tezdast	*Padshahnama*, I/2, pp.94–100 *Shah Jahan Nama*, pp.155–58	The capture of Orchha by imperial forces (October 1635)
36	176b	Khan Dawran receiving the heads of Jujhar Singh and his son Bikramajit (January 1636)	Payag (attribution)	*Padshahnama*, I/2, pp.116 *Shah Jahan Nama*, p.165	Khan Dawran receiving the heads of Jujhar Singh and his son Bikramajit (January 1636)
37	192b	Jahangir receives Prince Khurram (April 1616)	Abid (attribution)	*Tuzuk*, I, p.322 *Jahangirnama*, fols 132b–133a	The departure of Prince Khurram for the Deccan campaign (30 October 1616)
38	194b	Jahangir receives Prince Khurram on his return from the Mewar campaign (19 February 1615)	Murar	*Padshahnama*, I/I, pp.174ff *Shah Jahan Nama*, p.6 *Tuzuk*, I, pp.276–78 *Jahangirnama*, fols 107a–108a	Jahangir receives Prince Khurram on his return from the Deccan (12 October 1617)
39	195a	Jahangir presents Prince Khurram with a turban ornament (12 October 1617)	Payag	*Padshahnama*, I/2, pp.187–88 *Tuzuk*, I, pp.393–95 *Jahangirnama*, fol.162a	Jahangir receives Prince Khurram on his return from the Deccan (12 October 1617)
40	204b	The surrender of the Fort at Udgir to Khan Dawran (October 1636)	Dhola	*Padshahnama*, I/2, pp.217–19 *Shah Jahan Nama*, pp.192–94	The surrender of the fort at Udgir to Khan Dawran (October 1636)
41–42	205b–206a	Shah-Jahan visits the shrine of Khwaja Mu'inuddin Chishti at Ajmer (November 1654)	'Kashmiri Painter' (attribution)	*'Amal-i Salih*, III, p.193 *Shah Jahan Nama*, pp.502	Shah-Jahan visits the shrine of Khwaja Mu'inuddin Chishti at Ajmer (6 December 1636)
43	214b	Shah-Jahan honoring Prince Awrangzeb at Agra before his wedding (27 April 1637)	Payag	*Padshahnama*, I/2, pp.255ff *Shah Jahan Nama*, p.206	Shah-Jahan honoring Prince Awrangzeb at Agra before his wedding (27 April 1637)
44	217b	The arrival of Prince Awrangzeb at the court at Lahore (9 January 1640)	Murar	*Padshahnama*, II, p.174 *Shah Jahan Nama*, p.264	Uncertain
45	218b	Shah-Jahan honoring Prince Awrangzeb at his wedding (19 May 1637)	Bhola (attribution)	*Padshahnama*, I/2, p.268 *Shah Jahan Nama*, pp.207–208	Shah-Jahan honoring Prince Awrangzeb at his wedding (19 May 1637)
46	220b	Shah-Jahan hunting lions at Burhanpur (July 1630)	Dawlat (attribution)	*Padshahnama*, I/I, pp.310–11	Shah-Jahan hunting lions near Bari (26 July 1637)

1. Christian Era dates for events during Shah-Jahan's reign accord with those given in *Shah Jahan Nama*.

I

I~2

Double shamsa

FOLIOS 1B–2A, CIRCA 1657

At the birth of Prince Khurram, the future Shah-Jahan, a poet at the court of his grandfather, the Emperor Akbar, wrote a celebratory line of verse. According to Persian literary convention, letters hold numerical values, and this phrase was composed to add up to one thousand, the year of the Prince's birth according to the Muslim calendar.[1] The line written in honor of the occasion – 'A flash from the world-pervading Sun' – was praised for its appropriate evocation of the divine character of the event that had inspired it. The connection between light and royalty is made explicit by Akbar's historian, Abu'l-Fazl:

THE SHAMSA [image of the sun] ... is a divine light, which God directly transfers to kings, without the assistance of men; and kings are fond of external splendour, because they consider it an image of the Divine glory.[2]

The symbolism of light was used in many ways in Mughal India. For example, the placement of *shamsa*s such as these at the beginning of an especially important manuscript or album was a sign that it was intended for a royal patron.

1. Shah-Jahan's birthdate was 10 Rabi' II 1000 (15 January 1592).
2. *A'in-i Akbari*, I, p.52.

3~4

Timur and Shah-Jahan

FOLIOS 2B–3A, CIRCA 1657

Historical texts carefully recount Shah-Jahan's imperial lineage:

HIS MAJESTY IS the illustrious heir of *Jannat Makani* [he whose place is in the heavenly garden], the Emperor Nur al-Din Muhammad Jahangir; son of *Arsh Ashyani* [he whose nest is on the divine throne], the Emperor Jalal al-Din Muhammad Akbar; son of *Jannat Ashyani* [he who is nested in heaven], the Emperor Nasir al-Din Muhammad Humayun; son of *Firdaus Makani* [he whose place is in paradise], the Emperor Zahir al-Din Muhammad Babur; son of 'Umar Shaikh Shah; son of the martyr king Sultan Abu Sa'id Shah; son of Sultan Muhammad Shah; son of Mirza Miran Shah; son of the Great Lord and universal conqueror, the Pillar of the World and Religion – Emperor Amir Timur Gurgan, the first *Sahib-i-Qiran* [Lord of the Auspicious Planetary Conjunction].[1]

The link between Shah-Jahan and Timur is made even more explicit by one of the former's titles, *Sahib-i-Qiran-i-Sani* (Second Lord of the Auspicious Planetary Conjunction), which represents a specific claim that Shah-Jahan has inherited Timur's prestige and power. It is completely appropriate, therefore, that the *Padshahnama* opens with a frontispiece showing Timur presenting the imperial crown to Shah-Jahan.

1. *Shah Jahan Nama*, p.4.

3

5

Jahangir receives Prince Khurram on his return from the Mewar campaign

AJMER, DIWAN-I 'AMM, 20 FEBRUARY 1615.
FOLIO 43B PAINTED BY BALCHAND, CIRCA 1635

To help consolidate their power, the Mughal emperors sought continually to defeat the rulers of the various independent Rajput kingdoms that controlled important territories in northwestern India. The senior and most prestigious of these Hindu rulers was the Rana of Mewar, against whom Prince Khurram (the future Shah-Jahan) was sent to fight in 1614. One of several illustrations that describe Khurram's activities before his accession as emperor, this painting is placed to correspond with text describing the departure of Khurram for the Mewar campaign, but it actually shows the Prince's triumphant return to the court established at Ajmer by his father, the Emperor Jahangir, following the Rana's defeat. As the *Padshahnama* relates in a later passage on the events of 19 February 1615:

THE IMPERIAL RETINUE camped outside of Ajmer, and by imperial command all the amirs went out to greet [the Prince]. The next day, Sunday ... [Prince Khurram] attained the felicity of paying homage to his mighty father. His Imperial Majesty was so kind and appreciative that he took him in an embrace. His Highness the Prince offered a thousand *muhr*s [*mohur*, a coin] in fulfillment of his vow, another thousand as alms, a crystal chest made by the artisans of [?[1]] filled with precious gems, along with the Rana's famous large ruby, which gem connoisseurs have evaluated at 60,000 rupees, the Rana's best elephant, Alam-Guman,[2] along with the other elephants belonging to him and his followers that had fallen to the imperial troops during this campaign. The Emperor presented the Prince with an imperial goldspun robe with a gold-embroidered collar on which were placed flowers of gems, a bejeweled dagger, a bejeweled sword, two horses from the royal stables, one with a bejeweled saddle and the other with a golden saddle, an elephant from the royal stables named Bijai Kunjar with silver trappings, and a female elephant.[3]

The scene depicted here, however, corresponds to a specific incident during this welcoming ceremony as described in far more personal terms in the *Jahangirnama*, the memoirs of Jahangir:

ON SATURDAY the 10th of the month [19 February 1615], the successful Prince stopped on the outskirts of Deorani, which is near the city of Ajmer, and an order was given for all the amirs to go out to greet him and present gifts, each according to his rank and station. The next day, which would be Sunday the 11th, the Prince was to attain the felicity of being received. Upon the morrow, the Prince entered the Hall of Public and Private Audience with all splendor and magnificence, surrounded by all the victorious soldiers who had been assigned to him on this campaign. Two watches and two *ghari*s [a measure of time equivalent to 24 minutes] had elapsed of the day when the hours of reception came and he had the fortune to perform *körünüsh* [formal, 'kowtow'] prostrations, and salutes. He gave a thousand *ashrafi*s [a coin] and a thousand rupees in fulfillment of his vow and another thousand *mohur*s and a thousand rupees as alms. I summoned my son forward, embraced him, and kissed his head and face, singling him out for particular affection and kindness. When he had finished the rites of service, he had his vow and alms displayed and said that if it was ordered, Karan should be allowed to prostrate himself and perform *körünüsh*. I ordered him brought in. The *bakhshi*s [military officials] brought him in with the usual ceremonials.[4]

Karan Singh, the dark-skinned man in the yellow robe, was the eldest son of Rana Amar Singh. Sent to court as a mark of the Rana's capitulation to the imperialists, he is shown being greeted 'with the usual ceremonials' by imperial officials.

Sir Thomas Roe, the first English ambassador to the Mughal court, wrote of a *darbar* [official audience] which he attended at Ajmer the following year, on 10 January 1616:

...I WENT TO COURT at 4 in the euening to the *Durbar*, which is the Place wher the Mogull sitts out daylie, to entertayne ſtrangers, to receiue petitions and presents, to giue Commandes, to see, and to bee seene ... The Place is a great Court, whither resort all sorts of people. The king sitts in a little Gallery ouer head; Ambassidors, the great men and ſtrangers of qualety within the inmoſt rayle vnder him, raysed from the ground, Couered with Canopyes of veluet and silke, vnder foote layd with good Carpetts; the Meaner men representing gentry within the firſt rayle, the people without in a base Court, but soe that all may see the king. This sitting out hath soe much affinitye with a Theatre – the manner of the king in his gallery; The great men lifted on a ſtage as actors; the vulgar below gazing on ...[5]

Later that same year, on 15 March, Roe attended a *darbar* where Karan Singh was also present:

AT EUENING I WENT to the *Norose* [New Year's celebrations] and demanded of Asaph Chan a place. He bad me Choose; so I went within the rayle, and ſtood on the right hand of the kynge upon the rising of the Throne; the Prince and young Ranna [Karan Singh] on the other syde. So that I sawe what was to be seene, Presents, Elephants, horses, and many whoores.[6]

1. The word may be a mistake for *firang* (Frank, Europe). WT
2. For a painting of Alam-Guman, see S.C. Welch 1963, no.36.
3. *Padshahnama*, 1/1, pp.174ff.; translated by Wheeler Thackston.
4. *Jahangirnama*, fols 107b–108a; translated by Wheeler Thackston.
5. Roe 1967, pp.106ff.
6. Roe 1967, pp.149–50.

6~7

The submission of Rana Amar Singh of Mewar to Prince Khurram

GOGUNDA (MEWAR STATE), 5 FEBRUARY 1615.
FOLIO 46B PAINTED BY LA'LCHAND;
FOLIO 47A ATTRIBUTED TO THE 'KASHMIRI PAINTER', BOTH CIRCA 1640

Following the success of the imperial troops, Rana Amar Singh of Mewar himself appeared before Prince Khurram, the leader of the Mughal army, to acknowledge his defeat. Unlike the preceding illustration, which was incorrectly placed when the manuscript was assembled, this scene corresponds to the accompanying text:

IN A SHORT WHILE they had made things so difficult for [Rana Amar Singh] that he realized that his life and honor were in jeopardy, that his salvation lay in obedience, and that his only means of deliverance was to surrender. Appealing to Rai Sundardas, His Highness' [Prince Khurram's] majordomo, he pleaded his inability and brokenness and begged for mercy. With Hardas Jhala, his most important Rajput, and Subhkaran, his maternal uncle who was among his trusted men, he sent a petition to the aforementioned Rai, asking if he could come and pay homage and if Karan, his eldest son and successor, might join the Emperor's retinue. Although it had been the Emperor's intention to crush him, when Rai Sundardas asked for his forgiveness, the Emperor issued an imperial edict that his son should consider this event an act of outstanding service and not destroy him. The plea was granted. In accordance with Rana Amar Singh's wish, Mulla Shukrullah Divan and Rai Sundardas, who were later rewarded by the Emperor with the titles Afzal Khan and Rai Raian respectively, were dispatched to console the Rana and bring him to court. A letter adorned with the blessed handprint, which was a guarantee of his safety, was sent with the two. The Rana, who had given up all hope of escape and washed his hands of life, children, honor, and homeland, immediately upon receipt of the letter set out for the Prince's threshold. When he reached the imperial camp, the imperial Prince sent Abdullah Khan Bahadur, Raja Suraj Singh, Raja Sukhdeo Bundela, and Sayyid Sayf Khan Barha to receive him. After the eye of his hope was illuminated by the world-adorning brilliance [of Prince Khurram], he rubbed the forelock of politeness on the ground as servants do and executed the rituals of humility. With his own blessed hand the Prince lifted his head and ennobled him in form and content, soothing him with hope-increasing and fear-erasing words ... The Rana presented a famous large ruby, eight *tânk*s in weight, along with other gems and jewel-studded implements, nine horses, and nine elephants left over from the imperial troops' plunder. In turn, he was honored with the gift of a robe of honor, a jewel-studded *jamdhar* [a type of dagger], a jewel-studded sword, a fine horse from the royal stable with a golden saddle, and an elephant with silver trappings from the royal elephant stables ... The Rana was ordered to return to his seat, and as a sign of increased favor Afzal Khan and Rai Raian were ordered to escort him. Inasmuch as he had not brought Karan with him – it being the custom of the Zamindars not to bring their sons with them when they pay homage to rulers – and the imperial troops were scheduled to depart at the end of that day, it was settled that Afzal and Rai Raian would bring Karan to court that very day.[1]

This event, which evidently took place in a tent enclosure at Gogunda, northeast of Udaipur,[2] is also discussed by Jahangir in his memoirs:

IN THE MONTH OF Bahman there came pieces of good news one after the other. The first was that Rana Amar Singh had elected for obedience and service to the Court ... My lofty mind was always desirous, as far as possible, not to destroy the old families. The real point was that as Rana Amar Singh and his fathers, proud in the strength of their hilly country and their abodes,

had never seen or obeyed any of the kings of Hindustan, this should be brought about in my reign. At the request of my son I forgave the Rana's offences, and gave a gracious *farman* [decree] that should satisfy him, and impressed on it the mark of my auspicious palm.[3]

This description suggests that the Rana's prestige was respected when he appeared at court, but the illustration tells us otherwise. Rana Amar Singh, the proudest of all the Rajput rulers, is shown in a humiliatingly menial position, and the gesture of Abdullah Khan, whose hand is on the Rajput's back, leaves no doubt that he is being carefully controlled. The subsequent appearance of Karan Singh, the son of the Rana, before the Emperor Jahangir is depicted in the previous illustration (**5**).

1. *Padshahnama*, I/I, pp.171–72; translated by Wheeler Thackston.
2. G.N. Sharma 1954, p.120.
3. *Tuzuk-i-Jahangiri*, I, pp.273–74.

8~9

Jahangir receives Prince Khurram on his return from the Deccan

MANDU, DIWAN-I 'AMM, 10 OCTOBER 1617.
FOLIO 48B PAINTED BY RAMDAS;
FOLIO 49A PAINTED BY MURAR, BOTH CIRCA 1640

After his defeat of the Hindu Rana of Mewar, Prince Khurram was placed in charge of the military expeditions against various Muslim kingdoms, or sultanates, in the Deccan. It was because of his victories there that he was awarded the title of Shah-Jahan ('King of the World') by Jahangir in 1617.

This scene, inserted into the Windsor manuscript to correspond with a textual description of Shah-Jahan's return from the Mewar campaign, actually shows the Prince's arrival at the court established by his father in Mandu following the victories in the Deccan. According to the *Jahangirnama*, on Thursday, 10 October 1617:[1]

[KHURRAM] ATTAINED THE happiness of paying homage in the fortress of Mandu with auspiciousness and felicity. The length of our separation had been eleven months and eleven days. After he performed the ritual of *körünüsh* and kissed the ground, I summoned him up into the *jharoka* [balcony], and, out of sheer love and yearning, I involuntarily rose from my place and embraced him. The more he exaggerated the etiquette of humility, the more I increased my show of favor and affection by ordering him to sit near me. He presented a thousand *ashrafi*s and a thousand rupees in fulfillment of a vow and another thousand *ashrafi*s and rupees as alms. Since there was not enough time for him to have his offerings displayed for my view, at this time he showed only the elephant Sirnag, the finest of the elephants of the Adil Khan's offering, and a small chest of precious gems. After that, an order was given to the *bakhshi*s for the amirs who had accompanied my son to pay homage by rank. The first to pay homage was Khan Jahan. I summoned him up and allowed him to prostrate himself. He presented an offering of a thousand *mohur*s and a thousand rupees as a vow and a small chest of gems and jeweled objects. What was accepted of his offering was worth 45,000 rupees. After him, Abdullah Khan prostrated himself and gave his vow of a hundred *mohur*s. Then Mahabat Khan was honored to prostrate himself. He offered a hundred *mohur*s and a thousand rupees as a vow and a *gathri* [parcel] of gems and jeweled objects worth 124,000 rupees. Among the items was a ruby weighing eleven mithcals [253 carats] a Frank [European] had offered for sale last year in Ajmer. He had priced it at a lac [100,000] of rupees, but the jewelers valued it at 80,000 rupees. Therefore the transaction did not take place, and he was sent back to where he came from. When he reached Burhanpur, Mahabat Khan bought it from him for a lac of rupees ... The other amirs also paid homage by rank. After that, the Adil Khan's deputies were allowed to kiss the ground and present his message.[2]

1. 11 Shawwal 1026, corresponding to the 20th day of the Ilahi month of Mihr, the twelfth regnal year.
2. *Jahangirnama*, fols 155b–156a; translated by Wheeler Thackston.

10~11

Shah-Jahan receives his three eldest sons and Asaf Khan during his accession ceremonies

AGRA, DIWAN-I 'AMM, 8 MARCH 1628.
FOLIO 50B PAINTED BY BICHITR, CIRCA 1630;
FOLIO 51A ATTRIBUTED TO RAMDAS, CIRCA 1640

Shah-Jahan succeeded to the throne after the death of his father, the Emperor Jahangir, on 8 November 1627. Many events were held during the time of his formal accession the following spring, of which the ceremonial arrival in Agra – on 8 March 1628[1] – of his three eldest sons, who had been with their grandfather, Asaf Khan, in Lahore, has been depicted here.

THE PEARLS OF the casket of magnificence, the lights of the eye of fortune, the fruits of the garden of the sultanate, Prince Muhammad Dara-Shikoh, Muhammad Shah-Shuja', and Muhammad Awrangzeb, accompanied by the most magnificent of amirs Yaminuddawla Asaf Khan and other retinue and servants arrived from Lahore to the vicinity of Akbarabad [Agra] and camped by imperial command on the outskirts in a place known as Sikandara. Hearing the good news of the arrival of the princes, Her Majesty Mahd-i-Ulya [Mumtaz Mahal] set out by imperial command with the Sahib-Qiran's favorite child the Begum Sahib [Jahanara] and the other noble children to receive them ... Along the way, at the place where the imperial tents had been pitched for the encounter, Her Majesty Mahd-i-Ulya was delighted to behold her sons after a long period of separation, and the two parents' joy after such a long time was indescribable inasmuch as in the workshop of rhetoric there is no expression that can convey such joy. The entire day was spent by the parents admiring the beauty of their grand sons. The next day, which was a Thursday, all the nobles and grandees went out by royal command to greet the princes and escort them and Yaminuddawla to court. When the *jharoka* of the Hall of Public and Private Audience was illuminated by the rays of the Emperor's enthronement, the princes and Yaminuddawla were allowed to pay homage. First, the eldest pearl of magnificence, Muhammad Dara-Shikoh, attained the felicity of kissing the ground and offering a thousand *muhr*s as *nazr* [a vow] and another thousand as *nisar* [a gift]. After him, the apple of fortune's eye, Muhammad Shah-Shuja', kissed the ground and presented 1,500 *muhr*s, half as *nazr* and half as *nisar,* to the Emperor's gaze. The Emperor embraced the prince and kissed his forehead. After that, Imadussaltana Yaminuddawla Asaf Khan was exalted by kissing the imperial threshold and presenting a thousand *muhr*s in *nazr* and a tray full of various gems and another thousand *muhr*s as *nisar*. At the Emperor's signal he went to the *jharoka*, which is a stairway as lofty as heaven, and placed his head in servitude at the imperial foot. The appreciative Emperor, who nurtured and cherished his servants, raised with his own hands the head of this leader of his sincere servants and embraced him in favor. Through this regal act of compassion Asaf Khan's head was elevated to the skies. During that very assembly the Emperor gave him a robe of honor with a jewel-studded collar, a bejeweled dagger with a valuable *phul-katara* [an Indian thrusting dagger] a jewel-studded sword with a bejeweled scabbard that had been taken as booty in the conquest of Ahmadnagar and which His Majesty Arsh-Ashyani [Akbar] had given to His Majesty Jannat-Makani [Jahangir], and he had given as a prize to the Solomonic emperor [Shah-Jahan] for the conquest of the Deccan, and which was worth a lac of rupees ...[2]

1. 1 Rajab 1037, corresponding to 19 Isfandarmudh.
2. *Padshahnama*, 1/1, pp.177–82; translated by Wheeler Thackston.

12~13

The weighing of Shah-Jahan on his forty-second lunar birthday

AGRA, DIWAN-I KHASS, 23 OCTOBER 1632.
FOLIOS 70B–71A PAINTED BY BHOLA, CIRCA 1635

SINCE ALMS ARE beneficial for repelling bodily and psychic harm and for attracting spiritual and corporeal benefits, as all peoples, religions, and nations are agreed, His Majesty Arsh-Ashyani [Akbar] established the custom of weighing and had himself weighed twice [a year], once after the end of the solar year and the other after the end of the lunar year. In the solar weighing he was weighed twelve times, first against gold and then eleven other items, while in the lunar weighing he was weighed eight times, first against silver and then seven other items. The weighing of his sons was carried out once a solar year. The amounts from the weighings were given away in alms.

His Majesty Jannat-Makani [Jahangir] followed His Majesty Arsh-Ashyani's custom in putting this beneficial custom into practice. Inasmuch as it benefited the needy, His Majesty Jahanbani [Shah-Jahan] has his perfect self weighed twice, and in his generosity he has ordered that gold and silver be used each time. In the solar weighing he is first weighed against gold, then against silver, and then ten other items, while in the lunar weighing first against gold, then against silver, then against six other items. The princes are weighed once [a year].[1]

More specific aspects of these important ceremonies are described in another historical text, the *'Amal-i Salih*. Written by Muhammad-Salih Kanbo, this narrative covers Shah-Jahan's reign and includes the years following Abdul-Hamid Lahawri's death in 1654.

THE MANNER OF weighing is as follows. Since it is His Majesty's custom and habit to have beggars sought out, and his generous nature is always looking for a pretext to relieve those who are in need, therefore twice a year he sits, like the orient sun in majesty, in the pan of the scale of auspiciousness in the solar and lunar weighing ceremonies. Twice a year by solar and lunar calculation a magnificent celebration and a large-scale banquet is arranged by order of His Majesty. An amount equal to his weight in gold and silver is distributed among the destitute and the poor according to their deservedness and merits. Although this type of alms is not mentioned in the religious law, nonetheless since scholars of this country are all in agreement that such alms are the most perfect type of alms for repelling corporeal and spiritual catastrophes and calamities, therefore this pleasing method was chosen and established by His Majesty Arsh-Ashyani, whose personality was, like the world-illuminating sun, based upon pure effulgence. By this means the poor attained their wishes, and in truth the custom of *aqiqa* – which is an established custom in the law of the Prophet and his Companions, and in which on the seventh day after birth the equivalent weight of an infant's shaven hair in silver is given in alms, and a sacrificial animal is divided and distributed among the poor – has opened the way to making this custom permissible ... In short, from that time forward this praiseworthy act has become usual and customary in this mighty empire. The princes are also weighed once a solar year.[2]

While this illustration is placed within the manuscript to correspond to text describing the *Nawroz* (New Year) celebrations at Fatehpur-Sikri in 1629, it actually shows the weighing ceremony at Agra celebrating the Emperor's 42nd lunar birthday, on Sunday, 23 October 1632:[3]

... THE LUNAR WEIGHING ceremony for the end of the forty-third year of the Emperor's life was held. The Emperor, surrounded by a divine aura, was weighed against gold and the other usual things, and the skirt of the world was held out in expectation of gold and silver. On this auspicious day Muhammad-Ali Beg, the ambassador of Iran, was awarded a gold-embroidered

robe of honor, a jeweled belt, an elephant, a female elephant, and four large *ashrafi*s, one weighing 400 *tola*s [a measure of weight, slightly more than two mithcals], the second 300 *tola*s, the third 200 *tola*s, and the fourth 100 *tola*s, and four rupees also of the weights given above, and he was given leave to depart. From the time he paid homage until the time he set out to return he had been given 316,000 rupees in cash and nearly a lac of rupees in goods.[4]

An earlier weighing ceremony of the Emperor Jahangir, on 1 September 1617, was described by the always observant, and usually skeptical, first English ambassador to the Mughal court, Sir Thomas Roe:

— WAS THE KINGS Birth-day, and the solemnitie of his weighing, to which I went, and was carryed into a very large and beautifull Garden; the square within all water; on the sides flowres and trees; in the midst a Pinacle, where was prepared the scales, being hung in large tressels, and a crosse beame plated on with Gold thinne, the scales of massie Gold, the borders set with small stones, Rubies and Turkey, with Chaines of Gold large and massie, but strengthened with silke coords. Here attended the Nobilitie, all sitting about it on Carpets, vntill the King came; who at last appeared clothed, or rather loden with Diamonds, Rubies, Pearles, and other precious vanities, so great, so glorious! ... Suddenly hee entered into the scales, sate like a woman on his legges, and there was put against him many bagges to fit his weight, which were changed sixe times, and they say was siluer, and that I vnderstood his weight to be nine thousand *Rupias*, which are almost one thousand pound sterling. After with Gold and Iewels, and precious stones, but I saw none; it beeing in bagges might bee Pibles ...[5]

The Portuguese priest Sebastian Manrique also left an account of a later weighing ceremony, on 8 July 1641:

ON THE NIGHT preceding the following day, that of the festival commemorating that on such a day the Emperor Corrombo [Khurram, or Shah-Jahan] came into the world, before the lesser and unstable luminary had emerged to illuminate the shades of night with her dim beams, artificial thunder from the Imperial Palace began to fulminate with its awe-inspiring, deafening roar and announce the approaching feast.

Following a description of the ceremonial banquet, he continues:

... THE EMPEROR CORROMBO retired to a richly decorated private chamber adorned with all the most precious and valued products of the world.

In the centre of this wonderful room there hung, securely fastened by thick chains of gold, a pair of scales made of the same metal, their circular edges being set with many rich stones. The Imperial Majesty came forth to attend this solemn function dressed in a white satin robe covered with most precious stones of many colours ... He also wore round his neck very rich collars of most valuable jewels; so that when I mentally summed up everything estimated I could only think that all this adornment must serve rather as a troublesome burden than an elegant ornament.

On reaching the balance, dressed in this fashion, the Emperor squatted in one pan, and at once certain officials commenced to fill the other pan with bags full of silver, coined into rupees, until the weights were equalized and the balance was in equilibrium. This weighing over, they removed the silver and put it to one side. They then made a second weighing with other bags full of gold and precious stones. After adjusting this they made a third weighment with different kinds of cotton cloth, woven with gold, silver, and silk. They also added, in this third weighment, certain precious spices and various drugs. The fourth and last weighment was made against eatables, such as wheaten cakes, flour, sugar, *ghi* [ghee, clarified butter], and common kinds of cotton cloth.

All the articles included in the final weighing are, it is said, at once distributed amongst the poor Bramenes and Baneanes [brahmins, priests, and *bania*s, or merchants], but with much secrecy that, save those who actually gave the gifts or those who received them, no one saw them ...

This ceremony ... completed, the Emperor returns to his throne. As soon as he is seated on it the dish-bearing eunuchs advance with great trays of gold filled with artificial and imitation fruits, in silver, representing almonds, nuts, hazel-nuts, and many other kinds and species of fruit, all so delicately contrived ... The Emperor distributed this fruit among the Princes and Nobles ...[6]

1. *Padshahnama*, I/I, pp.243ff.; translated by Wheeler Thackston.
2. *'Amal-i Salih*, I, pp.321ff. I am grateful to Ebba Koch for this reference.
3. 8 Rabi' II 1042, corresponding to I Aban.
4. *Padshahnama*, I/I, pp.440–41; translated by Wheeler Thackston.
5. Roe 1967, pp.411–12.
6. Manrique 1927, I, pp.200–202.

14

The presentation of Prince Dara-Shikoh's wedding gifts

AGRA, DIWAN-I 'AMM, 4 FEBRUARY 1633.
FOLIO 72B PAINTED BY BALCHAND, CIRCA 1635

No adjoining text within the Windsor manuscript corresponds to this scene, which probably illustrates the presentation in the *Diwan-i 'Amm* (Hall of Public Audiences) at Agra of the elaborate wedding gifts given to Dara-Shikoh, the Emperor's eldest son and intended heir. This was an especially important event, at which honor was also paid to the memory of the Empress Mumtaz Mahal; she had died unexpectedly in 1631 after having initiated plans for the wedding ceremony. As related in the *Padshahnama* narrative, her daughter Jahanara Begam was eventually placed in charge when preparations for the wedding were resumed:

WHAT THE LATE Mahd-i-Ulya [Mumtaz Mahal] had put together while she was alive for her son and what was arranged in [Jahanara Begam's] establishment amounted in cash and goods to sixteen lacs of rupees, as follows. Seven lacs of precious gems and jeweled items; four lacs of gold and silver utensils and various sorts of brocades from Hindustan, Persia, Anatolia, China, and Europe; various carpets, rugs, canopies in goldspun and gold-embroidered velvets; and other things from the workshops suitable to mighty emperors; several elephants and female elephants with silver trappings; Arabian, Persian, Kachhi, and Turkish horses with jeweled saddles of gold and silver; one lac of rupees in cash. A total of four lacs of rupees will be listed among the offerings of the court servants and the princes and begams, and the amirs' robes of honor, et cetera. Since there was no place to spread all these items other than the portico in front of the public and private *jharoka* recently built by His Majesty, on Friday ... [4 February 1633],[1] the courtyard of the Hall of Public and Private Audience was emptied, and the servants of the harem by order of the Begam Sahib [Jahanara] and Satiunnisa Khanim laid out in that portico all the above-mentioned items from the beginning of sunrise until the end of the third watch in an arrangement that astonished jaded world-travelers.[2]

1. 24 Rajab 1042, corresponding to 'Mihr day', 16 Bahman.
2. *Padshahnama*, I/1, pp.454–56; translated by Wheeler Thackston.

15

A'zam Khan captures Fort Dharur

DHARUR (DECCAN), JANUARY 1631.
FOLIO 92B PAINTED BY BALCHAND, CIRCA 1633

Despite Shah-Jahan's successes in the Deccan as a prince, the area remained resistant to Mughal control during the early years of his imperial reign. In January 1631, however, the Emperor celebrated the defeat of the fortress at Dharur.

AT THIS TIME Multafit Khan was assigned ... to attack the town of Dharur and its *bitha* [market area], where once a week the people from the villages around Dharur and other places came to buy and sell all sorts of goods. Since the fortress of Dharur is renowned in the Deccan for its impregnability and ability to withstand siege, it being situated atop a hill with streams on two sides too deep and too steep for the imperial army to descend, it was decided that after the attack on the town and *bitha*, no attack on the fortress would be made but rather the 'Benizam's [Nizamulmulk's]' ill-starred forces would be pursued. After Multafit Khan entered the town with the other servants of the imperial court, he announced to A'zam Khan that a large contingent on that side of the ditch would begin firing. A'zam Khan and all the servants of the imperial court passed through the town and settled near the fortress inside a compound within cannon range. Multafit Khan and his comrades went to the edge of the ditch and fired musket shots to clear out the people of the town who had taken their possessions and families into the ditch to be clear of the cannons and gunfire. Then they entered the ditch nimbly and began to plunder and take prisoners. A'zam Khan dispatched a contingent of Raja Jujhar Singh Bundela's foot soldiers to capture several elephants that had wandered into the ditch. These men went boldly and bravely on their mission and brought back four elephants along with many horses, camels, and oxen, although several men were wounded and killed. Once again the heroes of the victorious army attacked, plundering the remaining goods in the ditch. This time three elephants and many horses fell into the hands of the men of Bundela. A'zam took the elephants for His Majesty and gave the horses back to the men. So brave and courageous was he that that night he entered the ditch, and, after inspecting the wall, it was obvious that on one side was a doorway that had been blocked up with plaster and stone. It could be penetrated by pick and shovel or blown up with gunpowder, and the fortress could be entered ... Since the defenders of the fortress had lost heart and given up hope in the face of the pugnacity and battle-readiness of the fortress-conquering heroes, on Monday morning ... [27 January 1631],[1] he mounted and headed for the fortress gate. He ordered that the imperial forces should bring their barricades forward all at once from the direction of the gate, which had no ditch and so facilitated a forward charge, and some should scale the walls by ladders and others by lassoes. These warriors considered arrows, bullets, and stones as the rain that makes the fields of success green: some of them attacked the moat, and others courageously gained the walls. Marhamat Khan and a few men entered the fortress and opened the door in the gate. A'zam Khan and the rest of the commanders entered. At the same time as he entered, nearly two thousand men came over the walls and entered the fortress, plundering many goods and an amount of jewels and gem-studded vessels ... So, through the good fortune of the adorner of the world throne and the good offices of the supporters of the eternal empire, such an impregnable stronghold was quickly and easily subdued.[2]

1. 23 Jumada II 1040, corresponding to 8 Bahman.
2. *Padshahnama*, I/1, pp.339–43; translated by Wheeler Thackston.

16

The death of Khan Jahan Lodi

KALINJAR, 3 FEBRUARY 1631.
FOLIO 94B PAINTED BY ABID, CIRCA 1633

Once a favorite of the Emperor Jahangir and the guardian of Prince Parvez, the elder brother of Shah-Jahan, Khan Jahan Lodi distrusted the new emperor and became openly rebellious early in the reign. He was hunted fiercely by imperial troops, by whom he eventually met his death:

PIRA [KHAN-JAHAN LODI] had given up all hope of salvation since the deaths of his sons, relatives, and cohorts, by whose assistance his brainless head had become a nest of false and demonic hopes and vain fancies. To his companions he said, 'I am sick of transient life, which is not worth all this humiliation. While the road of escape is open, any who want to leave may go.' Several remained firm, but many chose to take flight. In the meanwhile arrived Sayyid Muzaffar Khan, the vanguard of the victorious imperial forces, who had preceded the main body of the army with Madho Singh and a troop of mace-bearers and other servants of the imperial court. The unlucky Pira, with his dearest son Aziz, Emal, and a group of Afghans, put the two elephants remaining to them in front and set out in the direction of Sayyid Muzaffar Khan. The Khan and his comrades attacked. When the misfortunate saw that the approaching army would in no way leave him alone, he dismounted from his horse and began hand-to-hand combat with Madho Singh and the group of mace-bearers and others, *sayyid*s [lineal descendants of the Prophet Muhammad] of Barha, et cetera, who formed Sayyid Muzaffar Khan's vanguard. During the struggles Madho Singh struck him with a lance, and before Sayyid Muzaffar Khan arrived, the imperial warriors had attacked and torn to shreds the miscreant, his son Aziz, and Emal. Most of his companions, who were nearly a hundred in number, were killed, and the mace-bearers cut off the heads of those wrong-headed rebels, although some escaped ... Just at this time Abdullah Khan Bahadur arrived from the rear and sent the heads of Pira, Aziz, and Emal to court with Khwaja Kamgar, his nephew. Farid, Pira's other son who was captured alive, was put into chains. Since Bahadur Khan's mother was residing in Sihenda, which was her son's fief, and Pira's son Jan-i-Jahan had escaped and taken refuge with her, Abdullah Khan sent his eunuch Iqbal to bring him. He brought Jan-i-Jahan and Bahadur Khan's younger brother Jalal before the Khan, who entrusted him to Yakataz Khan for delivery to court.[1]

1. *Padshahnama*, I/I, pp.350–51; translated by Wheeler Thackston.

17

Shah-Jahan receives the Persian ambassador, Muhammad-Ali Beg

BURHANPUR, DIWAN-I 'AMM, MARCH 1631.
FOLIO 98B ATTRIBUTED TO THE 'KASHMIRI PAINTER', CIRCA 1633

The arrival of an embassy from the Persian shah, the most immediate and powerful of Shah-Jahan's imperial rivals, on 22 March 1631,[1] and the exchange of gifts four days later, were occasion for grand celebrations.

... AFZAL KHAN AND Sadiq Khan, by imperial command, who had gone to greet Muhammad[-Ali] Beg, the ambassador of Iran who had arrived in the vicinity of Burhanpur, escorted the ambassador to pay homage to the imperial presence. After performing the rites of etiquette, *körünüsh*, and kissing the ground, he presented a letter Shah Safi had written in congratulations for the accession and conveyed the Shah's wishes. He also conveyed some of the words he had been ordered to deliver. The grateful Emperor took the letter into his own blessed hand and turned it over to Afzal Khan. After that, at Muhammad-Ali Beg's request, a time for the presentation of gifts was fixed, the 22nd of the month [26 March 1631], and the ambassador was favored with the gift of a gold-brocaded robe of honor, a jewel-studded Qizilbash crown, a bejeweled *jigha* [turban ornament], and a valuable jewel-studded dagger. Since during the days of *Nawroz* betel-leaf and aromatics in bejeweled and golden trays are given to the nobles of court and others attending assemblies, a jeweled betel-leaf box with a lid and a golden tray worth together nearly 20,000 rupees were given to Muhammad-Ali Beg. Mu'taqid Khan, the *Subadar* [Governor] of Malwa who had escorted the ambassador from Ujjain to court, attained the felicity of kissing the imperial threshold. After that, Abdul-Haqq, Afzal Khan's brother who had gone from Akbarabad to Panipat to greet the ambassador and brought him from there to Akbarabad and from Akbarabad to Burhanpur, entertaining him all along the way, was honored to make *körünüsh*.[2]

Sir Thomas Roe was present at the court of Jahangir when an earlier embassy arrived from Persia. In his account for 19 October 1616, he records the gifts presented to the Emperor:

HE BROUGHT FOR Presentes 3 tymes 9 horses of Persia and Arabia, this beeing a Ceremonius Number among them, 9 mules very fayre and lardg, 7 Camells laden with veluett, two Sutes of Europe Arras (which I suppose was Venetian hanginges of veluett with gould, and not Arras), two Chestes of Persian hanginges, on Cabinett rich, 40 Muskettes, 5 Clockes, one Camell laden with Persian Cloth of gould, 8 Carpettes of silke, 2 Rubyes ballast, 21 Cammelles of wyne of the Grape, 14 Camelles of distilld sweet waters, 7 of rose waters, 7 daggers sett with stones, 5 swordes sett with stones, 7 Venetian looking glasses, but these so faire, so rich that I was ashamed of the relation.[3]

Roe was particularly alert to these ceremonies, for he had been acutely conscious that his own gifts had not been successful. As he wrote back to the East India Company in January 1616:

... HE [JAHANGIR] accepted your presents well; but after the English were come away he asked the Iesuyte whether the King of England were a great Kyng that sent presents of so small valewe ...[4]

1. 18 Sha'ban 1040.
2. *Padshahnama*, 1/1, pp.365ff; translated by Wheeler Thackston.
3. Roe 1967, pp.296–97.
4. Roe 1967, p.119.

18

The siege of Qandahar

QANDAHAR (DECCAN), MAY 1631.
FOLIO 102B PAINTED BY PAYAG, CIRCA 1633

WHEN NASIRI KHAN was appointed, along with a group of servants of the imperial court, to subdue the realm of Telengana, he made his foremost concern the conquest of the Qandahar fortress, one of the renowned strongholds of that area, famed for its strength and impregnability, and then held by Sadiq, Yaqut Khudawand Khan's son-in-law. Having set out for there, he camped within one league of the fortress on Saturday … [28 December 1630].[1] The next day, together with Raja Bharath, Shahbaz Khan, and other officers and soldiers, he mounted, determined to take the town … In the meanwhile, Randola, Muqarrab Khan, Bahlul the Blitherer, and other Adilkhanis and Nizamis arrived and attacked Nasiri Khan's station. Taking no account of the numbers of these foxy types, that lion of the forest of courage began to do battle, performing manly deeds in the face of cannon and gun fire from the fortress and strikes by the enemy [who] turned away. Having given up a few to death, they pulled back three *kos* [a measure of distance, approximately two miles] and sat. Encouraged by the enemy's withdrawal, the imperial forces began to make even more earnest endeavors to take the fortress … While preparations were being made to take the stronghold, A'zam Khan, who had come to help Nasiri Khan, arrived in the vicinity … Nasiri Khan went out to greet him and then took him to watch the fortress being blown up, and in his presence the three mines that had been filled with gunpowder were lit. One went out, but the other two took: the *sher hajji* [outer fortification wall] was blown up along with half of the fortress tower, and the men who were in the tower and at the base of the wall were dispatched to hell. Although from within the fortress cannons, guns, mortars, and rockets were being lit and fired, the imperial soldiers ran forward on foot. From noon until dusk there was pitched battle. Since the fortress wall had not been completely blown away and the defenders were using artillery, the imperial forces could not get in. After sundown they made barricades where they were and busied themselves with lighting the other mines. When the defenders of the fortress realized their terrible predicament, they requested amnesty … Nasiri Khan granted their request and dispatched Raja Bharath to the gate with a written pledge of amnesty. Sadiq, the warden of the fortress, … opened the gates to the representatives of the empire, and in so doing they opened the gates of fortune to themselves. The next day, at an auspicious hour, [Sadiq] went before Nasiri Khan and presented ten elephants. On behalf of the Emperor, Nasiri Khan presented him with a robe of honor and a horse. It should not remain hidden that the locus of divine generosity and inspiration [the Emperor] had written a letter to Nasiri Khan, saying, 'God willing, the fortress of Qandahar will be conquered at the end of the month of Urdibihisht.' Four months and nineteen days after the initial besiegement, on … [17 May 1631],[2] the fortress was taken. A'zam Khan, Nasiri Khan, and other representatives of the Emperor went to inspect the fortress and were astonished by [its] strength and fortitude …[3]

The Qandahar referred to here is a site in the Deccan, northwest of Hyderabad, rather than the important fortress of the same name now in Afghanistan, which was soon to absorb Shah-Jahan's attention as it had that of Jahangir.

1. 23 Jumada I 1040.
2. 15 Shawwal 1040, corresponding to 7 Urdibihisht.
3. *Padshahnama*, I/I, pp.374–77; translated by Wheeler Thackston.

19

Europeans bring gifts to Shah-Jahan

AGRA, DIWAN-I 'AMM, JULY 1633.
FOLIO 116B, PAINTED CIRCA 1650

The Jesuit missionary John Cabral described the Portuguese trading colony at Hoogly, as he himself saw it in 1632:

OF THE VARIOUS BANDELS or trading ports which the Portuguese had in the kingdom of Bengal during the last fifty years, the richest, the most flourishing, and most populous was that of Hugli. To it, as to the common emporium, vessels of India, China, Malacca, and Manila repaired in great numbers. Not only the natives of the country, but also the Hindustanis, the Mogols, the Persians, and the Armenians came there to fetch goods ....[1]

Because it threatened their own trading activities, the Mughals attacked Hoogly late in 1632 (see the next illustration, **20**), with the result that some Europeans were imprisoned and taken to the court at Agra. Their arrival on 18 July 1633 is described in the *Padshahnama*:[2]

... OF THE FRANKISH PRISONERS, four hundred men and women, young and old, along with the idols of those erroneous infidels, were brought before the Islam-nourishing *padshah*'s [Shah-Jahan's] view. The religion-propagating infidel-eradicating Emperor gave a command to the lords of the religious law first to invite the miserable group onto the straight path of the Muhammadan nation and the highway of the Ahmadi religion and inculcate in them the precepts of Islam. Some who found it worthwhile to be thus ennobled by accepting the correct religion were shown imperial favor. Most of them, out of implacable ignorance and erroneousness, turned their heads away from attaining this good fortune. They were divided among the amirs, and it was ordered that this miserable group be held in prison and tortured. Any who inclined to Islam were to be reported to the Emperor, and any who failed to accept the honor were to be kept in prison. Most of them perished in prison. Of their idols, those that were figures of the prophets were ordered to be thrown into the water of the Jumna, and the rest were smashed.[3]

Another missionary, Sebastian Manrique, gives an account of the capture of his compatriots:

... THEY DEFENDED THEMSELVES with such valour as to oblige the Nababo himself to send them several embassies promising them on behalf of the Emperor, that if they surrendered all would be given their lives, and liberty to return to Portuguese territory, together with whatever was necessary for the journey. The besieged accepted these terms as Catholics should ... In regard to these terms, none were adhered to except that of sparing their lives, as they were made prisoners and taken to the Court of the Great Mogol at Agra. They spent about eleven months on this journey, owing to the sick people and the women and children who were with them, as well as to the endless toils and difficulties they met with. They suffered unbearable hardships ... they were all brought into the presence of the Mogol Monarch, who ordered them to be distributed, both men and women, amongst the various Princes and Nobles of the Court, he himself retaining the two Augustinian friars and the two Priests ....[4]

This episode is out of sequence; historically it occurred a year later than the attack depicted in **20**. Moreover, both scenes have been placed within spaces intended originally for one double-page composition.

1. Manrique 1927, I, pp.392–93.
2. 11 Muharram 1043.
3. *Padshahnama*, I/1, pp.534–35; translated by Wheeler Thackston.
4. Manrique 1927, II, pp.325–26.

20

The capture of Port Hoogly

HOOGLY (BENGAL), JUNE–OCTOBER 1632.
FOLIO 117A, PAINTED CIRCA 1634

The Mughal attack on the Portuguese trading colony at Hoogly was a major event that hastened the decline of Portuguese influence in India. The arrival of Portuguese prisoners at court subsequent to the battle depicted here is shown in the preceding illustration (**19**), which is now out of its correct historical sequence.

The battle is described at length in the *Padshahnama*:

DURING THE REIGN of the Bengalis, a group of Frankish [European] merchants ... settled in a place one *kos* from Satgaon on the banks of the [?] and, on the pretext that they needed a place for trading, they received permission from the Bengalis to construct a few edifices. Over time, due to the indifference of the governors of Bengal, many Franks gathered there and built dwellings of the utmost splendor and strength, fortified with cannons, guns, and other instruments of war. It was not long before it became a large settlement and was named Hoogly .... The Franks' ships trafficked at this port, and commerce was established, causing the market at the port of Satgaon to slump ... Of the peasants of those places, they converted some to Christianity by force and others through greed and sent them off to Europe in their ships ...

Since the improper actions of the Christians of Hoogly Port toward the Muslims was accurately reflected in the mirror of the mind of the Emperor before his accession to the throne, when the imperial banners cast their shadows over Bengal, and inasmuch as he was always inclined to propagate the true religion and eliminate infidelity, it was decided that when he gained control over this region he would eradicate the corruption of these abominators from the realm.[1]

John Cabral, who was present at the siege of Hoogly, provides an account of the same event from the European viewpoint:

HUGLI CONTINUED AT peace all the time of the great King Jahangir. For, as this Prince, by what he showed, was more attached to Christ than to Mohammad and was a Moor in name and dress only ... Sultan Khurram was in everything unlike his father, especially as regards the latter's leaning towards Christianity ... he declared himself the mortal enemy of the Christian name and the restorer of the law of Mohammad ... he sent a firman [*farman*] to the Viceroy of Bengal, commanding him without reply or delay, to march upon the Bandel of Hugli and put it to fire and the sword. He added that, in doing so, he would render a signal service to God, to Mohammad, and to him ...

Consequently, on a Friday, September 24, 1632 ... all the people [the Portuguese] embarked with the utmost secrecy ... Learning what was going on, and wishing to be able to boast that they had taken Hugli by storm, they [the imperialists] made a general attack on the Bandel by Saturday noon. They began by setting fire to a mine, but lost in it more men than we. Finally, however, they were masters of the Bandel ...[2]

1. *Padshahnama*, 1/1, pp.433–36; translated by Wheeler Thackston.
2. Manrique 1927, 1, pp.392ff.

21~22

The delivery of presents for Prince Dara-Shikoh's wedding

AGRA, THE RIVERFRONT, NOVEMBER–DECEMBER 1632.
FOLIOS 120B–121A ATTRIBUTED TO BISHANDAS, CIRCA 1635

Preparations for the wedding of Prince Dara-Shikoh with his cousin, Nadira Banu Begam, were suspended at the unexpected death of his mother, Mumtaz Mahal, on 17 June 1631. Late the following year, ceremonies were resumed when elaborate gifts (*sachiq*) were sent to the home of the bride's mother, Jahan Banu Begam (the widow of Prince Parvez, Dara's uncle). The *Padshahnama* text describes the ceremony on Sunday, 21 November 1632:[1]

... AFTER THE ELAPSE of eleven *gharis*, two lacs of rupees, one lac in cash and one in goods, half of which were precious gems and the other half fine textiles, were taken from the royal treasury as *sachiq* by Allami Afzal Khan the chief *Diwan* [finance minister], Sadiq Khan the *Mir Bakhshi* [chief fiscal administrator of the empire], Musavi Khan the head comptroller, Mir Jumla the chief steward, the chaste ladies of Her Majesty Mahd-i-Ulya [Mumtaz Mahal]'s establishment, her elder sister, their paternal aunts, and Satiunnisa Khanim, as befitted the status of grand rulers, to the domicile of the chaste Jahan Banu Begam, the daughter of Prince Murad and wife of Prince Parvez.[2]

1. 8 Jumada I 1042, corresponding to the day of Aniran, 3 Aban.
2. *Padshahnama*, I/I, p.453; translated by Wheeler Thackston.

23-24

The wedding procession of Prince Dara-Shikoh

AGRA, PALACE COURTYARD, 12 FEBRUARY 1633.
FOLIOS 122B PAINTED BY MURAR, CIRCA 1635;
FOLIO 123A PAINTED BY AN UNKNOWN ARTIST, CIRCA 1640

To initiate the actual wedding ceremony, on 12 February 1633,[1] the imperial princes, senior ministers, and major nobles escorted Dara-Shikoh from his residence in Agra to the palace.

... BY THE EMPEROR'S command, the Princes Muhammad Shah-Shuja', Muhammad Awrangzeb, and Muradbakhsh, along with Yaminuddawla and other renowned *amirs* and dignitaries of state, went to the Prince's exalted domicile and held a celebration at which the exalted officials presented suitable gifts in congratulations for this happy ceremony. Then they mounted the pearl of the crown of the sultanate [Dara-Shikoh] on a mountainous, heavenly-moving rose-colored horse and, according to their stations, some mounted and others on foot, set out in his entourage. In the Hall of Public and Private Audience they attained the great fortune of making *körünüsh* and kissing the ground before the Emperor of the seven climes, the lord of throne and crown. He who sits on the throne of good fortune presented [the Prince] with a special robe of honor, a jeweled *jamdhar*, a jeweled sword with a valuable jeweled strap, a rosary of pearls that had precious rubies in it, two horses from the royal stable, one with a jeweled saddle and the other with a golden saddle with enamel work, an elephant from the royal elephant stables with silver trappings, a female elephant, the total value of which was four lacs of rupees.[2]

1. 2 Sha'ban 1042.
2. *Padshahnama*, 1/1, pp.457ff.; translated by Wheeler Thackston.

21

25~26

Shah-Jahan honoring Prince Dara-Shikoh at his wedding

AGRA, DIWAN-I KHASS, 12 FEBRUARY 1633.
FOLIOS 124B–125A PAINTED BY BULAQI, SON OF HOSHANG, CIRCA 1635

After the procession reached the palace, Dara-Shikoh greeted his father outside the *Diwan-i Khass*, the Hall of Private Audience:

IN ACCORDANCE WITH the custom of Hindustan, for good luck His Majesty fastened around the Prince's head the *sehra* [veil], which consists of strands of lustrous pearls with brilliant rubies and emeralds, and which His Majesty Jannat-Makani [Jahangir] had fastened with his own blessed hand around His Majesty's [Shah-Jahan's] head on the eve of his marriage … When the day had passed in celebration and night had fallen, the servants of the court at imperial command lit lamps in the garden below the private hall and on the ground beneath the *jharoka-i-darshan* [balcony in the wall of the royal palace for the use of the emperor] and brought boats on the Jumna and arrayed them with so many candles, torches, and lamps that the stars of the firmament were jealous. Fireworks from the imperial establishment were lit and lighted up the earth like the heavens.[1]

By an extraordinary coincidence, the English trader and traveler Peter Mundy observed these ceremonies and also described them in his journal:

IN FEBRUARY [1633] the eldest and second sonns to the kinge were both married within 8 dayes one of the other, first Daroo Shuckoore [Dara-Shikoh], the eldest to the Sultan Parvaez [Parvez] his daughter, hee about 17 and shee about 14 years of age … each Couple had two nights of fireworkes, longe beforehand fittinge and prepareinge, the first night when hee goes to see her, and the next night when shee is brought home to him, being married in the Interim. In briefe the manner was thus.

On the Strand by the River side, under the castle wall and the Kinges windowe, there was a place Rayled in, about half a mile in Compasse att least. In it were placed the fireworkes, *vizt.*, first a ranck [row] of great Eliphants, whose bellies were full of squibbs, Crackers, etts. Then a ranck of Gyants with wheeles in their hands, then a ranck of Monsters … All theis being fired (although not att one tyme) innumerable were the Rocketts, reports, squibbs, and Crackers that flewe about and alofte in the Ayre, makeinge the night like day. The noyse was as terrible. Also I think there were noe lesse than a million of lights burninge in the meane tyme … Lanthornes, Lampes, etts. Fastned and placed in rancks one above another on the Castle wall … Mee thought it a brave and pleasant shew … Here was Cost and Labour enough, but it wanted it may bee the Arte wee have in Europe of those kinde of workes.[2]

1. *Padshahnama*, I/I, pp.458ff.; translated by Wheeler Thackston.
2. Mundy 1907–24, II, pp.201–202.

27-28

The wedding procession of Prince Shah-Shuja'

AGRA, PALACE COURTYARD, 4 MARCH 1633.
FOLIOS 126B–127A ATTRIBUTED TO BHOLA, CIRCA 1635

The marriage of Shah-Shuja' was held about three weeks after the ceremonies for his elder brother Dara-Shikoh, on 5 March 1633:[1]

... AT THE END OF the day His Imperial Majesty illuminated the *jharoka* of the Hall of Public and Private Audience with the light of his presence. When the starry carpet of night spread over the firmament and the world was lit by many camphor candles and torches, at His Majesty's command the Princes Muhammad Awrangzeb and Muradbakhsh and Yamin-uddawla Asaf Khan, accompanied by the other amirs, went to the exalted Prince's quarters on the banks of the Jumna, which had been His Majesty's domicile when he was a prince, and in the same manner as described before for the first banquet, escorted the Prince along the banks of the river to the palace to pay homage. His Majesty presented a sumptuous robe of honor, a jeweled sword, a jeweled *jamdhar* with a *phul-kattara*, two horses from the royal stable, one with a jeweled saddle and the other with a gold enameled saddle, a personal elephant with silver trappings, and a female elephant, worth a total of two lacs of rupees. Then, with his own hand, he fastened around the prince's head the *sehra* of pearls with a ruby and an emerald that he had bound around the eldest prince's head, and the prince made obeisances in return for the imperial bounty. The quality and quantity of fireworks are beyond description. When it came time for the marriage, Qazi Muhammad Aslam pronounced the address of marriage in the Shah Tower in His Majesty's presence. A *kábín* [dowry] of four lacs of rupees was set.[2]

1. 23 Sha'ban 1042.
2. *Padshahnama*, I/I, pp.461–64; translated by Wheeler Thackston.

29

Prince Awrangzeb facing a maddened elephant named Sudhakar

AGRA, THE RIVERBANK, 7 JUNE 1633.
FOLIO 134A, PAINTED CIRCA 1635

The Englishman Peter Mundy wrote during his stay in Agra in 1632:

IN INDIA ARE USED many other fightings of beasts, as of Eliphants, wild Buffaloes etts. The fighting of Eliphants is seldome seene but where the King is, and there often used, sometimes twice a weeke, *vizt.*, Tuesdaies and Sattardaies in the afternone at Agra. The manner thus, partly as my selfe sawe, and partly by report. First the Elephants appoynted for the day, which are usually one Couple, other tymes there may be two and some tymes three Couple. The King cometh to the Jarooca [*jharoka*] or windowe, that looketh into the River, upon whose strand, right before the said Windowe, being the place appoynted, they are brought; with each a guide sitting on his Neck. Att the word given they are lett goe, and soe runninge one against the other with their Truncks aloft they meete head to head.[1]

In June of 1633, however, an exciting episode provided variety to this rather routine form of entertainment:

ON MOST DAYS the Emperor enjoyed watching elephant fights, except on auspicious Mondays, the day of accession, on which felicitous day the imperial mind is focused upon enjoyment and no harm should come to any animal, without which elephant fights cannot be held. Occasionally some spectators are killed beneath the feet of these demonic, mountainous creatures, and sometimes in their rage the elephants kill each other. By chance, on Tuesday … [7 June 1633],[2] two enormous elephants, one with tusks named Sudhakar, and the other without tusks named Surat Sundar, renowned elephants of the imperial establishment, were made to fight each other by imperial order beneath the *jharoka-i darshan* of His Majesty's princehood quarters. These two creatures of demonic aspect could rip out an opponent's foundation with one movement like a blowing wind or churning sea or blow out an enemy's lamp of life with one puff, grew heated in the field of combat and set the pillars of the earth to shaking with their granite-breaking collisions. During their fighting they went off a bit out of the Emperor's sight and clashed beneath the *jharoka-i darshan* of the palace outside the fortress. The Emperor, desirous of watching, mounted and set off in that direction with the princes, who rode several paces forward, Prince Muhammad Dara-Shikoh to the right of Sudhakar and Prince Muhammad Shah-Shuja' and Prince Muhammad Awrangzeb to the left, where they began watching the amazing battle. When the two behemoths separated, they took several steps backward, and there was little distance remaining. Sudhakar, seeing his opponent at a distance, made mighty charges and violent motions and ran in the direction of Prince Awrangzeb. This heroic stalker of elephants held firmly the reins of his fleet-footed steed and valiantly stood his ground. 'In manliness he budged not an iota; he moved not an inch from before such an onslaught.' When the elephant drew near, with heavenly assistance and through the problem-solving attention of the Emperor he raised his valiant arm and wounded the demonic beast on the forehead with a *birchha* [lance]. The angels and celestials recited Koranic charms on the Prince's arm; those fast asleep in the slumber of negligence woke at the cries of bravo; spectators were left in amazement. Inasmuch as in his youth the Emperor His Majesty Sahib-Qiran II had given a raging lion a taste of his bloodthirsty sword, as will be reported shortly, why should one think it strange that such feats should be performed by his offspring? 'A lion is he who is born of a lion.' Now wounded, the enraged elephant came even closer and headed straight for the Prince, and even though the face of the earth had been turned into a fiery globe with fireworks and rockets, it was to no avail. The elephant gored the Prince's horse and sent it rolling. Then the lion of the forest of courage leapt from the saddle onto the ground nimbly and stood erect, his hand on the hilt of his sword. Seeing this, Prince Muhammad Shah-Shuja' Bahadur, although access had been blocked by the crowd of people, the fireworks, and the smoke, charged his steed to get closer and wield his spear. During his charge suddenly a *charkhi* [firework mounted on a stick] hit his horse on the forehead, the horse reared, and the Prince fell. Raja Jai Singh ..., a hereditary servant of the court and successor to Raja Man Singh, also attacked the elephant, but his horse shied and would not come face to face with the elephant. Coming in from the right side, he threw a spear. At this point, when everyone was excited, the Emperor himself headed in that direction, ordering that the mace-bearers and other servants of the stirrup advance. When Surat Sundar found his opponent otherwise occupied, taking advantage of the opening, he charged. Unable to retreat, Sudhakar attacked, and Surat Sundar retreated. Now both elephants were running wild, and the princes, enveloped by divine protection and seeing signs of the Emperor's love, gained a new lease on life. The Emperor first drew Prince Muhammad Awrangzeb into the bosom of affection and kissed him with emotion, awarding him all sorts of favors and the title Bahadur [champion]. Then he showed Prince Muhammad Shah-Shuja' favor and praised his action.[3]

The Mughal emperors' exclusive right to stage elephant fights was asserted, among other royal prerogatives, by Jahangir:

... I ORDERED THAT the Bakhshis should circulate orders, to be obeyed amongst the Amirs of the borders, that hereafter they should not interfere in such things, which are the private affair of kings. The first thing is this, that they should not sit in the *jharokha* (private window), and should not trouble their officers and captains of the auxiliaries with keeping guard or saluting them, and should not have elephant fights, and should not inflict the punishment of blinding, and should not cut off ears and noses, and should not force Islam on anyone, and should not confer titles on their servants, and should not order the royal servants to do *kúrnish* or prostration, and should not force singers to remain on duty in the manner customary in (royal) darbars, and should not beat drums when they go out, and when they give a horse or elephant to anyone, whether to the king's attendants or to their own servants, they should not place reins or elephant's goads on their backs and make them perform obeisance. In going in procession they should not take with them on foot in their retinue the royal attendants. If they write anything to them they should not put a seal on it.[4]

1. Mundy 1907–24, II, p.127.
2. 29 Dhu'l-Qa'da 1042.
3. *Padshahnama* I/1, pp.489–92; translated by Wheeler Thackston.
4. *Tuzuk-i-Jahangiri*, I, p.205. I am grateful to Ebba Koch for this reference.

30

Prince Khurram attacking a lion

BARI, LATE 1610.
FOLIO 135B PAINTED BY BALCHAND, CIRCA 1640

In the *Padshahnama* narrative for the preceding illustration (**29**), the bravery of Awrangzeb evokes a parallel episode from Shah-Jahan's life before his accession, when 'in his youth the Emperor ... had given a raging lion a taste of his bloodthirsty sword.' That earlier event is further described in the *Padshahnama*:

ONE DAY HIS MAJESTY Jannat-Makani [Jahangir] was hunting with cheetahs in the *pargana* [administrative district] of Bari. All of a sudden, Anup 'the Unprincipled' – who had been promoted through great service at the end of His Majesty Arsh-Ashyani [Akbar]'s reign to the rank of chief of the servants who are called *khawass* [the elite], and who had also continued a while in that same service at the beginning of His Majesty Jannat-Makani's reign – who was leading the *bara,*[1] which is the group that accompanies the monarch when he is mounted to hunt, at some distance behind, received word of a huge, powerfully-built lion in the vicinity and charged off after it. Together with the men of the *bara,* he surrounded the lion and sent word to His Majesty Jannat-Makani. So eager was His Majesty to hunt the lion that, although the day was near its end and there were no elephants, which are necessary for taking such a ferocious animal, he set out on his horse in that direction. After spotting the lion, he dismounted and shot several times. Since the wound was not fatal, [the lion] went off and sat on a patch of sloping ground. Inasmuch as the sun had set and enough time had elapsed that prayer could have been performed,[2] and since the Emperor was determined to hunt the lion, His Majesty, accompanied only by the World-Lord [Shah-Jahan],[3] Raja Ramdas Kachhwaha, Anup, I'timad Rai, Hayat Khan the superintendant of imperial drinking-water, and Kamal Qaravul, took several steps forward and fired another musket shot. By chance, this time too the wound was insufficient to prevent the animal from attacking, and the lion, roaring from anger and rage, ran in His Majesty's direction. Another shot was fired. Although it hit the mark, it too was not fatal, and the angered, rampaging animal reached Anup, who was holding the musket stand, and knocked him to the ground. At this point the lion's head was to His Majesty's right. Anup put one hand on the lion's mouth and put the other around the lion's shoulder. The Emperor of the Age [Shah-Jahan], who had been standing forward and to the left of His Majesty, unsheathed his lightning-raining sword from its scabbard and was about to deliver a blow to the lion's shoulder when he saw Anup's arm. So worried was he lest an injury be done to Anup that he decided to strike the lion's waist and break its back. So nimbly and swiftly did he sheathe his blade that only Hayat Khan, who was standing to His Majesty Jannat-Makani's right with I'timad Rai, was aware of it. Ramdas, who was to the Emperor's left, also gave a blow with his sword. Hayat Khan struck it several times with a stick. The lion let Anup go and ran off. In the meantime men came running from all directions and finished off the lion. When Hayat Khan reported the world-lord [Shah-Jahan]'s bravery, resolve, and nimbleness to the Emperor, His Majesty drew his blood-stained sword from its scabbard with his own blessed hand and inspected it, praising and extoling the hand and arm of the regal lion-hunter who had been nurtured by divine power. Anup was awarded the title of Ani Rai Singhdalan, which means 'lion-crusher,' and an increase in station.[4]

The episode was also discussed at length, as a first-person account, in Jahangir's memoirs. The Emperor gives the date as 10 December 1610:

ANUP RAI, ONE OF my intimate servants, had led a troop accompanying me on the hunt a bit forward of me when he came to a tree on which were sitting a couple of kites. When he spied the kites he shot them with a bow and a few arrows and proceeded. By chance, he saw a half-devoured cow in the vicinity of the tree, and immediately thereafter a large, powerful lion emerged from some near-by bushes and started forward. Although there were no more than two *gharis* of daylight remaining, since he knew my avidity for hunting lions, he and several of those with him surrounded the lion and sent a messenger to tell me. As soon as I got the news I set out at a gallop. Baba Khurram Shah-Jahan, Ramdas, I'timad Rai, Hayat Khan, and one or two others were with me. As soon as I arrived I saw the lion sitting in the shade of a tree. I wanted to shoot while on horseback, but I realized that the horse was exhausted, so I got off, set up the gun, and fired. Since I was on a rise and the lion was in a depression, I didn't know whether it had been hit or not. I took another shot. I thought this one had hit the mark, but the lion leapt up and charged. A falconer with a falcon on his arm happened to be standing right in the lion's path. It mauled him and returned to its place. I had just placed another musket on the tripod and taken aim. Anup Rai was standing holding the tripod, and he had a sword in his belt and a wooden staff in his hand. Baba Khurram was standing a short distance to the left, and Ramdas and the others were behind him. Kamal Qaravul had filled the musket and handed it to me. Just as I was about to fire, the lion set out roaring in our direction and attacked. I fired the musket, and the ball passed right by its mouth and teeth. The sound of the musket startled it. A group of servants who had charged could not withstand its attack. They all collided against one another, and in the rush I was knocked back one or two paces. I know for certain that two or three of them stepped on my chest getting over me. I got up with the help of I'timad Rai and Kamal Beg Qaravul. Now the lion was headed for the men who were to the left. Anup Rai let go of the tripod and went for the lion, which turned around on him as nimbly as it had attacked, but he faced the lion and, grabbing with both hands the staff he was holding, struck it hard on the head. The lion opened its mouth and got both of Anup Rai's hands in its mouth. Then it bit so hard that its teeth went straight through both of his hands, but the staff he was holding and the several rings he had on his fingers helped keep him from losing his hands. Knocked down by the lion, Anup Rai fell on his back between the lion's paws so that his head and face were directly opposite the lion's chest. In the meantime Baba Khurram and Ramdas had come to help Anup Rai. The prince struck the lion on its waist with his sword, and Ramdas got in two sword blows. The one that fell on the lion's shoulder really cut in. Hayat Khan struck it on the head with the club he was holding. Anup Rai managed to pry his hands out of the lion's mouth and hit it several times on the head. Then, rolling on his side, he got himself erect. As he was pulling his hands out of the lion's mouth, since the lion's teeth had gone all the way through, his hands were a bit torn, and the lion's claws had scraped his shoulder. While he was getting up the lion also reared and scratched his chest with its claw. His wounds hurt him for a few days. On the downward slope it looked as though two wrestlers had grappled with each other and rolled down. In the place I was standing the ground was perfectly level. Anup Rai said, 'God gave me the sense to take the lion in the other direction on purpose. Otherwise, I don't remember anything.' Now the lion left Anup Rai and headed off. In his dazed state Anup Rai chased after it, brandishing his sword, and struck it on the head. When the lion turned its head around, he struck it again in the face, cutting both its eyes. The flesh of the eyelids, which had been severed by the sword, hung down over its eyes. Just then, as it was time for lamps to be lit, a lamp-holder named Salih stumbled onto the scene. By chance, he bumped into the lion. With one swipe of its paw the lion sent him flying. He fell to the earth dead. Other men arrived and finished off the lion. For having performed such a service, and because his valor was witnessed, after he recovered from his wounds and came to pay homage, I gave Anup Rai the title of Ani Rai Singhdalan. In the Hindi language the commander of a troop is called *ani rai*, and *singhdalan* means lion-crusher. I also gave him a royal sword and increased his rank a little.[5]

1. The word is probably the Hindi *bara* (twelve). WT
2. That is, enough time had elapsed after sunset for the evening prayer to be said. WT
3. Here and elsewhere various epithets, all of which mean 'world-lord,' are used for Shah-Jahan, which itself means 'king of the world'. See the commentary on **8–9** above. WT
4. *Padshahnama*, I/I, pp.493–95; translated by Wheeler Thackston.
5. *Jahangirnama*, fols 74b–76a; translated by Wheeler Thackston.

31

The siege of Daulatabad

DAULATABAD (DECCAN), APRIL–JUNE 1633.
FOLIO 144A PAINTED BY MURAR, CIRCA 1635

Military activities during the first decade of Shah-Jahan's rule concentrated on the Deccan, and no fortress in that area was more daunting than Daulatabad, which embodied the power of the Nizamshahi dynasty of Ahmadnagar.

THE FORTRESS OF Daulatabad is the basis of the Nizamulmulk's fortune and the gubernatorial seat of that realm. In impregnability and fortification it is renowned throughout the world and is unique among all the bastions of Hindustan, for this mighty fortification is in reality nine fortresses situated [on] a splendid granite mountain of the utmost magnitude and height in the midst of a flat plain. Four of the fortresses are on top of the mountain, one above the other, and all around the mountain is a wide, deep, water-filled moat carved from the granite. The other five mighty fortresses and two deep moats adjoin the mountain.[1]

After an extensive campaign, it fell to the imperialists in June 1633, as described in the *Shah-Jahannama*:

ON... [19 APRIL 1633],[2] the royalists charged a mine ... with 70 maunds [5600 lbs] of gunpowder. Their strategy was for Raja Pahar Singh Bundela – with his brothers, Nazar Bahadur Khwishgi, and some others of the royalists, and 600 horsemen of Nasiri Khan's – to proceed under arms and assemble at this trench before dawn. As soon as it became light and the commander-in-chief arrived, the mine should be sprung, and simultaneously with the explosion this force should storm the breach supported by Khan Zaman in the rear.

However, it chanced that the officer in charge of the mine made a mistake and fired the mine before daybreak and before the arrival of the commander-in-chief and the storming party. The explosion made a good-sized breach by demolishing 28 yards of the curtain wall and 12 yards of a bastion. Unfortunately, the force assembled for the assault had not yet assembled; and since Khan Zaman, Nasiri Khan, Luhrasp, and Diler Himmat were all outside the camp ready to repulse any sally that the garrison might undertake just as the mine was about to be sprung, there was consequently no one to storm the place.

Khan Khanan [Mahabat Khan] was on the point of dismounting and leading on the storming in person, when Nasiri Khan prevented him. The latter averred that such a duty was unbecoming Khan Khanan's high rank, and that he would instead place himself at the head of the party and attempt to force an entrance .... Fortune eventually smiled on the imperial troops ...

On Monday ... [27 June 1633], Fath Khan, in company with Husain, the youthful Nizam al-Mulk and his own relatives and dependents, evacuated the fortifications and delivered over to the servants of the crown the whole nine forts of Daulatabad – five of which are on level ground and four on the summits of hills – with about 1,000 cannon of different calibre and the whole of the military stores. Khan Khanan entered the fort with all the royalists, and after making a tour around the defenses, he caused the *khutba* [part of the Friday congregational prayer in which the name of the ruler is announced] to be read in the august name and titles of His Gracious Majesty, the asylum of the Caliphate, the Shadow of God.[3]

1. *Padshahnama*, 1/1, p.484ff.; translated by Wheeler Thackston.
2. 9 Shawwal 1042.
3. *Shah Jahan Nama*, pp.105ff. For a corroborating description of the fortress by a European observer, see Tavernier 1925, p.116.

1

32

The departure of Prince Shah-Shuja' for Kabul

AGRA, DIWAN-I 'AMM, 16 MARCH 1638.
FOLIO 147B PAINTED BY MURAR, CIRCA 1640

In 1638, the Emperor sent troops north to regions in present-day Afghanistan, where control of the fortress of Qandahar was disputed by the Mughals and the Persian shah. Placed in conjunction with text describing Shah-Shuja's departure in August 1633 for a campaign in the Deccan, this scene seems instead to show the Prince's leave-taking at Agra on 16 March 1638,[1] before his expedition to Kabul in support of the Qandahar expedition.

AT AN AUSPICIOUS HOUR …, Prince Muhammad Shah-Shuja' Bahadur was awarded a regal robe of honor with a gold-embroidered *charqab* [garment given as a gift by the emperor], a jeweled aigrette, a jeweled *jamdhar* with *phul-katara*, a jeweled sword, an increase of 1,000 horses to his rank of 12,000/8,000,[2] a hundred horses, of which two were from the royal stable, one with a jeweled saddle and the other with a gold enameled saddle, an elephant from the royal elephant stables with silver trappings and one female elephant, and ten lacs of rupees in cash as a bonus, and he was given leave to depart with 20,000 mounted warriors. As he was being given leave, the Emperor said, 'If Shah Safi comes to Qandahar himself, my exalted son will go thither with the army and engage him in battle. If he sends his army, let him assign a contingent of our blood-thirsty warriors under Khan Dawran Bahadur *Nusrat-Jang* to chastise those misguided offenders.' Each of the amirs accompanying the Prince was ordered to trample the ill-starred Qizilbash and received regal favors. Raja Jai Singh was awarded a regal robe of honor, a horse from the royal stable with a gilded saddle, and a royal elephant. Najabat Khan was awarded a robe of honor and a horse with a gilded saddle. Raja Gaj Singh's son Amar Singh and Rao Rattan's son Madhu Singh were both awarded robes of honor, horses with silver saddles, and the privilege of drums. Mahabat Khan's son Luhrasp was awarded a robe of honor, a horse, and the title of khan … Diyanat Khan was made *diwan*, and Mu'tamad Khan's son Dostkam was made *bakhshi* [military official] and recorder for this expedition, and they were awarded robes of honor. Khan Dawran Bahadur *Nusrat-Jang*, who had been summoned by a court decree, attained the honor of paying homage several days after the felicitous Prince was given leave to depart. He was awarded a robe of honor, a jeweled *jamdhar* with *phul-katara*, a jeweled sword, two horses from the royal stable, one with a golden saddle and the other with a gilded saddle, and an elephant from the royal stable with silver trappings and a female elephant. He was made commander and put in charge of battles, and he was given leave to depart along with Sayyid Hizhabr Khan and Nazar Bahadur Khweshgi, who were awarded robes of honor and horses. They were ordered to join the Prince as soon as possible. An imperial order was issued for Bahadur Khan, Mubariz Khan, and several other servants to come from their estates and join the expedition. Wazir Khan, the governor of the Punjab, was ordered to undertake provisioning and assign his men to gather grain in the Punjab and send it to Kabul so that provisions would reach the army along the way. He was also ordered to proceed himself from Lahore to Kabul.[3]

1. The commencement of Dhu'l-Qa'da 1047.
2. These numbers refer to the ranking system for Mughal officers. The first denotes the number of horsemen for which the officer is paid a stipend and the second the actual number of cavalry that he is required to make available for service in the imperial armies whenever needed.
3. *Padshahnama* II, pp.35–37; translated by Wheeler Thackston.

33

Shah-Jahan hunting

PERHAPS NEAR RUPBAS, AFTER 1640.
FOLIO 165A, PAINTED CIRCA 1645

The elderly appearance of the Emperor indicates that this scene does not illustrate the adjacent text, which concerns a hunt at Palam (near Delhi) in February 1634. Nonetheless, since the exact event depicted remains unidentified, the narrative for that episode, which is typical of the many hunts described in the *Padshahnama*, is given here:

ON ... [26 FEBRUARY 1634][1] the procession left Nurgadh and reached the hunting grounds of Palam, where descent was made in the imperial buildings that had been built in that area. For four days in that delightful spot [the Emperor] indulged in hunting and bagged much game. In one day he himself shot forty black antelope with the royal gun called Khassban, and not a one needed a second shot. This of course occasioned astonishment in all. The following words were heard to cross the royal lips, 'One day while I was Prince I downed sixteen antelope with my own hand in Jalesar, the royal hunting preserve near the capital Akbarabad [Agra]. When the news reached the hearing of His Majesty Jannat-Makani [Jahangir], he said, "We too have shot eighteen antelope in one day." ' Taliba Kalim sent the following quatrain to the royal presence and received a reward: 'When Shah-Jahan Padshah the lion hunter chanced to hunt in Palam, / In one day, with the gun Khassban, he shot forty antelope without a second shot.'[2]

1. 27 Sha'ban 1043.
2. *Padshahnama*, I/2, pp.6–7; translated by Wheeler Thackston.

34

A royal procession

UNIDENTIFIED CITY, CIRCA 1655.
FOLIO 166B ATTRIBUTED TO THE 'KASHMIRI PAINTER',
CIRCA 1655

The adjoining text describes the procession of the imperial party, sometimes termed the 'Sublime Cortege,' as it traveled between Lahore and Agra:

ON THE 7TH [26 JANUARY 1635],[1] at the hour chosen by the astrologers, the sublime banners set out .... On the 28th, shadows of glory and magnificence were spread over the plain of Sirhind, and the Hafiz Rakhna garden, which had been graced by the advent of the imperial procession on the way from the capital Akbarabad [Agra] to the Punjab, was honored by the imperial advent .... Several amazing and far-sighted modifications suggested themselves to the prescient imperial mind, and Mir Ali-Akbar, the *crori* [tax-gatherer] of Sirhind, was ordered to execute them according to this new plan. On the 11th [of Ramadan], the vicinity of Kairana, the birthplace of Muqarrab Khan the surgeon, was illuminated by the imperial descent, and the Emperor inspected in detail the building and garden he had made on the banks of the Jumna River. Muqarrab Khan had paraded before the imperial view gems, jeweled utensils, et cetera, as *payandaz* and *nisar* [offerings and gifts]. The Emperor honored his old servant by accepting some of them .... On the 15th, the victory-inscribed banners arrived in the outskirts of the city of Delhi, the imperial tents were unfurled in Nurgadh. The next morning the Emperor made a visit to the blessed tomb of His Majesty Jannat-Ashyani [Humayun], and for the repose of his soul he bestowed 5,000 rupees upon the servants of that paradisiacal threshold and the unfortunate and deserving who were occupied with divine worship in the tomb.[2]

Peter Mundy left a very detailed account and a drawing of the processional entry of Shah-Jahan into Agra on 1 June 1632, following his stay at Burhanpur:

MYSELFE, WITH SUNDERDAS, went towards Darree ca bagh [the garden Dehra-ka-bagh] to see the Kinge comeinge thither. By the way, before wee could gett forth of the Cittie, wee were stopped and hindred by a great number of Eliphants, Cammells, Carts and Coaches laden with lumberment [baggage], which came from the laskerre, or Campe, also many Coaches, Palanqueenes and doolees with woemen ...

Att length wee were informed whereabout the king was himselfe; for all the face of the earth, soe farr as wee could see, was coverred with people, troopes of horses, Eliphants, etts., with innumerable flaggs small and great, which made a most gallant shew; for it is the Custome of every perticular great man to goe with a great many of theis flaggs carried before him ... Then thousands of horsemen going breadthwise; then came about 19 or 20 great Eliphants of State with coverings and furniture; most of them of Cloth of gold, the rest rich stuffe, velvetts, &c.; some of them carryinge a flagg with the kings Armes, which is a Tygar couching with the Sunne riseinge over his backe ... Then came the kinge himselfe mounted on a darke gray horse, and with him Mohabutt Ckaun [Mahabat Khan] ... A litle distance behinde rode his eldest sonne Daroo Shuchir [Dara-Shikoh] all alone, All the rest of the Amrawes or Lords on foote, before and behinde, and on each side of him ... All theis moveinge in one, on soe many huge Eliphants seemed like a fleete of shipps with flagg and streamers ... Soe that all theis together made a most majesticall, warlike and delightsome sight ...[3]

1. 7 Sha'ban 1044.
2. *Padshahnama*, I/2, p.70; translated by Wheeler Thackston.
3. Mundy 1907–24, II, pp.188ff. For his drawing, see the discussion of **36** in the catalogue below.

35

The capture of Orchha by imperial forces

ORCHHA, OCTOBER 1635.
FOLIO 174A PAINTED BY TEZDAST, CIRCA 1637

The account of the rebellious activities of Raja Jujhar Singh Bundela of Orchha, in central India, is presented in the guise of a moral lesson.

AN ILL-FATED PERSON who is guided by adverse fortune will hasten off the right path onto the crooked road of abomination and make camp in the bramble patch of ignorance, his eyes closed to the favors of his metaphorical lord and true god, and shake with his idolatrous hand the chain of rebellion, which bequeaths death and destruction. Quickly his fortune turns dark, and the eye of his expectation is dazzled. 'When a man's fortune turns dark, everything he does goes against him.' A good example of this assertion is the disastrous story of the good-for-nothing Jujhar, for as long as that wayward one knew service and dedication to this heavenly court to be the highway of his salvation and had not entered the path of error, the star of his fortune was on the rise, but the instant he turned the reins of his conduct over to adverse fortune and placed his foot in the wrong path to the desert of misery, the gates of eternal perdition were opened in the face of his fortune as his life, property, honor, and family were plundered by untoward events and trampled under the feet of misfortune.[1]

Following a period of trouble and rehabilitation, Raja Jujhar Singh had been given an assignment in the Deccan, which he deserted, leaving his son Bikramajit (also called *Jagraj*) in charge. He then seized imperial territories for himself, and refused to obey the infuriated Emperor's demand that they be returned. Not unexpectedly, Jujhar and his son were pursued through their home territories of Orchha and Dhamuni by imperial troops.

BIKRAMAJIT WAS ALMOST caught in the jaws of death during the fray, but since he was fated to survive for a while longer, he escaped from the field wounded and got himself and a few of his comrades into the thick forests and impassable mountains, through the secret ways of which only one thoroughly acquainted with the terrain could pass, and joined his father in the *pargana* of Dhamuni ... When this news reached the court, twenty thousand cavalrymen were appointed under the leadership of three outstanding warriors: Abdullah Khan Bahadur *Firoz-Jang*, who had recently received permission to return to Patna, was ordered to turn around and exterminate the wretch; second was Sayyid Khan Jahan, and third was Khan Dawran, who had stopped in Malwa after pursuing Bikramajit and was awaiting orders. These three were ordered to uproot the corruption of the wretched miscreants and set an example to all other short-sighted individuals who did not appreciate their own well-being.[2]

In this illustration, the three commanders, Khan Dawran, Abdullah Khan, and Sayyid Khan Jahan Barha are shown with their troops during the campaign in the Deccan. The slaying of Jujhar Singh and his son is depicted in the following illustration (**36**).

1. *Padshahnama*, 1/2, pp.94–96; translated by Wheeler Thackston.
2. *Padshahnama*, 1/2, pp.96–100; translated by Wheeler Thackston.

36

Khan Dawran receiving the heads of Jujhar Singh and his son Bikramajit

DECCAN, JANUARY 1636.
FOLIO 176B ATTRIBUTED TO PAYAG, CIRCA 1638

In January 1636, following the seizure of the fortresses at Orchha and Dhamuni, the rebellious Raja Jujhar Singh and his son were captured and beheaded.

KHAN DAWRAN ARRIVED from behind and put to the sword a group who were fighting as viciously as they could. Jujhar's son Durga Bhan and Bikramajit's son Durjan Sal were captured ... Udai Bhan and his younger brother Syam Diva, who had fled to Golconda, were also captured a little later, details of which will be given. At Khan Dawran's order the warriors raised Rani Parbati and the other wounded women from the dust and took them, along with the elephants, jeweled utensils, coins, and other booty to *Firoz-Jang*. The leaders of the imperial forces camped beside a lake in the vicinity to rest the horses, count the booty, and search for the other belongings of and find out what had happened to Jujhar and Bikramajit, since it was the primary object of this campaign to kill them or take them prisoner. Just then, through divine assistance, the sound of victory drums arose, and the good news that Jujhar and Bikramajit had been killed was received. In mortal fear of attack by the bloodthirsty imperial warriors, the two had taken refuge and hidden themselves in a jungle nearby, and a tribe of Gonds who lived in the vicinity put the ingrates to death. Hearing this news, Khan Dawran praised God and giving thanks for the Emperor's fortune, rode to where the traitors' bodies were, and ordered their heads, which were beehives of sedition and corruption, severed from their bodies. [Abdullah Khan] *Firoz-Jang* took the heads, along with their rings, horses, and arms, and sent them to court by Bahadur Beg, the supervisor of the imperial muskets.[1]

1. *Padshahnama*, 1/2, p.116; translated by Wheeler Thackston.

37

Jahangir receives Prince Khurram

AJMER, DIWAN-I 'AMM, APRIL 1616.
FOLIO 192B ATTRIBUTED TO ABID, CIRCA 1635

This illustration is placed in relation to text that describes, in flashback, Prince Khurram's departure for the Deccan campaign in 1616. Because it is the only scene in which Jahangir wears a jeweled armlet, however, it may instead illustrate the following episode described in the *Jahangirnama*:

AS A SIGN OF divine grace and favor, something happened around this time [*circa* 23 April 1616] that is not a little strange. After defeating the Rana, my son Khurram presented a fine, brilliant ruby to me in Ajmer. It was worth 60,000 rupees. It occurred to me that I should wear it on my arm. However, two rare and lustrous matched pearls worthy to sit with this ruby were necessary. Then, when Muqarrab Khan obtained one superb pearl worth 8,000 rupees and offered it for *Nawroz*, it occurred to me that if a mate to this pearl could be found it would make a perfect armband. Khurram, who had been in my exalted father's service from childhood and had spent day and night with him, said, 'I have seen a pearl of this size and shape on an old headband.' A big old turban band was brought, and upon examination it was found to contain a pearl of exactly the right size and shape. It didn't differ an iota in size. The jewelers were astonished that it matched in size, shape, and luster. You'd say they had been cast in the same mold. I had the pearls set on either side of the ruby and put it on my arm, placing my head in gratitude upon the ground to render thanks to the Lord.[1]

In this scene, several of the courtiers are shown with small gold medallions bearing a portrait of the Emperor worn in their turbans or hanging around their necks. The English ambassador Sir Thomas Roe discussed these in an account dated just four months after the episode that may be depicted here:

AUGUST 17 – I went to visitt the King, who, as soone as I came in, called to his woemen and reached out a picture of him selfe sett in gould hanging at a wire gould Chaine, with one pendant foule pearle, which hee deliuered to Asaph chan [Asaf Khan], warning him not to demand any reuerence of mee other then such as I would willingly giue, it being the Custome, when soever hee bestowes any thing, the receiuer kneeles downe and putts his head to the ground (which hath been exacted of the Embassadores of Persia). So Asaph Chan came to mee, and I offered to take it in my hand; but hee made signe to putt of my hatt, and then putt it about my neck, leading me right before the king ... This giuft ... was five tymes as good as any hee giues in that kynd, and held for an especiall fauour, for that all the great men that weare the kings Image (which none may doe but to whom it is giuen) receiue noe other than a meddall of gould as bigg as sixpence, with a little chayne of 4 inches to fasten it on their heads, which at their owne Chardg some sett with stones and garnishe with pendant Pearles.[2]

1. *Jahangirnama*, fols 132b–133a; translated by Wheeler Thackston.
2. Roe 1967, pp.244–45.

37

38

Jahangir receives Prince Khurram on his return from the Mewar campaign

AJMER, DIWAN-I 'AMM, 20 FEBRUARY 1615.
FOLIO 194B PAINTED BY MURAR, CIRCA 1640

While placed within the manuscript to correspond with text that recalls Shah-Jahan's return from his triumphant campaign in the Deccan when he was still a prince, this scene contains no specific visual reference identified with that event. The illustration instead shows the reception for Prince Khurram and Karan Singh of Mewar following the Prince's defeat of the Mewar Rana. Another version of this episode is seen in an earlier plate (5), where the text passage appropriate to both illustrations is quoted.

39

Jahangir presents Prince Khurram with a turban ornament

MANDU, DIWAN-I 'AMM, LATE 1617.
FOLIO 195A PAINTED BY PAYAG, CIRCA 1640

This scene may show the occasion on which Jahangir gave to his son, who has just been titled Shah-Jahan, a dynastically important jewel:

ON THURSDAY THE 9TH [30 November 1617], a drinking party was convened. On this day I gave my son Shah-Jahan a flawless ruby weighing nine *tánk*s and five *surkh*s [184 carats] worth 125,000 rupees set with two pearls. This was the ruby Her Majesty Maryam-Makani, His Majesty Arsh-Ashyani's [Akbar's] mother, gave me as a present at my birth. For years it was in his [Akbar's] turban band, and thereafter I too kept it in my turban band for good luck. Aside from its monetary value, since it has been a good luck charm for this eternal dynasty, it was given to my son.[1]

An alternative identification of this scene would be the ceremony on 12 October 1617 at which Khurram was awarded the title Shah-Jahan, following his victories in the Deccan.[2]

1. *Jahangirnama*, fol.162a; translated by Wheeler Thackston.
2. See *Tuzuki-i-Jahangiri*, I, pp.393–95.

40

The surrender of the fort at Udgir to Khan Dawran

DECCAN, OCTOBER 1636.
FOLIO 204B PAINTED BY DHOLA, CIRCA 1650

Despite earlier battles and temporary victories, the Mughals did not finally defeat the Deccani kingdom of Ahmadnagar until 1636, when Khan Dawran was sent against the fortresses of Udgir and Ausa.

DURING THESE AUSPICIOUS days [October 1636], the stronghold of Udgir and the fortress of Ausa were conquered through the Emperor's great good fortune. This came about as follows. When Khan Dawran Bahadur, at imperial order, was assigned to conquer the above-mentioned fortresses from the Adil Khan, he first sent word to the castellans of both fortresses saying, 'Now that all the Benizam's [Nizamulmulk's] strongholds and provinces have been subdued by the Emperor's good fortune, and the Adil Khan has withdrawn his improper claim to these two, it would be best to turn over the fortresses to imperial agents; otherwise soon both fortresses will be taken by force, and your lives and property will be lost.' Those short-sighted men, disregarding the inevitable end of their actions, set about reinforcing their towers and ramparts. On Sunday ... [29 June 1636][1] Khan Dawran Bahadur set up camp on the outskirts of the town of Udgir and assigned battle stations after inspecting the perimeter of the fortress .... The brave warriors courageously and valiantly moved their positions forward, and when the mines had neared the fortress, the defenders lost their nerve and Sayyid Miftah Habashi, the warden of the castle, sent a terrified message to Khan Dawran Bahadur, saying, 'If you will enroll me among the court servants, I will surrender the fortress.' When Khan Dawran accepted his proposal, the short-sighted Sayyid Miftah put forward other, unacceptable demands, and Khan Dawran set fire to the mine that had been run to the foot of the *sher hajji*. Although the entire tower, which was nearly a hundred ells [275 feet] round, was blown up along with cannons, catapults, and other implements of war that were on top of it, since it did not do any damage to the foundation of the citadel, Khan Dawran ordered his warriors to break off the attack and sent a message to Miftah, saying, 'If you have the foresight and wisdom to turn the fortress over to the imperials, you will be given quarter. Otherwise you will soon taste the fire-raining swords of my victorious warriors.' Sayyid Miftah was so terrified that he could do nothing but acquiesce and come out to Khan Dawran Bahadur. On ...[8 October 1636],[2] which had been a little over three months from the beginning of the siege, he surrendered the fortress ...[3]

1. 25 Muharram 1046.
2. 8 Jumada I 1046.
3. *Padshahnama*, I/2, pp.217–19; translated by Wheeler Thackston.

41–42

Shah-Jahan visits the shrine of Khwaja Mu'inuddin Chishti at Ajmer

AJMER, NOVEMBER 1654.
FOLIOS 205B–206A ATTRIBUTED TO THE 'KASHMIRI PAINTER', CIRCA 1656

The text accompanying this illustration describes an imperial visit to Ajmer in 1636, but the elderly appearance of the Emperor suggests that the scene must refer to a later visit, in 1654:

SINCE THE EMPEROR was extraordinarily devoted to visiting the shrines of the great men of religion, particularly that of Khwaja Mu'inuddin, and it had been a long time since he had attained that felicity, he promoted Fazil Khan to the rank of 2500/2600[1] and made him governor and garrison commander of the capital, and on Friday ...[6 November 1654][2] he entered the palace at the dam on Ana Sagar Lake, a beautiful pleasure park. At the end of the day he betook himself to the saint's blessed shrine. In accordance with the custom of his noble ancestors, he dismounted at the shrine gate and performed the rites of visitation. Having acquired heavenly grace and eternal blessings, he made offerings, which are obligatory upon visiting the likes of this charitable institution. Again on the 27th, out of his great devotion, he visited the shrine, and after reciting the Fatiha, he returned to the palace. On ...[19 November 1654][3] he visited the shrine again, performed the afternoon prayer in the marble mosque he himself founded, and returned to the palace. On this day, by imperial order, 140 Shah-Jahani maunds of *nilgai* meat from the royal hunt, oil, and spices were put in a large kettle that had been endowed to the shrine, cooked, and given away to the general populace. On Wednesday the 14th ... reins were turned in the direction of the capital.[4]

When Shah-Jahan's grandfather, the Emperor Akbar, had prayed for the birth of his first son, he had vowed that he would walk from his capital at Agra to the shrine of Shaykh Mu'inuddin Chishti at Ajmer to give thanks for the saint's intervention. Jahangir's birth was then recorded in his father's official history, the *Akbarnama*:

WHEN SUCH A night-gleaming jewel of the casket of the Caliphate [Jahangir] arrived at the shore of hope, he [Akbar] recognised his obligation and set out on foot from Agra ... He spent several days there [at the shrine] in devotion and good works. He distributed gifts among the attendants of the shrine.[5]

Jahangir, in turn, lived in Ajmer for almost three years, while Prince Khurram, the future Shah-Jahan, was conducting his campaign against the Rana of Mewar, and in his memoirs he records that he made nine visits to the shrine of Shaykh Mu'inuddin during this period.[6] On one of these occasions, during an extensive illness early in 1614, Jahangir 'went to the mausoleum of the revered Khwaja, and in that blessed abode prayed to God Almighty for recovery.'[7] Once his health was restored, he vowed:

INASMUCH AS I WAS inwardly an ear-bored slave of the Khwaja [Mu'inuddin] and was indebted to him for my existence, I should openly make holes in my ears and be enrolled among his ear-marked slaves. On Thursday ... I made holes in my ears and drew into each a shining pearl ...[8]

This mark of distinction was followed by many courtiers and can be seen in certain paintings in the *Padshahnama* such as **5**, which depicts an event that took place just after this episode. It is clear, therefore, that a visible link with the Ajmer shrine had become a desirable imperial symbol.

Following Jahangir's death outside Lahore in 1627, Shah-Jahan had also stopped at Ajmer on his procession from the Deccan to Agra to establish himself as successor. Then he too visited the shrine of Shaykh Mu'inuddin Chishti. The *Shahjahannama* records:

ON HIS ARRIVAL AT Ajmer, His Majesty paid a visit to the radiant mausoleum of the venerable saint Shaikh Mu'in al-Din Chishti (may God illumine his soul), and was pleased to offer alms and charity and assign salaries for the pious recluses and the devotees of the august shrine. At the time of his earlier conquest of the Rana, His Majesty had made a vow that he would build a mosque near that illuminated tomb but its fulfillment was obstructed due to disorder in the country. He now fulfilled that vow and ordered a mosque of marble to be constructed to the west of the dome of the mausoleum of His Holiness.[9]

At almost the same moment, Shah-Jahan's only surviving brother, Shahriyar, as well as the sons of his uncle Prince Daniyal, and his two nephews (the sons of Prince Khusraw) were murdered by his supporters. Since these men were potential rivals, this was a more immediate means to assure Shah-Jahan's safe ascension to the position of emperor.

1. For the meaning of these numbers, see above, **32**, note 2.
2. 25 Dhu'l-Hijja 1064.
3. 9 Muharram 1065.
4. Since the *Padshahnama* text does not extend to so late a date, this passage is taken from *'Amal-i Salih*, III, p.193; translated by Wheeler Thackston.
5. *Akbarnama*, II, pp.510–11.
6. *Tuzuk-i-Jahangiri*, I, p.341.
7. *Tuzuk-i-Jahangiri*, I, p.267.
8. *Tuzuk-i-Jahangiri*, I, p.267.
9. *Shah Jahan Nama*, p.15.

43

Shah-Jahan honoring Prince Awrangzeb at Agra before his wedding

AGRA, DIWAN-I ʻAMM, 27 APRIL 1637.
FOLIO 214B PAINTED BY PAYAG, CIRCA 1640

In 1637, Prince Awrangzeb returned to Agra from the Deccan, where he was stationed, for his marriage to the daughter of Shahnawaz Khan.

ON [26 APRIL 1637][1] Prince Muradbakhsh, together with Yaminuddawla, Khan Dawran Bahadur *Nusrat-Jang*, Allami Afzal Khan, Sayyid Khan Jahan, Shayista Khan, and Jaʻfar Khan, went as ordered to bring the Prince [Awrangzeb] to perform *körünüsh*. The Emperor embraced that princely pearl of the crown in the bosom of affection and awarded him a regal robe of honor and other kingly tokens of favor. The Prince presented a thousand *ashrafi*s as a vow and the same amount as a gift.[2]

Shahnawaz Khan was the son of Mirza Rustam Safavi, and his daughter, Dilras Banu Begam, was Awrangzeb's chief wife. When she died at Awrangabad in 1657, Awrangzeb built for her a mausoleum modeled on the Taj Mahal. It remains a principal monument of the city.

1. 1 Dhu'l-Hijja.
2. *Padshahnama*, 1/2, pp.255ff.; translated by Wheeler Thackston.

44

The arrival of Prince Awrangzeb at the court at Lahore

LAHORE, DIWAN-I 'AMM, 9 JANUARY 1640.
FOLIO 217B BY MURAR, CIRCA 1645

While placed to correspond with text describing Prince Awrangzeb's appointment as Viceroy of the Deccan in 1637, the scene actually depicts his return to court – then located at Lahore – on 9 January 1640:[1]

BY IMPERIAL COMMAND, the exalted Prince Muradbakhsh, Alimardan Khan, Islam Khan, Ja'far Khan, and the great *bakhshi*s Salabat Khan and Asalat Khan went out to greet the 'cypress of the brook of the caliphate,' Prince Muhammad Awrangzeb Bahadur, and escorted him into court. The exalted prince attained the felicity of paying homage and presented a thousand *mohur*s in fulfillment of a vow and an elephant worth 50,000 rupees with gold trappings, and three female elephants as a *peshkash* offering [a present.] He was given an Arabian horse named Tamam-Ayar from the royal stables – one of the horses Hakim Masihu'z-Zaman had brought from Arabia – with a golden saddle, and nine Persian and Turkish horses.[2]

1. 15 Ramadan 1049.
2. *Padshahnama*, II, p.174; translated by Wheeler Thackston.

45

Shah-Jahan honoring Prince Awrangzeb at his wedding

AGRA, DIWAN-I KHASS, 19 MAY 1637.
FOLIO 218B ATTRIBUTED TO BHOLA, CIRCA 1640

ON THE EVE OF Tuesday, ...[May 18, 1637],[1] four *gharis* before the appearance of dawn, the hour chosen for the marriage by the astrologers, at His Majesty's order, the pearl of the casket of might, Prince Muradbakhsh, Yaminuddawla, and other notables went to Prince [Awrangzeb's] quarters, in which the Emperor had dwelt while he was a prince, on the banks of the Jumna, and which [Shah-Jahan] had given [Awrangzeb] after the accession to the throne. From there they escorted [Awrangzeb] along the banks of the river to the palace and brought him to perform *körünüsh*. The Emperor of the Age honored the Prince with a robe of honor with a goldspun *charqab*, two rosaries of precious pearls, a jeweled *jamdhar* with a *phul-katara*, a jeweled sword with a jeweled strap, two Arabian and Persian horses from the royal stables with a jeweled and gold-enameled saddle, an elephant from the royal stables with silver trappings, goldspun velvet cloth, and a female elephant. Then, with his own hand, the Emperor fastened on the prince's head a *sehra* of lustrous pearls among which were strung rubies and emeralds, and the prince expressed gratitude for these favors. It was ordered that Prince Muradbakhsh, Yaminuddawla, and the other high-ranking amirs should mount and accompany the Prince to the house of Shahnawaz Khan. As they were crossing the field, the varieties of fireworks that lit up the night are beyond description. Toward the end of the night the Emperor himself got into a royal ship and honored Shahnawaz Khan's house, and in his presence the two were wed ... Congratulations and rejoicing rose from the earth to the highest celestial sphere, and singing and merriment engulfed time and space.[2]

1. 23 Dhu'l-Hijja 1046.
2. *Padshahnama*, I/2, p.268; translated by Wheeler Thackston.

46

Shah-Jahan hunting lions at Burhanpur

BURHANPUR, JULY 1630.
FOLIO 220B ATTRIBUTED TO DAWLAT, CIRCA 1635

The corresponding text describes a hunt at Bari in 1637, but the illustration relates more closely to an event at Burhanpur in July, 1630:

SEVERAL LIONS SEEN in the hunting grounds were reported to the Emperor. By imperial order the ferocious beasts were surrounded with a *bâdar*,[1] which is a very strong net 10,000 imperial cubits long and six cubits high, set up like a *sarâparda* [canvas wall], with thick poles, and all sorts of wild animals are collected in it and hunted. The lions were then driven to outside the Zaynabad garden, where the lion-hunting Emperor mounted on an elephant entered the hunting ground and shot a lion with his gun. Four lion cubs were caught by the mace-bearers.[2]

François Bernier, who traveled in India between 1656 and 1668, left a lengthy account of Mughal hunting practices, beginning with a description of the land held by the Emperor as a private hunting preserve:

IN THE NEIGHBOURHOODS of *Agra* and *Dehli* [*sic*], along the course of the *Gemna* [Jumna], reaching to the mountains, and even on both sides of the road leading to *Lahor*, there is a large quantity of uncultivated land, covered either with copse wood or with grasses six feet high. All this land is guarded with the utmost vigilance; and excepting partridges, quails, and hares, which the natives catch with nets, no person, be he who he may, is permitted to disturb the game, which is consequently very abundant ....

Of all the diversions of the field the hunting of the lion is not only the most perilous, but is peculiarly royal; for except by special permission, the King and the Princes are the only persons who engage in the sport. As a preliminary step, an ass is tied near the spot where the gamekeepers have ascertained the lion retires. The wretched animal is soon devoured, and after so ample a meal the lion never seeks for other prey, but without molesting either oxen, sheep, or shepherds, goes in quest of water, and after quenching his thirst, returns to the former place of retirement. He sleeps until the next morning, when he finds and devours another ass, which the gamekeepers have brought to the same spot ... When information is received of the King's approach, they fasten at the spot an ass where so many others have been sacrificed, down whose throat a large quantity of opium has been forced ... The next operation is to spread, by means of the peasantry of the adjacent villages, large nets, made on purpose, which are gradually drawn closer, ... Everything being in this state of preparation, the King appears on an elephant protected in places with thin plates of iron, and attended by the Grand Master of the Hunt ....

You must know, then, that as it is considered a favourable omen when the King kills a lion, so is the escape of that animal portentous of infinite evil to the state. Accordingly, the termination of the hunt is attended with much grave ceremony. The King being seated in the general assembly ... the dead lion is brought before him, and when the carcass has been accurately measured and minutely examined, it is recorded in the royal archives that such a King on such a day slew a lion of such a size and of such a skin, whose teeth were of such a length, and whose claws were of such dimensions, and so on down to the minutest details.[3]

1. *Bâdar* is probably a variant of the Hindi word *bâdal* (cloud), an apt description of the overall effect of such a vast net.
2. *Padshahnama*, I/I, pp.310ff.; translated by Wheeler Thackston.
3. Bernier 1891, pp.375ff.

The Padshahnama in context

THE SUN OF THE HEAVEN of good fortune and the caliphate, the emperor who gives asylum to the world, Shadow of God, the splendor of whose brow illuminates the eyes of the world from the Public and Private *Jharoka*, which is so great, high, and vast in its beauty that it vies with the Sphere of Spheres, and honors the servants of the court at which sultans take refuge with the felicity of *körünüsh*. As soon as the sun appears, as is customary, the kettledrums of gladness and rejoicing and the drums of success are sounded loudly from the *naqarakhana* opposite the *jharoka*, and the sound travels for leagues. First, in accordance with the rule of the great sultanate, fleet-footed horses with jeweled and enameled trappings and renowned line-breaking elephants, mountainous in body, with jeweled, enameled, and gold trappings, as well as female elephants, are arrayed, and on either side and behind elephants with drums and all sorts of ornaments pass before the imperial gaze. After that, on every side rises a tumult of God-given splendor, and a commotion of might and majesty occurs in the public and private arena with a crowd of people of all classes and castes beyond enumeration. Renowned princes, exalted offspring of kings, and felicitous sultans have permission to sit rank by rank near the throne of the sultanate; *khan*s, sultans, *mir*s, and *mirza*s of Iran and Turan; exalted viziers, lords of the pen and the sword, high-ranking amirs, amirs' sons in service, and all sorts of swordsmen from the ever-victorious army, triumphant elite servants, nimble quiver-bearers, sword-wielding arms-bearers, granite-splitting yeoman, sword-wielding warriors, rank-shattering mace-bearers, mountain-shattering musketeers, and other respected men like high-ranking *sayyid*s, great shaykhs, noble learned men, skilled physicians, and boon companions of various classes of Arabs, Persians, Turks, and Tajiks, Kurds, Tatars, Russians, Ethiopians, Circassians, *et alii* from Anatolia, Egypt, Syria, Iraq, Arabia, Persia, Fars, Gilan, Mazanderan, Khurasan, Transoxiana, the Quipchaq Steppe, Turkistan, Georgia, and Kurdistan, various groups and castes of learned men from Hindustan, people of the sword and pen, true *sayyid*s, warrior *shaykhzada*s [the offspring of a religious personage], Afghan tribes like Lodi, Rohela, Khweshgi, Yusufzai, et cetera, and classes of Rajputs – ranas, rajas, raos, and rais – like Rathor, Sisodia, Khachhwaha, Hada, Cauhan, Jhala, Chandrawat, Jadun, Tonwar, Baghela, Gujar, Bwar, Bhadauriya, Singhi, Bundela, Sagarwal, and the rest of the people of India and other tribes from [the ranks of] 7000 to 1000 and from 1000 to 100, and from 100 to *ahadi* [cavalry-man], *zamindar*s [landholders] from plain and hill from the realms of the Carnatic, Mugg, Assam, Udaipur, Srinagar, Kumaun, Bandhu, Tibet, Kishtwar, et cetera. From all the protected imperial realms, class by class, and group by group, they are honored to kiss the threshold of Saturnian splendor, and they stand in magnificence in accordance with their ranks, posts, and stations, unable to move or speak beyond their appointed stations. At the *kathara* [railing] and inside and out, on every side *mir tüzük*s [officers in charge of order and precedence] with plumes in the turbans and bejeweled staffs, guards, nimble sergeants-at-arms, huge bodyguards, and other intrepid foot soldiers keep such guard that no living soul is able to budge from his place. Sincerely loyal servants, their gazes fixed in perfect servitude on the Emperor's countenance, which is more brilliant than the sun, feast their eyes on his unparalleled beauty. Despite their nearness and proximity to each other, they cannot speak, and with seals of silence on their lips, they stand with loins girded in service and fully armed, their ears and minds focused upon the Emperor's orders which they execute with the swiftness of lightning and the wind. Likewise

Fig.7 Detail of 13.

glib ambassadors from the caesar of Rum [the Ottoman sultan] and the rulers of Iran and Turan who have come bearing letters and gifts in order to maintain the continuation of cooperation have permission to stand in a place appropriate to their stations, while stewards and rulers of the Deccan, like the Adil Shah, Qutbulmulk, and the people of the Carnatic, bearing petitions and gifts, are granted audience and display their servitude. The great mercantile class of merchants, traders, and the wealthy of every region from Persia and Khurasan, Anatolia, and Syria, Chinese Turkistan and beyond, Cathay, and Khotan, Turkistan and Firangistan [Europe], and other far-away countries, islands, fabled ports come to court bearing expensive gems, valuables, rarities, and goods and display their wares at court, receiving great compensation and profit and bearing the good name and repute of this eternal empire to the corners of the world. The court of majesty and splendor is the focus of the hopes of the world. Masters of the sword and the pen, men of learning and perfection, the wise and the expert, the eloquent, and masters of all other arts and crafts from all over the world, like Istanbul, Aleppo, Egypt, Basra, Baghdad, Hamadan, Shirvan, Shemakha, Gilan, Mazanderan, Astarabad, Ganja, Bards, Tabriz, Ardabil, Qazwin, Qum, Tehran, Yazd, Isfahan, Simnan, Damghan, Bistam, Sabzavar, Nishapur, Merv, Mashhad, Tus, Tabas, Qayin, Tun Isfarayin, Jam, Herat, Khwaf, Bakhtar, Selstan, Qandahar, Balkh, Badakhshan, Bukhara, Samarkand, Andijan, Tibet, Kashgar, and other regions of Turkistan, as well as the farthest lands – all classes of men come full of hopes to [the] court that gives asylum to the world, the center of the circle of the four corners of the world and enlist among the servants of the mighty threshold. Since this eternal empire gathers together all good people, on the one hand rarities and novelties are offered as gifts to the transmuting gaze of the Emperor; on the other hand the servants of the court [are] rewarded, as befits their stations, with ranks, increase in rank, fiefs, *charqab*s, robes of honor, horses, elephants, gems, jeweled utensils, auspicious likenesses [of the Emperor], turban ornaments, swords, *jamdhar*s, daggers, *tümän tugh*s, banners, and drums.[1]

1. *Chahar Chaman*, fols 20a – 21a (translated by Wheeler Thackston).

THE PADSHAHNAMA
AND MUGHAL HISTORICAL MANUSCRIPTS

by Milo Cleveland Beach

By the mid-sixteenth century, painters in India were hard at work inventing a visual equivalent for the extraordinary verbal descriptions with which historians were recording events at the court of the Mughal emperor. Few passages are as vivid as these words by Chandarbhan Brahman, a Hindu historian writing of the court of Shah-Jahan, yet the world he describes is easily recognizable in the illustrations to the Windsor Castle copy of the *Padshahnama*, the official history of Shah-Jahan's reign.

The manuscript is clear proof of the success of the painters' efforts, the result of which was a brilliant kind of illustration that quickly distinguished itself from existing traditions of book production both in India and in the Mughals' physical and cultural homeland in central Asia. The Mughals, whose name is a variant of 'Mongol,' had arrived in India in 1526 from family territories in Central Asia now divided among Uzbekistan, Afghanistan, and Iran – a world dominated by Persian culture. They quickly gained control of important territories in the northern regions of the Indian subcontinent, bringing together formerly separate kingdoms with varied Hindu and Muslim religious and cultural affiliations. The Mughals, themselves Muslim, also brought to India the traditionally Muslim respect for books as both sources of secular and religious knowledge and embodiments of wealth and power.

The personal interest in history of Babur, the first Mughal emperor, was the basis of his successors' concern for minute and accurate documentation of the events that filled their lives. His grandson, Akbar, established specific offices to keep precise records of the happenings at court, and these became sources for official historical accounts compiled by court poets and historians.[1] In the *A'in-i Akbari* ('Annals of Akbar'), the historian Abu'l-Fazl wrote: 'Keeping records is an excellent thing for a government; it is even necessary for every rank of society. Though a trace of this office may have existed in ancient times, its higher objects were but recognized in the present reign. His Majesty [Akbar] has appointed fourteen zealous, experienced, and impartial clerks, two of whom do daily duty in rotation, so that the turn of each comes after a fortnight ... Their duty is to write down the orders and doings of His Majesty and whatever the heads of the departments report; what His Majesty eats and drinks; when he sleeps, and when he rises; the etiquette in the State hall; the time His Majesty spends in the Harem; when he goes to the general and private assemblies ... the acts of His Majesty as Spiritual guide of the nation; vows made to him; his remarks; what books he has read out to him ...'[2] In two cases, those of Babur and his great-grandson, Jahangir, actual imperial memoirs have also survived – the *Baburnama* and *Jahangirnama* – as have the personal accounts of family members, court officers, and servants.[3]

No book illustrations or other paintings made for Babur are known, but Mughal artists working for his successors gradually developed a special expertise in the accurate depiction of historical events, probing analyses of human personality, and visual studies of the flora and fauna with which they were surrounded – interests which also make descriptions of India in the imperial memoirs so enthralling. When these historical accounts were illustrated by court painters, the documentary elements that they contained were vastly increased. Artists provided images of details that were seldom included in written texts, such as styles of dress and personal ornamentation, architectural decoration, or human appearance and interaction. Thus, the illustrations were not merely recreations of the narrative, but provided additional insights and information. Artists employed in the imperial studios were also responsive to the tastes and interests of the emperors who controlled their employment, to the extent that this was important to each individual patron. But from Humayun, Babur's son, to the fifth emperor, Shah-Jahan, all painters were encouraged to be inventive in their compositions and receptive to the lessons of unfamiliar artistic traditions, whether of India itself or of those foreign cultures introduced to the subcontinent by missionaries and traders in this period of unprecedented international contact. When the *A'in-i Akbari* discusses Akbar's interest in painting, for example, it asserts his personal involvement in the work of the artists, and then continues: 'Most excellent painters are now to be found, and masterpieces, worthy of a *Bihzad* [the most famous painter of Iran], may be placed at the side of the wonderful works of the European painters who have attained world-wide fame.'[4] The statement is as much a testament to the cosmopolitan artistic atmosphere of the court as is the description of Chandarbhan Brahman to its physical internationalism. Through both subject matter and style, therefore, manuscripts illustrating Mughal historical texts are among the richest sources of information about India and its international contacts during any period.

The text for the Windsor Castle copy of the *Padshahnama*, the only known version with illustrations dating to Shah-Jahan's reign, was written out by the scribe Muhammad-Amin of Mashhad in 1656–7. Since pre-existing illustrations made at vari-

Fig.8 Unknown artist, *Yusuf is presented with a mirror by the traveler from Canaan*, mid-16th century.

Fig.9 Unknown artist, *Audience scene*, dated 1567–68.

ous times throughout the period of Shah-Jahan's rule were eventually inserted into spaces deliberately left blank within the text area,[5] the Windsor manuscript actually provides – despite its date – a survey of painting throughout Shah-Jahan's lifetime. By 1628, the year of Shah-Jahan's accession as emperor, artists were working within a style developed and brought to maturity over the preceding three-quarters of a century. There are no stylistic traits that immediately distinguish the later works commissioned by Jahangir from the first paintings made for his son, Shah-Jahan, although after a few years innovations responsive to this new patron emerged. By the end of the reign, however, Mughal painting had fundamentally changed: with the exception of deliberately archaistic illustrations, few images made after Shah-Jahan's death could be confused for earlier works. It is especially important to note that serious and sustained Mughal interest in illustrated official histories and individualized portrait studies also died with Shah-Jahan, depriving Mughal painting of its unique contribution to Indian and Islamic artistic traditions.

Early evidence of Mughal interest in historical narratives is found not only in the *Baburnama* ('History of Babur'), the first distinctively Mughal literary work, but also in the earliest identifiably Mughal paintings, made during the reign of Humayun. (No illustrations for the *Baburnama* are known from that emperor's lifetime, but several illustrated copies of a translation into Persian made for the Emperor Akbar have survived.)[6] Humayun spent fifteen years in exile from India, but was an active patron during this period.[7] Once the power of the court was restored, in 1555, and then solidified under his son and successor, Akbar, the imperial workshop of craftsmen and artists to make and decorate books was quickly enlarged. From Akbar's reign into that of Shah-Jahan, the output of the Mughal painting studios was impressively prolific.

After about 1580, following initial exploration of a wider range of subjects, the illustration of historical narratives and portraits of important personalities became an increasingly dominant focus of Mughal painting. The first manuscripts had recounted popular fables and stories, and were illustrated with paintings that stressed the excitement of the narrative – an appropriate artistic response to Akbar's youth, energy, and intellectual curiosity.[8] This interest culminated in specially-commissioned Persian translations of the *Mahabharata* (known as the *Razmnama* or 'Book of Wars') and the *Ramayana* ('Story of Rama'), two great Hindu epics written in Sanskrit.[9] The focus of these projects was outward, into the often unfamiliar religious and cultural traditions by which the Mughals were surrounded in their new homeland. Many artists were enthusiastic about the new ideas and different attitudes to painting to which they were introduced by the freedom of contact between Hindus and Muslims in the imperial workshops at this time and by their access to such novelties as the European images brought to India by missionaries and traders.[10] For Mughal painters, this period under the Emperor Akbar was a phase of exploration and assimilation, and choices were constantly being made as to which subjects, techniques, and ideas were appropriate for development.

The Mughal emperors, however open to new ideas initially, nonetheless constantly affirmed their own links with the world of Persian culture; Persian was the court language, and Akbar hired painters from the employ of the Iranian Shah Tahmasp Safavi (r. 1524–1576) to head his painting workshops in India.[11] Since the Safavid rulers of Iran were their political rivals as well as their contemporaries, however, the Mughals were also eager to distinguish themselves by asserting their descent from the great Turko-Mongol warrior Timur (Tamerlane), whose Timurid dynasty had preceded the Safavids. The Mughals referred to themselves as Timurids, and collected manuscripts that had belonged to, or revived the styles associated with, Timur's patronage.[12] In the early sixteenth century, many artists trained under the Timurids had taken refuge with Uzbek chieftains at Bukhara (in present-day Uzbekistan), and from there they provided illustrated books for neighboring courts as well as for a thriving commercial market; others took employment in Mughal India.[13]

An example of a Bukhara manuscript is the *Tuhfat al-ahrar*

Fig.10 Unknown artist, *The throne of prosperity*, dated 1570–71.

Fig.11 Unknown painter, *The court of the queen of Sheba*, dated before 1512.

('Gift of the Free') of Jami copied in 1515–16, although its illustrations were added about four decades later. This volume was once in the imperial Mughal library, for it bears an inscription by Jahangir as well as a seal dated 1658–9.[14] One of its illustrations, *Yusuf is presented with a mirror by the traveler from Canaan* (Fig.8), is compositionally similar to a page from the *Dewal Devi Khizr Khan* ('The Romance of Dewal Devi and Khizr Khan') of 1567–8, the earliest specifically dated manuscript in a distinctively Mughal style (Fig.9). Both show a ruler seated in a pavilion with rich and carefully detailed architectural decorations, a foreground courtyard thronged with nobles whose figures are placed against strongly patterned flooring, and a background landscape. While this is a basic, oft-repeated compositional type within Islamic painting, the Indian work already reveals a distinctively Mughal taste for lively activity and emphatic gestures; strong color; an abundance of detail; the importance of narrative – for even if we do not know the story, it is instantly clear that a story is being told; and the use of landscape to further that narrative, rather than to serve simply as an attractive backdrop. More understated in its presentation of any specific episode, the concern of the Bukhara painter is to create an artistically harmonious, decorative image, rather than to encourage excited involvement with the story being told. The formulas repeated in the Bukhara illustration nonetheless represent a general compositional prototype for the dated Mughal page, and the scheme for the audience scenes of the *Padshahnama* is rooted in both works.

In the later sixteenth century, regions of India outside immediate Mughal control had quite different political and cultural affiliations, and as a consequence different artistic interests. The separate kingdoms of the Hindu Rajputs tended to be highly conservative and local in their aesthetic approach, and maintained strong links with the art of Indian villages,[15] while Muslim kingdoms such as Bijapur maintained closer cultural ties to Ottoman Turkey than to Iran. The great *Nujum al-Ulum* ('The Stars of the Sciences') manuscript made in Bijapur in 1570–71 (Fig.10), for example, bears little visual relation to the Mughal *Dewal Devi Khizr Khan* (Fig.9), although both were painted at virtually the same time for Muslim rulers in India; instead, it closely follows the compositional format and style of an early sixteenth-century *Sulaymannama* ('Story of Solomon') manuscript (Fig.11), almost certainly made in Turkey for the Ottoman Sultan Bayazid II (r. 1481–1512). This is only one example, of course, of the extensive vocabulary of artistic styles that may have served as examples to the Mughals at the time they were formulating their own visual language. The rejection of this particular style, when placed alongside the fascination with and acceptance of European techniques and imagery noted by Abu'l-Fazl, helps to define the intentions of Mughal patronage. The affiliation of the Mughal arts with the traditions of Iran and Europe was a conscious decision and not the result of a lack of alternatives.

Early in the 1580s, Akbar defined the new direction for Mughal painting when he commissioned factual accounts of both Islam during its first millennium, the *Tarikh-i Alfi* ('History of a Thousand'), and of Timur, the *Timurnama* ('History of Timur').[16] Both works included discussion of the descendants of Timur, including the young Emperor Akbar and his predecessors in India. While the extant *Tarikh-i Alfi* pages and most of the *Timurnama* deal with distant history and their illustrations could not pretend to accuracy of appearance, the figures in the *Timurnama* certainly include portraits of Akbar and other Mughals. The depictions of Timur sometimes show him in situations exactly repeated in episodes from Akbar's life, which reveals a conscious parallelism in the accounts.[17] This kind of manipulation of an otherwise staunchly accurate historical depiction is occasionally found in later Mughal paintings, including the *Padshahnama* – especially in the allegorical scene where Shah-Jahan meets the mythical Khwaja Khizr (**41–42**). The book, in fact, opens with an allegory, a double page in which Shah-Jahan is depicted as if receiving his crown from Timur (**3–4**), but homage to Timur is not expressed there simply by the inclusion of his portrait. When seen in relation

Fig.12 Husayn Naqqash and Kesu, *Akbar receiving prisoners-of-war from Gujarat* (right half), circa 1596.

to the portrayal of Shah-Jahan, the symmetry and relative flatness of the composition depicting Timur, the prominently calligraphic lines used to define the edge of the ruler's skirt, the indifference to a convincingly portrait-like characterization, and the relatively flat background together indicate that the painter was also paying tribute to Timurid ideals in painting. (The presentation of Timur here is in many ways closer stylistically to the prince in the Bukhara painting, Fig.8, than to the facing portrait of Shah-Jahan.) This reference to Timurid style demonstrates that Mughal painters were highly conscious of the distinctive character of the tradition within which they worked, and realized that because references to other styles would be recognized, they could be made expressively significant. Scholarly studies continually remind us of the down-to-earth realism of Mughal painting, so it is important to remember that even very specific depictions could include allegorical references to higher truths and distant historical relationships. Occasional non-historical details were therefore in no way contradictory to the purpose of the paintings. Indeed, manuscript illustrations often had a carefully planned political significance that demanded flexibility with historical accuracy.

Among the most important illustrated historical manuscripts produced in the sixteenth century were two copies of the *Akbarnama* of Abu'l-Fazl. The earlier in date was closely related in its compositions to the *Timurnama* project,[18] while the second – far more refined and mature in style – may have been the official copy made at Lahore for presentation to the Emperor in 1596.[19] *Akbar receiving prisoners-of-war from Gujarat* (Fig.12), from the earlier book, no longer draws on the generic architectural structures seen in the *Dewal Devi Khizr Khan* (Fig.9) but very specifically suggests the Hall of Public Audience at Fatehpur-Sikri, where the episode took place.[20] The figure of Akbar, too, is an individualized portrait. (Because successful portraiture was still beyond the reach of many painters, particularly competent master artists would often add portraits of the most important figures to scenes for which they had no other responsibilities.) Vivid gestures and strong movement suggest that a lively throng is gathered before the Emperor, and the artist skillfully evokes the conversations and curiosity of the crowd. All these features testify to the ability of Mughal painters to respond quickly to imperial interests, of course, for Akbar's concern for historical accuracy preceded his artists' ability to describe events and details of appearance with documentary realism. Both *Akbarnama* manuscripts are also full of scenes that confirm Akbar's active personal participation in the events that established and consolidated Mughal control of north India; he is shown, for example, hunting, in battle, passing judgment on rebels, building his new capital at Fatehpur-Sikri, and celebrating the births of his sons. The illustrations reinforce the dynamism of these subjects by the lively visual activity of their compositions and colors.

Certain paintings from the *Jahangirnama* reveal the next stage in the development of historical audience scenes. The *Jahangirnama*, written as a memoir by the Emperor, is an extraordinary narrative. Whereas the *Akbarnama* stresses Akbar's physical activity and his personal role in conquering and bringing together the various independent territories of north India, Jahangir's memoirs are set in the powerful, wealthy, well-ordered empire that resulted from his father's effort. Jahangir could therefore describe a quite different life, and was primarily interested in having himself depicted in the accompanying illustrations not in battle or at the hunt, but as a man of peace, interested in learning about the world which he ruled and concerned with spiritual matters. He is shown receiving exotic animals (Fig.15), examining and measuring a peculiarly shaped tree, watching the fight between a snake and a spider, talking with a naked Hindu ascetic, and visiting the Muslim shrine at Ajmer.[21] An interest in highly individualized portraiture, exploring the personalities by which Jahangir was surrounded, is also an important ingredient of the painting of this period, for unlike the court scenes in the two major *Akbarnama* manuscripts, almost every courtier in the *Jahangirnama* illustrations is recognizable as a precisely described individual. Because these are subjects and attitudes often without precedent in either Islamic or Indian painting traditions, they demanded compositional innovations and direct observation by the artists. The encouragement of artistic spontaneity and originality initiated by Akbar therefore continued unabated under Jahangir's rule.

In the text of the *Jahangirnama* for the summer of 1618, the Emperor actually refers to the making of an illustrated copy of the narrative: 'On this date Abu'l-Hasan the artist, who had been awarded the title *Nadiruzzaman* ('rarity of the time'), presented a painting he had made on the opening page of the *Jahangirnama*. Since it was worthy of praise, he was shown limitless favor. Without exaggeration, his work is perfect, and his depiction is a masterpiece of the age.'[22] His text for one month later incidentally informs us of the existence and purposes of multiple copies: 'When a fair copy of the events of the twelve years in the *Jahangirnama* had been made, I had ordered the supervisors of the royal library to make one volume of the twelve years and produce multiple copies for me to award to particular servants and to be

Fig.13 Abu'l-Hasan, *Ceremonies at the accession of Jahangir*, dated 1605 (left half) and circa 1618 (right half).

Fig.14 Attributed to Manohar, *Prince Khusraw brought to Jahangir as a captive*, circa 1610.

Fig.15 By Bulaqi, son of Hoshang, *Jahangir receiving Muqarrab Khan with exotic animals from Goa*, circa 1625.

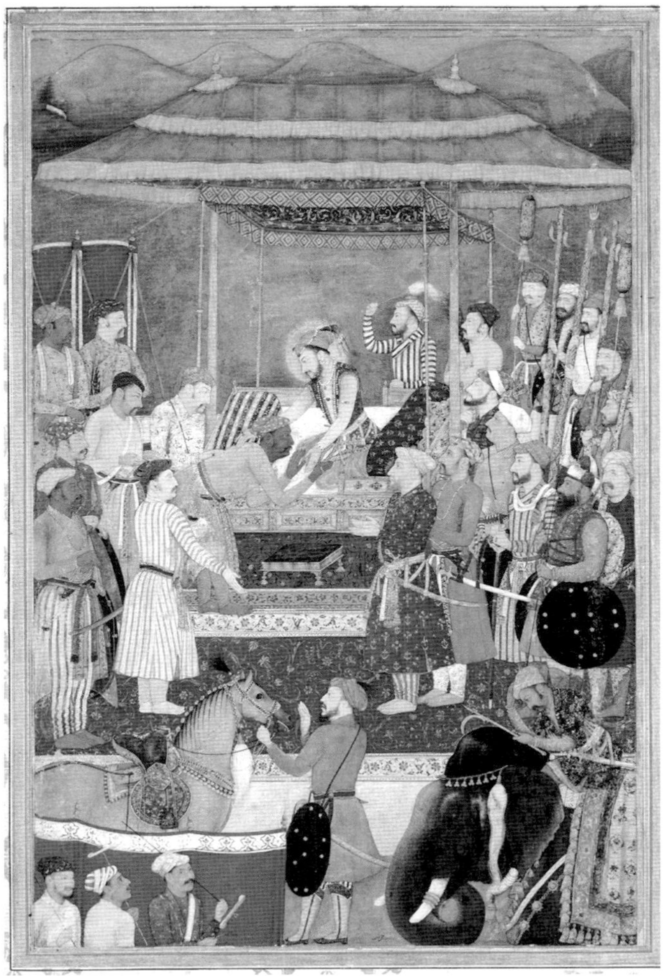

Fig.16 La'lchand, *The submission of Rana Amar Singh of Mewar to Prince Khurram* (see **6**).

Fig.17 Nanha, *The submission of Rana Amar Singh of Mewar to Prince Khurram*, circa 1625.

sent to other countries to be used by the rulers as a manual for ruling. On ... [20 August 1618] one of the recorders brought me a copy completely written out and bound. Since it was the first copy to be made, I gave it to my son Shah-Jahan, whom I consider the first of all my sons in everything.'[23]

While the connection is not explicit, it can be assumed that the frontispiece would have been made for this first formal copy of the text. And it is equally likely that this scene was *Ceremonies at the accession of Jahangir* (Fig.13), for this is the first event described in the narrative.[24] The placement of the Emperor, enthroned within a carefully defined architectural framework, above a densely organized and rather sedate assembly of nobles is the format that becomes standard for the *Padshahnama*. The left-hand illustration of the throngs attending the ceremonies, however, is among the most spirited of all Mughal court scenes; no mere repetition of established formulas, it is full of direct observation of the behavior of humans in crowds. Dated to the year of the accession (1605–6) and signed by the same artist, Abu'l-Hasan, it is a perceptive and elegantly executed development of the style of the *Akbarnama* audience scenes. Whereas Akbari illustrations are especially effective for evoking the general brouhaha of court gatherings – a quality understood by Abu'l-Hasan – the accession *darbar* seems more intent on celebrating the personalities of the courtiers shown as well as, more perfunctorily, documenting who was there. Many of the figures are even accompanied by tiny inscriptions giving their names and titles. The stylistic differences in the two halves of the scene correspond to their apparently different dates of execution. Other pictures had been painted earlier for eventual inclusion in the manuscript; the passage of Jahangir's memoirs that records the arrival of the animals from Goa, for example, is dated 1612 and continues, 'Since these animals looked so extremely strange to me, I both wrote of them and ordered the artists to draw their likenesses in the *Jahangirnama* so that the astonishment one has on hearing of them would increase by seeing them.'[25] Paintings were therefore occasionally commissioned in anticipation of their future inclusion in some specific volume. Among other things, this implies that the imperial workshops would have had a stock of pictures continually on hand for intended projects; and that if plans for these projects changed, as they did, the purpose anticipated for any single painting would also change. Knowledge of this procedure is important to an understanding of the *Padshahnama* and the many single paintings of historical events made for Shah-Jahan.

Unfortunately, no bound illustrated copy of the *Jahangirnama* has survived, although many separate paintings that illustrate episodes described in the text are known. (Unlike the historical paintings remaining from Shah-Jahan's reign, however, there is no duplication of subject, raising the possibility that only one copy of the text was intended to contain illustrations by imperial artists.) The remaining illustrations seem to form two groups: those in the first relate to what we term early-Jahangiri style (and were perhaps intended for the copy presented to the Emperor in 1618); the others, more numerous, are in styles that merge more easily with the paintings made later for the *Padshahnama*. The early group includes a scene in which the Emperor confronts his rebellious son Khusraw and his followers in a garden laid out in Lahore by Mirza Kamran, the brother of Humayun (Fig.14). Although the painting is badly worn, the landscape shows an unprecedented naturalism in the depiction of the trees, the sense of free space, and the light-filled sky. Impossible (because of these traits) to confuse with works made for either Akbar or Shah-Jahan, this is a fine example of the distinctive style developed under Jahangir's patronage.

Among those later Jahangiri works that are closely allied to the *Padshahnama* in style, *Jahangir receiving Muqarrab Khan with exotic animals from Goa* (Fig.15) is the compositional prototype for two wedding scenes in the *Padshahnama* (**25** and **45**), while an even more direct correspondence occurs between the depictions ostensibly made for both books showing *The submission of Rana Amar Singh of Mewar to Prince Khurram*.[26] The illustration in the *Padshahnama* (**6** and Fig.16) is compositionally identical to a page thought to have been intended for the *Jahangirnama* and recently acquired for the Victoria and Albert Museum, London (Fig.17).[27] The scenes are by different painters, and stylistic distinctions include the more graceful proportions of the figures and the increased psychological subtlety of the interactions in the Windsor Castle painting – especially the relationship between the Rajput ruler who is bending in homage, the imperial henchman Abdullah Khan who controls his movements, and the Mughal prince (see also the further discussion on page 163). The Windsor illustration, by La'lchand, also has fewer figures; each group of men is smaller in number, and the composition is thereby clarified and strengthened. A potentially more significant difference concerns the placement of Afzal Khan Allami, one of the men assigned to escort Amar Singh into Khurram's presence. In the painting in London, which is by Nanha, an artist who had been active under Akbar's patronage, Afzal Khan stands inconspicuously among the men directly behind the Rajput whereas La'lchand places him very prominently to the future emperor's immediate left – the place where he regularly stood after his appointment as *Diwan* (Finance Minister) to Shah-Jahan in 1629.[28] By this move, Afzal Khan pushes aside Sundardas, whose placement in the earlier version proclaimed him the most prominent noble in attendance. In fact, for his role in the Mewar campaign Shah-Jahan granted Sundardas, who was then his *Mir Saman* (Imperial Steward), the title Rai Raian (later increasing his rank and changing the title to Raja Bikramajit, described as the highest honour in India.)[29] La'lchand may have been responding to Afzal Khan's position when the later painting was made, rather than his rank at the time of the event. Raja Bikramajit had died in 1622–3, when Shah-Jahan was still a prince, and recognition of his former status had become of less importance. Such adjustments to the positioning of the officials in Mughal paintings were made continually, revealing that they were intended not merely as documentary records, but could be determined by artistic or political purposes.

Even with these differences, the scene by La'lchand is an unusually exact copy of the presumably earlier work,[30] and it makes us aware that existing paintings were important sources of information for artists. The *Padshahnama* and other historical texts do not give complete lists of the men shown in the various *darbar*s, for example; that information had to be gathered for painters from archival sources, from memory, from guesswork (sometimes wrong), and from the sketches and illustrations made by other artists in attendance.[31] This suggests that painters had access to historical archives and other sources of information about which we know nothing.[32]

In comparison to *Ceremonies at the accession of Jahangir* (Fig.13), the accession portrait from the *Padshahnama* (**10–11**) impresses us by the staidness and decorum of the event, and the contrast corresponds to contemporary accounts of the opposing personalities of Jahangir and his son. Sir Thomas Roe had already perceived these differences during his embassy, when Shah-Jahan was still a prince: 'Breefly, I stand on very fickle termes, though in extraordinarie Grace with the King [Jahangir], who is gentle, soft, and good of disposition; yet on Poyntes and disputes with an insolent and Proud sonne of his, into whose handes he hath remitted all Power, which hee is neyther woorthy nor able to manage. He is Lord of our Port, and by his folly giues mee much trauell; so sordidly ambitious, that hee would not haue mee acknowledge his father King, nor make any addresses, nor diliuer any Presentes nor Complementes of honor, but to hymselfe; which I will neuer yeild too[33] ... I neuer saw so settled a Countenance [as that of Shah-Jahan], nor any man keepe so Constant a grauety, neuer smiling, nor in face showeing any respect or difference of men; but mingled with extreame Pride and Contempt of all.'[34]

Illustrations for the *Jahangirnama* stress the Emperor's intellectual curiosity about the physical world and his sympathy with sages and men of religion. The passages chosen for illustration in the *Padshahnama*, however, most often confirm Shah-Jahan's bravery and military prowess as a prince, and then celebrate major events of his rule as Emperor – confirmation of the pride and ambition noted by Roe, rather than a revelation of more personal interests. (And, of course, it is the spaces initially intended for illustrations, rather than the scenes actually inserted into the book, that reveal the original programmatic intentions of this manuscript.) The use of double-page illustrations – scenes with far greater physical impact on the viewer than single-page images, which would have faced pages of text – also establishes a more specific hierarchy among the painted scenes: two successive double-pages were intended to relate Shah-Jahan's defeat, while a prince, of the Rana of Mewar and thus demonstrate his bravery (**6–7**); a 'flashback' depiction showing Jahangir granting the title *Shah-Jahan* to the prince indicates the future emperor's right to succeed (**38**, **39**, now separate scenes, although the space was originally intended for a double-page presentation of a single episode); the accession ceremonies establish him as Emperor in fact (**10–11**); the defeat of the Portuguese at Hoogly proclaims Shah-Jahan's world-dominion (**19**, **20**, also originally intended as a double-page opening showing a single episode but now divided between two separate scenes); three sequential double-page illustrations of the wedding of Prince Dara-Shikoh make an undeniable statement about dynastic succession, family strength, and the specific importance of Dara (**21–26**); and the Emperor's visit to the shrine of Shaykh Mu'inuddin Chishti at Ajmer, and his allegorical encounter there with Khwaja Khizr, establish his credentials as a spiritual leader (**41–42**). Around these key illustrations are placed scenes that expand on these themes and confirm Shah-Jahan's official status without offering those revelations of the private human being that make the memoirs of Babur and Jahangir, and even the official history of Akbar, such engrossing and satisfying texts.

Despite the somewhat haphazard grouping of images, and the inconsistency of their historical sequence, the illustrations still tell us a great deal about the evolution of the arts under Shah-Jahan. Above all, the craftsmanship of the best pages is superb: the scenes are minutely detailed; the descriptions of people, the costumes they wear and the textiles, decorative objects, and architecture by which they are surrounded are not only convincingly accurate, but are full of life and verve. This extends even to the portraits of the imperial family. The portraits of the Emperor and his sons that appear in the various *darbar*s are intensely sensitive to physical appearance. The scene recorded in *Shah-Jahan receives his three eldest sons and Asaf Khan during his accession ceremonies* (**10–11** and Fig.168), for example, is dated 1628, while *Shah-Jahan honoring Prince Dara-Shikoh at his wedding* (**25–26** and Fig.169) is 1633; *Shah-Jahan honoring Prince Awrangzeb at Agra before his wedding* (**43** and Fig.170), 1637; and *The arrival of Prince Awrangzeb at the court at Lahore* (**44** and Fig.171), 1640. During the years represented in the four illustrations cited, Shah-Jahan's beard thickens and turns grey, while Dara-Shikoh, who first appears as a beardless youth of twelve, is seen also as a seventeen-year-old with the beginnings of a mustache, a mature twenty-two-year-old, and a fully bearded man in his thirties – with similar changes visible in the appearances of his brothers. From this point of view, images by the best painters can often serve as objective documentation of the people and events shown, and the age of the subjects can be used as partial evidence for identifying those scenes which have no accompanying textual referent. Furthermore, just as the appearances of the courtiers have been carefully studied, so have their personalities been interpreted. With few exceptions, and even when we realize that individual painters have recognizably distinct ways of depicting character, we see these men mature and age over time in all their portraits. This is incisive description, not simply the repetition of established pictorial formulas. And the inclusion of several painters in *Padshahnama* scenes (for example, in **5** and **39**) suggests both that artists were socially acceptable members of the court community, and that they were sometimes present at the events they depicted.

THE PADSHAHNAMA AND LATER INDIAN PAINTING

From the earliest known paintings made for the Emperor Humayun in the 1550s to the middle years of Shah-Jahan's reign almost a century later, the development of Mughal painting followed a sequence in which increasing attentiveness to the precise appearance of people and the world in which they lived was combined with continuous refinement of techniques and physical materials. The artistic and political interests of each of the emperors were also reflected in the images produced during their individual reigns, although shifts of style at the accession of a new emperor were neither radical nor immediate. During this century-long period, the size and organization of the imperial workshops were periodically adjusted, but artists were given continuous support, and their assignments remained predominantly historical scenes, portraits (including flora and fauna), and illustrations of Muslim poetic texts. In the early 1640s, Sebastian Manrique was able to report that the Imperial Library held 'twenty-four thousand separate volumes, mainly the works of the oldest and most important Writers ... with their rich and valuable bindings ...'[35] This situation altered as the elderly Shah-Jahan withdrew from his earlier personal involvement with the painters and the generation of senior artists became inactive. The changes were even more drastic following the imprisonment of Shah-Jahan and the accession of his son, Awrangzeb.

The rivalry of the four sons of Shah-Jahan had been particularly intense. The eldest, the intended heir, was Dara-Shikoh, so obviously his father's favorite that his brothers' hatred was inevitable. An intellectual, he was interested in the various religious movements of his time, and translated and wrote commentaries that showed his sympathetic understanding of Hindu texts. His interest in the world around him recalled to many the personality of his great-grandfather, Akbar. But this liberalism placed him in conflict with orthodox followers of Islam and encouraged his younger brother, Awrangzeb, to present himself in an opposing way, as a pious upholder of Islam against what he claimed to be the apostasy of Dara. In 1658, Awrangzeb boldly seized the throne by imprisoning his father, who was declared to be ill and confined to rooms in the fort at Agra until his death in 1666. The assembly of an *'Alamgirnama*, an official chronicle of Awrangzeb's rule – *'Alamgir* ('Seizer of the Universe') was his formal title – was begun almost automatically, as it had been for his predecessors. A few historical paintings, indistinguishable from those made for Shah-Jahan, were executed early in Awrangzeb's reign and must have been intended for this project.[36] Once the chronicle of the first ten years was completed, however, the Emperor withdrew his support, ostensibly for financial reasons. More importantly, he closed the offices that had gathered and maintained documentary and

Fig.18 Attributed to Hunhar, *Awrangzeb in a shaft of light*, circa 1660.

historical information about each imperial reign, dispersing the elaborate structure formed to record in minute detail events at court.[37] This lack of interest in historical chronicles represents an important change within the Mughal world. Moreover, as reported in the *Ma'asir-i-'Alamgiri*, an historical account commissioned after his death as a completion of the *'Alamgirnama*: 'As the emperor had no liking for pleasure, and his devotion to duty left him no time for festivity, he ordered that the chief musician Khush-hal Khan ... and others might come to Court, but must not make music. Gradually (music) was entirely forbidden.'[38]

By withdrawing his support for both the accurate recording of historical events and the pleasures which the arts could bring, Awrangzeb effectively destroyed the basic incentives for Mughal imperial painting. One result is that there are few images related to the events of Awrangzeb's reign, and those that exist are generally uninterested in either recording or creating an artistic equivalent for the opulence and physical *richesse* of court life. Nor is there interest in the kind of acute study of individual appearance and character that had been the most innovative aspect of Mughal imagery within either Indian or Islamic traditions.

It is possible to place the paintings made during and immediately following Awrangzeb's rule into several categories. The first includes paintings made for Awrangzeb and his circle that continued the character of painting under Shah-Jahan, supporting those artists who had previously worked in the imperial studios. The majority of these images can be dated soon after the accession.[39] The second group consists of works made in a more severe, simplified style that lasted throughout the reign and became indicative of Awrangzeb's personal taste and interests. It may be that the Emperor's financial austerity forced a decline in the physical mate-

Fig.19 Bichitr, *Shah-Jahan with Asaf Khan*, circa 1640.

Fig.20 Unknown artist, *Shah-Jahan with Asaf Khan*, early 18th century.

Fig.21 Unknown artist, *Bahadur Shah enthroned*, circa 1710.

Fig.22 Unknown artist, *Prince Azimushshan enthroned with Khwaja Khizr*, circa 1713.

rials used by painters (for example, the pigments), and perhaps less time was allotted to the portraits and other studies that were produced. In comparison to the works made during the first half of Shah-Jahan's reign, these illustrations for Awrangzeb are far less opulent in style and materials. They do, however, represent a new development, and the basis for the majority of later illustrations for Mughal patrons (Fig.18). The third category includes copies of earlier paintings (Fig.20), and the fourth, evocations of earlier images made in recognizably later styles (Figs 21–23).

Awrangzeb in a shaft of light (Fig.18) can be dated to the early years of the Emperor's reign, and it may include a portrait of Kushhal Khan, the musician mentioned above.[40] The explicit visual depiction of an emperor illuminated by a ray of light sent down from the heavens was only established in Mughal painting under Shah-Jahan and its source in European iconography can be seen in *Shah-Jahan with Asaf Khan*, signed by Bichitr (Fig.19) – an image in which the light emanates directly from God the Father.[41] In the later illustration, the effect of light is more theatrically realized, although the overall workmanship is less finely detailed than in either *Shah-Jahan with Asaf Khan* or any other *Padshahnama* illustration. There is also less interest in evoking the volumes, textures, or physical richness of the material world. This change of emphasis, which defines the new direction of Mughal painting, is seen also in an early eighteenth-century version of *Shah-Jahan with Asaf Khan* (Fig.20). While compositionally identical, it completely lacks the subtlety of portraiture and the technical skill found in Bichitr's earlier image (Fig.19), qualities that made earlier imperial works so distinctive.[42]

Early in the eighteenth century there was a revival of concern at court for historical records and for illustrations that evoked the sumptuousness of imperial life and ceremony. Several important paintings relating to the Emperor Bahadur Shah (r. 1707–1712), his son Azimushshan (1664–1712), and the Emperor Farrukhsiyar (r. 1713–1719) are known.[43] *Bahadur Shah enthroned* (Fig.21) consciously evokes the compositional format of *Padshahnama* illustrations. While he never formally acceded to the throne, Azimushshan did proclaim himself emperor in 1712, and a remarkable painting shows him aureoled, enthroned, and accepting the imperial regalia from Khwaja Khizr (Fig.22). He died about a month later, however, when his elephant disappeared into quicksand during a battle. This accession portrait, which must have been commissioned by Azimushshan and completed soon after his proclamation, the portrait of Bahadur Shah, and a closely related posthumous painting of the aged Awrangzeb,[44] are the closest equivalents found in the eighteenth century to the style of Shah-Jahan's painters: they revive interest in matters to which Awrangzeb had seemed indifferent – the richness and intensity of architectural and textile patterns, or careful technical accomplishment, for example. Even these later works, however, lack individualized characterizations, and the figures have no mass. These traits, and the increased concern for the decorative brilliance of flat surfaces, denies them the sense of physical substance so important to the naturalistic effects achieved in earlier times.

The revivalism that centered around the imperial personalities of Bahadur Shah and his descendants was mirrored at courts such as Lucknow, from where the Windsor Castle manuscript traveled to England. The local rulers there were eager to proclaim their own importance and wealth by evoking the world of Shah-Jahan, which now represented a model of achievement to which they aspired. Images such as *The wedding procession of Dara-Shikoh* (Fig.23), for example, are in a distinctively Lucknow style.[45] The painting's inspiration would have come from a variant of the wedding procession found in the *Padshahnama* (**23–24**), but here no one individual – including the prince and his father – is a recognizable portrait, and there is no psychological interaction among the attendants. These figures are so generalized that they have become virtual cartoons.

In the sixteenth century, when the Mughals first established their power, the rulers of several of the varied kingdoms that together controlled north India also patronized painters. The resulting styles were defined both by local traditions and by the cultural affiliations of the rulers, and the latter were inevitably affected by religion – Hinduism, Jainism, or Islam. As the political power of the Mughals grew, however, and their artistic aspirations were made clear, Mughal ideals in the arts profoundly affected these existing traditions. In some cases, local rulers tried to emulate the Mughals, while others refused consciously to change. In each case, however, Mughal arts were central to the definition of local styles.

Partly as a political tactic, Akbar had arranged marriage alliances with the daughters of several important Rajput rulers, men whose political reliability was vital to the stability of Mughal rule. Jahangir's mother was Hindu, a Rajput princess of Amber (Jaipur), and he, too, arranged alliances for himself with Amber, as well as with the neighboring states of Jodhpur, Jaisalmer, Bikaner, and Bundelkhand.[46] By blood, Shah-Jahan – whose mother was the daughter of Mota Raja Udai Singh of Jodhpur – was actually three-quarters Hindu. And while a mother's family had little significance (and was rarely even recorded) within official imperial lineages, these links did have important artistic results. Residence at and involvement with the Mughal court inevitably affected the social habits and artistic taste of the Rajputs. The *Padshahnama* records the first appearance at court of Jaswant Singh Rathor, the great-grandson of the Mota Raja (see pages 190–91), and when he returned home he directed his artists to paint with the attention to realism in portraiture and polished technique that he had seen at Agra.[47] Madho Singh Hada, son of the Rajput ruler of Bundi, is shown prominently in *The death of Khan Jahan Lodi* (**16**). It was Madho Singh who actually killed the rebel leader, and for this act the Emperor awarded him a portion of his father's kingdom to hold and rule as the separate state of Kota. Many paintings executed for the rulers of Kota in the early and mid-seventeenth century are also adaptations of Mughal style,[48] as are works from Bikaner.[49] The relationship of the architectural depictions in the *Padshahnama* and a series of paintings attributed to the Rajput state of Mandi, in the Punjab Hills, is also important; compare Fig.24, *Krishna celebrates Holi festival*, with Fig.144, for example. Skill at depicting grand imperial spaces in which small-scale figures convincingly enact court ceremonials, and for which a ground-plan could easily be made, is a result of Mughal interest in historical and spatial recording. Earlier Rajput works were not concerned with material documentation or the expressiveness of empty space – these were Mughal contributions to the arts of sixteenth- and seventeenth-century India.

The Rajput (and several other) states drew from different aspects of the imperial Mughal style according to the interests and tastes of their own rulers, and the development of each of these Rajput schools forms a separate topic that cannot be explored here. It is important to realize, however, that while imperial Mughal painting after the middle years of Shah-Jahan's reign did not maintain either the quality or inventiveness of its first productions, its continued influence was assured through its role in expanding the range and meaning of painting at Rajput and other workshops across north India.

THE ASSEMBLY OF THE PADSHAHNAMA

The revival of historical illustration that occurred at the Mughal court, and in its provincial counterparts, in the early eighteenth century may have encouraged renewed interest in the *Padshahnama*. The fate of this manuscript can only be fully understood, however, if we return to the problem of how such illustrated books were assembled.

The Windsor Castle *Padshahnama* of Abdul-Hamid Lahawri is the major imperial manuscript relating to the period of Shah-Jahan's rule.[50] Few other illustrated manuscripts of any text that can be dated to his reign are known, in fact, and it may be that the different format provided by *muraqqa*s (albums) was more appealing to the Emperor than the illustration of literary texts or historical narratives. Albums were elaborately designed volumes that alternated facing pages of – often independent – illustrations with paired pages of calligraphic specimens. The pages of the best albums were enclosed within superbly designed borders, with the right and left halves of each opening composed as mirror images of

Fig.23 Unknown artist, *The wedding procession of Dara-Shikoh*, mid-18th century.

Fig.24 Unknown painter, *Krishna celebrates Holi festival*, circa 1740.

each other, and with no two openings exactly alike.[51] Unlike manuscripts, there was no narrative or literary sequence linking the passages of calligraphy in albums; they were chosen purely for aesthetic reasons. The images, however, tended to be gathered into categories such as imperial portraits, depictions of nobles, scenes with women, or flora and fauna. Illustrations were drawn from individual paintings already in the imperial collections, new commissions, pages from manuscripts that were disassembled or unused, images looted in battle, diplomatic and New Year's gifts, and so on. Mughal albums were, in effect, imperial scrapbooks, and often several originally distinct images were artfully combined on one page. The willingness to move illustrations from one context to another indicates, therefore, that the combination in the *Padshahnama* of text with pre-existing illustrations intended originally for some other use was in no way unusual.

The first known illustrated version of the official history of Akbar's reign, the *Akbarnama* written by Abu'l-Fazl, provides a theoretical precedent for the re-use in the *Padshahnama* of images made for earlier historical manuscripts. Recent research has revealed that Abu'l-Fazl's words have occasionally been written on separate pieces of paper and superimposed (glued) over an earlier, different, and presently unidentified text.[52] It has been concluded that a previous chronicle was made unacceptable by the new official history, and that this new narrative was then substituted in an existing manuscript with such skill that the change remained unnoticed, at least by European and American scholars, until very recently. In the case of the *Padshahnama*, however, the revision was handled differently: illustrations removed from (or never placed in) their original contexts were inserted into empty spaces within the new copy of Abdul-Hamid Lahawri's text. As noted above, the availability in the imperial workshops of quantities of such independent images awaiting assignment to specific projects seems to have been normal.[53] So many other examples of re-assigned paintings exist that there can be no doubt about the mobility of paintings among the varied projects of the Mughal painting workshops.[54]

At the time that the *Padshahnama* was put together, there existed a group of separate paintings large enough for adequate matches to be found for the spaces left open for illustration among the text pages. Robert Skelton has observed (in conversation) that the existence of distinct patterns of worm holes in the illustrations as a group and in the text pages suggests that these two elements of the present volume were each stored separately for some length of time before they were combined. It is most probable that such a quantity of diverse material would have been kept within the inventory of the imperial workshops, where all these pages were originally made and housed.

It is unlikely that the *Padshahnama* was assembled in Shah-Jahan's lifetime. Only one year remained between the writing of the text and the end of his reign in 1658; his usurpation by Awrangzeb provides the most convincing explanation of why the project was never completed as planned. While the manuscript could have been put together during this short period, its inaccuracies would indicate an indifference to historical fact that completely contradicts the reasons for commissioning the history in the first place. Moreover, the opening *shamsa*s lack the central imperial inscription or seal that was normally placed in volumes made for the imperial library.[55] It is also extremely unlikely that the

Fig.25 Attributed to Payag, *Officers and wise men*, circa 1640. From the Ardeshir Album.

book would have been assembled under Awrangzeb, with his known antipathy to the arts and his hatred of Dara-Shikoh (who is so clearly celebrated in the book). A date between Awrangzeb's death in 1707 and the sack of Delhi by Nadir Shah in 1739 therefore seems most likely. During this period the imperial treasuries and storerooms were constantly being drained to provide funds for the varied factions attempting to seize power. To cite only one instance, in December 1712 the palace at Delhi was looted of precious materials that had belonged to Babur, gold vessels assembled by Akbar, the unused stock of gold and silver remaining in the workshops, jewels, and carpets. Even the gold plating on the palace ceiling was removed.[56] It is at such times that manuscripts and paintings would also have been seized, to be sold to whomever was interested.

The period following Awrangzeb's death was one of nostalgia for earlier Mughal ceremonial, as was noted in discussion of the portrait of the ill-fated Azimushshan (page 125, above, and Fig.22), and this provides a possible historical context for renewed interest in the remains of the unfinished imperial *Padshahnama*. The broad outer borders presently framing each page also point to these years. On different paper than that of either the text or the illustrations, they are decorated with repeat motifs of Shah-Jahani origin (Fig.65). The designs are within the general decorative repertoire that served Mughal and Indian art into the nineteenth century, however, so no specific date for them can be proven. The closest equivalent is the broad border decoration in a volume known now as the Ardeshir Album (Fig.25).[57] This is a group of images and calligraphies – several of Shah-Jahani date – that was almost certainly assembled during the period of Muhammad Shah's rule (1719–1748).[58] The craftsmanship in the assembly of the pages and in the execution of the border designs is less fine than that of the *Padshahnama*, however.

There is therefore no evidence presently available to date exactly the formation of the great *Padshahnama* manuscript. Its combination of a group of imperial paintings made for a variety of projects and almost certainly kept in stock within the Emperor's workshops, with a separate text written for the Emperor but never completed with its own planned illustrations, was almost certainly the result of a quirk of fate, which in itself offers us more insight into Mughal history than Shah-Jahan's chroniclers had ever intended.

1. The structure and purposes of the Mughals' news-gathering system is briefly discussed by Jadunath Sarkar in the introduction to his translation of a history of Awrangzeb's reign; see *Ma'asir-i-'Alamgiri*, pp.iii–iv.
2. *A'in-i Akbari*, I, pp.268–69. This particularly informative section of the text is entitled 'Regulations Regarding the Waqi'a-Nawis [Newswriters].'
3. The most important of these are Gulbadan Begam's *Humayunnama* and Jouher's *Tezkereh al Vakiat*, both assembled at the request of Akbar to serve as sources for his history projects.
4. *A'in-i Akbari*, I, p.113.
5. As discussed in the Introduction, pages 17–19.
6. For a discussion of these manuscripts, see Smart 1973 and Beach 1981, p.77. The translation, from the original Turkish, was completed in 1589. Illustrations made for the initial copy presented to the Emperor have been published in Suleiman 1969.
7. An important painting in the Jahangir Album in Berlin that shows Humayun and his family in the mountains has recently been identified with a specific incident; see Wescoat 1991. For the historical account of an interesting bird which Humayun had painted, see *Tezkereh al Vakiat*, p.43.
8. These included the *Tutinama* (*circa* 1560), *Hamzanama* (*circa* 1562–77), *Dewal Devi Khizr Khan* of Amir Khusraw (dated 1567–8), and *Anwar-i-Suhaili* (dated 1570–71). For illustrations and discussions of these works, see especially Beach 1987a; Chandra 1976; and Seyller 1990.
9. For discussion of the historical attitudes that produced these volumes, see Beach 1981, pp.68–71 and 128–35.
10. See Beach 1978, pp.155–58, where further references are given.
11. For discussions of the early formation of the Mughal style and its relationship to painting in Iran, see especially Beach 1987a and Chandra 1976.
12. For a survey of Timurid art, see Lentz & Lowry 1989.
13. Paintings in Bukhara style are found in both the *Hamzanama* and *Anwar-i-Suhaili* manuscripts; see Beach 1987a, figs 45 and 48.
14. Lowry & Beach 1988, no.179.
15. See Beach 1975.
16. These and other Akbari historical manuscripts are discussed further in Beach 1981.
17. See Beach 1987a, pp.115–17.
18. See Sen 1984. For a discussion of the links with the *Timurnama*, see Beach 1987a, pp.115–22.
19. See Beach 1981, pp.102–23, where bibliographic references and a list of dispersed pages are given. For a proposed revision of the date of the work, see Seyller 1990.
20. For a discussion of the accuracy of architectural depictions in the *Akbarnama* manuscripts, see Beach 1987b.
21. A list of dispersed *Jahangirnama* paintings is given in Beach 1978, pp.63–64.
22. Translated by Wheeler Thackston.
23. Translated by Wheeler Thackston.

24. The right-hand page may be the work referred to by Jahangir in his memoirs. It was identified in its recent exhibition at the Metropolitan Museum of Art, New York, as the celebration of *Nawroz* (New Year) in 1616, but the presence of Raja Man Singh (who died in 1614) and the inclusion of the young Khurram (b. 1592) standing by his father contradicts this theory.
25. Translated by Wheeler Thackston.
26. But see also the comparisons cited on page 181.
27. An eighteenth-century adaptation of the composition is found in the Polier (Hamilton) Albums in Berlin; see Enderlein & Hickman 1979, no. 27.
28. See the key to participants on page 163.
29. *Ma'athir ul-Umara*, I, p.413.
30. The painting by Nanha is also within the range of sizes known for *Padshahnama* folios. Given its closeness to La'lchand's illustration, there is little reason to be certain that it was intended specifically for Jahangir's memoirs – subject matter alone is not always adequate proof, especially since this was an event included in the *Padshahnama* as well.
31. In Fig.17, Nanha can be seen at work just behind the Emperor at the Rana's ceremony of submission.
32. B.N. Goswamy (1975, pp.29–35) has shown that Pahari illustrations were checked regularly for their correspondence to related texts, and it is likely that a similar system was in place in the Mughal studios to assure historical accuracy in the official manuscripts of the emperors' lives.
33. Roe 1967, p.310.
34. Roe 1967, p.329.
35. Manrique 1927, II, p.294.
36. See the *Darbar of Awrangzeb*, attributed to Hashim (S.C. Welch 1978, pl.37); *Awrangzeb in darbar*, an illustration almost certainly intended for the original *'Alamgirnama* (S.C. Welch 1963, no.58); and a scene of *Awrangzeb hunting nilgais* (S.C. Welch 1978, pl.38). The latest scene in this category is *Awrangzeb hunting lions* (Arnold & Wilkinson 1936, III, pl.91), which shows an older Emperor, and can be dated to about 1680.
37. This is discussed briefly by Sarkar 1972–4; see *Ma'asir-i-'Alamgiri*, pp.ii–iii.
38. *Ma'asir-i-'Alamgiri*, p.45. (This is included at the opening of the account for the eleventh regnal year, 1668–9).
39. See note 36.
40. The identification of the subject matter has been made by Terence McInerney and Dr Ellen Smart. For a related portrait by Hunhar, which seems to show the younger seated figure, who is singing, see Falk & Archer 1981, p.406 and fig.93.
41. For related images, especially a portrait of Jahangir, whose titles included *Nuruddin* ('Light of the World'), see Beach 1981, no.17b. There, however, the light is more typically depicted as a radiant aura, rather than light sent down to earth from the heavens.
42. In its most extreme form, this led to a real decline in illustrative powers that is most evident in a series of illustrated *Padshahnama* and *Shahjahannama* manuscripts datable to about 1800, in which the workmanship is crude, the details muddy, and the personalities undifferentiated. These were presumably made to sell to the English traveling in India. See Losty 1982.
43. For another important portrait of Bahadur Shah, see Binney 1973, no.71; and for Farrukhsiyar, see Enderlein & Hickman 1979, pl.55.
44. Chester Beatty Library, Dublin, attributed to Bhavani Das; reproduced in Leach 1981, fig.164. This work and the portrait of Azimushshan are so similar in style and size that they may have been intended for a later imperial history.
45. A second copy of this composition is in the National Museum of India; see Carroll 1972, p.80. It is signed 'work of Hajji Madani.'
46. For a recent discussion of Jahangir's wives, see Findly 1993, pp.123–27.
47. One result is a drawing, *Jaswant Singh and courtiers*, a work that is almost completely in Mughal style; see Beach 1992a, fig.93.
48. See Topsfield & Beach 1991, no.14 and fig. 2; and Beach 1992a, fig.126, a drawing that derives its style from the background scenes found in such *Padshahnama* illustrations as **40**.
49. The earliest paintings from Bikaner, datable to the beginning of the seventeenth century, are rough in technique, with angular forms and strong colors that betray their derivation from the pre-Mughal styles popular over much of north India. By the middle years of the century, however, a number of artists had developed more precise and controlled techniques and gained access to finer materials, especially pigments. They began to depict delicately featured women and men with detailed clothing that evoked the actual texture of cloth, and these physical types are very close to those encountered in the *Padshahnama* (see **25–26**). Since texts relating to Krishna were especially popular at Bikaner, artists were challenged by the need to adapt figure types invented for one context to another that was quite different. This challenge gave Bikaner painting in the second half of the seventeenth century extraordinary dynamism.
50. The earliest extant commission for an official history of the reign was given to Muhammad-Amin Qazwini in 1634, and he finished his narrative (titled the *Padshahnama*) in 1646. Qazwini's text, which covered the first ten years of the reign, was then incorporated into Abdul-Hamid Lahawri's more extensive history. Earlier historical accounts for specific periods or events have been reported (see *Shah Jahan Nama*, pp.xv-xxvii), but these seem to have been part of ongoing documentation rather than polished literary history. One historical painting that is dated prior to Qazwini's commission (Fig.142) therefore seems to have been made on speculation, with no specific project in mind.
51. Many of these albums are listed and discussed in Beach 1978, pp.74–77. See also S.C. Welch *et al.* 1987, which is a more recent, full publication of the Kevorkian Album, now divided between the Metropolitan Museum of Art, New York, and the Freer Gallery of Art, Washington, DC.
52. Seyller 1990. A detached page from the manuscript in the Freer Gallery of Art (Beach 1981, no.9) may provide further important evidence about this volume. The text clearly describes Babur's capture of Kabul in 1504, although the illustration equally incontrovertibly shows Humayun enthroned; moreover, in this case, both text and illustration share a single sheet of paper. This page therefore suggests problems with the manuscript's production beyond those cited by Seyller.
53. This principle may well have been established by Akbar, and considerations of the dates of the various *Akbarnama* manuscripts should take this working method into account. Since only dispersed pages for an imperial *Jahangirnama* manuscript are known, it cannot now be determined whether a final bound copy covering the entire reign was ever completed for the imperial library.
54. The *Tuhfat al-ahrar* mentioned on pages 116–17 above (and see Fig.8) is one of many manuscripts in which text was only later enhanced by paintings. Among further examples of image mobility in the Shah-Jahan period is the *Masnavi* of Zafar Khan (Royal Asiatic Society, London), in which paintings of the 1640s were placed in a manuscript with text dated 1663 (see Beach 1978, pp.110–11).
55. Note S.C. Welch *et al.* 1987, no.1 (with an inscription naming Shah-Jahan) and no.5 (stamped with a seal of Awrangzeb).
56. Irvine 1971, I, pp.219–21.
57. I am grateful to Robert Skelton for this observation. For a publication of one section of the album, see Sotheby & Co. 1973. The volume is also discussed in Khandalavala 1953–5.
58. Because its reverse is blank, a portrait of the Emperor Muhammad Shah must have served as either the opening or closing image of the volume, a prominence that suggests either his patronage or that of someone close to the court at that time. See Sotheby & Co. 1973, lot 38; Binney 1973, no.76.

THE HIERARCHICAL PRINCIPLES OF SHAH-JAHANI PAINTING

by Ebba Koch

SHAH-JAHAN'S ART AS STATE ART

Under the rule of Shah-Jahan (r. 1628–1658) the Mughal Empire entered its classical phase of greatest prosperity and stability. From the earliest conquests of Babur (1526), which were later enlarged and consolidated under Akbar (r. 1556–1605) and Jahangir (r. 1605–1627), the Mughal Empire joined those of the Ottomans and Safavid Persia as one of the three leading powers of the Islamic world, and was known in East and West as the Empire of the 'Great Moghul.' Under Babur the Mughal court was somewhat informally organized, with Babur as *primus inter pares* to his Central Asian followers, but it developed – through the efforts of Akbar, in particular – into the court of an absolute ruler and head of a centralized state who personally and diligently oversaw the administration of his vast empire, supported by his amirs and *mansabdar*s, the militarily structured ruling elite. Distinction in the hierarchy of the bureaucratic apparatus had to be acquired through personal achievement, as acknowledged by the emperor; even the succession to the throne was not regulated by primogeniture.[1] This fused the ruling class – whose members belonged to diverse ethnic and religious communities reflecting the heterogeneous components of the empire – into a coherent administrative body which, by the time of Shah-Jahan, depended entirely on the emperor. This applied more strongly to the Muslim nobles than to the indigenous Rajput chieftains who were able to retain some autonomy, including a hold over their ancestral dominions, when incorporated into Mughal service. Since imperial authority represented the main guarantee for the functioning of this hybrid political system, the emperor's appeal to his subjects' loyalty depended largely on his personal charisma, which was supported and magnified by the myth created around his kingship. Here the Mughals drew with their own disarming lack of inhibition on all autocratic notions of kingship they could lay their hands on, and which they deemed to serve their cause. The ideas and systems they exploited included the Muslim caliphal, Koranic prophetic and mystical Sufi; Achaemenian and Sasanian Persian, and Perso-Islamic; Turko-Mongolian; Hindu and Islamic Indian; and Christian-Messianic, as well as recent European concepts of universal monarchy. Upon this eclectic plethora each ruler set his own accent. Shah-Jahan tried even more consistently than his predessors to live up to his self-created image, and architecture, art, poetry, historiography and court life during his reign all served to manifest the imperial ideal.

The hierarchic relationship between Shah-Jahan and his subjects was confirmed and acted out symbolically in rigid court ceremonial, repeated daily, focusing on the Emperor. Hand in hand with the regulation of every detail of court life went an ever-increasing formalization of the court arts, which were represented as a necessary instrument to rule. The monarch who exerted himself in the sphere of artistic representation was sure of the loyalty, obedience and recognition of his subjects, as well as the esteem of his rival rulers:[2] '[Such matters] may belong [to the category] of beautiful and external things the existence of which is not so necessary [in the context] of overall rule, but they must be [present] to give full distinction and spectacular display – the more so since it becomes a matter of increase of pomp and power, magnificence and elegance .... It is evident that the increase of such things creates esteem for the rulers in the eyes [of the people] and augments respect [for the ruler] and [their own] dignity in [their] hearts. In this form the execution of divine injunctions and prohibitions and the enforcement of divine decrees and laws which is the ultimate aim of rulership and kingship are carried out in a better way.'[3]

The political role assigned to the arts, and the emperor's function as the supreme administrator of his court-led state, meant that Shah-Jahan sought to assert as close a control over his artists as over his amirs and *mansabdar*s. The personal tastes of the first six rulers of the dynasty, the 'great' Mughals, dominated the arts of the Empire, to the extent that art historians have used the spans of their reigns to indicate artistic periods, but Mughal court art was never so strictly regulated as under Shah-Jahan. All his historians agree that Shah-Jahan made the personal overseeing of his artists a part of his daily routine, thus acting as his own artistic director in a manner typical of Shah-Jahani perfectionism. The focus of his attention – and the best known aspect of Shah-Jahan's patronage – was architecture, the imperial art *par excellence*.[4] Less well known – because only mentioned by Shah-Jahan's first court historian, Qazwini, who was replaced by Lahawri in about 1638, after covering the first ten years of Shah-Jahan's reign – is that the Emperor's daily morning session with his artists in the *Dawlat Khana-i Khass* or *Diwan-i Khass*, the Hall of Private Audiences, also included the close inspection of the work of his painters: 'Part of the time [His Majesty] spends in seeing gems [*jawahir*] and precious objects [*nafa'is*]. And part of the time he examines with care and in detail the masterpieces of artists [*karnamaha-yi arbab-i sana'i'*], such as painters [*musawwir wa naqqash*], carvers and engravers [*naqqar*], goldsmiths [*zargar*], enamelers [*minakar*] and [illegible]. The

Fig. 26 Detail of **31**.

superintendants of the imperial buildings [*daroghaha-yi 'imarat-i sarkar-i khassa-i sharifa*] together with the masterful architects of excellent abilities [*mi'maran mahir nadira-kar*] bring architectural designs [*tarha-yi 'imarat*] before the exalted sight [of the Emperor]. And since his most pure mind is inclined entirely towards building ['*imarat*] – which is the cause of the flourishing of the state and the adornment of the apparent world [*kargah-i zahir*] and which gives value and splendor to the material world [*karkhana-i surat*] – he attends to it fully by creating most of the designs himself and also by carrying out appropriate changes to whatever the architects have thought out.'[5]

This representation of Shah-Jahan as being above all involved in the creation of architecture certainly reflects his personal preference – he was one of the most tireless builders in Mughal history – but it accorded also with the view of architecture as the most appropriate artistic form of expression for a ruler. As the most prestigious and useful art, it had the capacity to represent the ruler and his state in the eyes of a wider public, and to provide a lasting memorial to his fame.[6]

Even when confined to the miniature format, painting's status was always more controversial than that of architecture and the applied arts, because of the theological injunction against the depiction of human and animal life. This seems to be the reason why Lahawri, who revised Qazwini's first volume of Shah-Jahan's history, omits the explicit reference to the painters in his description of the daily meeting of Shah-Jahan with his artists.[7] This purging of the text to play down the Emperor's involvement with painting reflects the increasingly orthodox concerns of Shah-Jahan, which might eventually have had an impact on his patronage of painting.[8] Lahawri's account, which was the version finally approved by the Emperor, has not failed to make its desired impression on posterity. Until recently, art historians have preferred to see Shah-Jahan as a patron who was not personally involved in painting:[9] the evidence strongly suggests, however, that in the first ten years of his reign at least Shah-Jahan exerted as decisive an influence on the painted illustrations to his official history as he did on the text itself, which he edited personally in lengthy sessions with his historian and which is thus a far more personal statement by the Emperor than its formal style would lead us to believe: 'Sometimes, the writer of these pages [Qazwini] enters the assembly[10] by imperial command and reports on the content of each and every narrative that has recently been written. If a slip in the contents or an error in the expression should have occurred, His Majesty corrects it and guides this worthless speck of dust to the exalted words and pleasing turns of phrase that occur to the royal mind and the inclusion of which in this history would occasion felicity of expression, indeed which are necessary concomitants to this art. This honor lasts two or three *gharis*,[11] or longer when there is more work.'[12]

The Emperor did not want to leave historiography to the historians, and the same was certainly true of its illustration. Shah-Jahan's highly personal involvement in the work of his painters is borne out by the pictorial evidence of the artists' direct access to the ruler. The artists not only included their self-portraits in the *darbar* scenes (see **5**, **9**, and **39**) but also left written statements or signatures on their depictions of the wall below the *jharoka*, or of the platform or stool on which the *'Arz-i Mukarrar* (Reviser of Petitions) or the *Mir Bakhshi* (Head of Personnel) would step up to present a petition (see **5**, **9**, **10**, **38** and **43**). The painter Murar, for instance, openly expressed his hopes for generous remuneration, for which he promised, in return, his best artistic effort (**38**). By bringing themselves to the attention of the Emperor in such a prominent place, the painters – several of whom were *khanazad* (born in the imperial household, and therefore second-generation servants) – demonstrated their special relationship to him.

The text and the illustrations of the *Padshahnama* emerge as an immediate expression of the close collaboration between the imperial patron, his historians and his painters. Their objectives were the creation of appropriate images in an official style, both in writing and in painting, and the maintenance of a uniform standard of quality. Artistic freedom was suppressed almost as thoroughly as political freedom. There is a certain contradiction in this: the Emperor had the most creative and skilled artists at his disposal but their talent was subject to the strictest form of control. This left only a narrow margin of licence for the artists, who nevertheless managed to produce work that was both of high quality and innovative.

Several painters in Shah-Jahan's atelier had already worked under Jahangir and had illustrated his history, the *Jahangirnama*.[13] While closely patterned on the painterly achievements of the previous period, the *Padshahnama* illustrations systematized and transformed trends set by the paintings in the earlier chronicle. The fundamental components of miniature painting, in terms of composition and figure arrangement, as well as antithetical stylistic modes, were systematically explored to political ends, to create programmatic statements of order and hierarchy, the basic tenets of Shah-Jahan's ideology. The themes of art were also affected. The focus was on static court assemblies or *darbar* scenes, which evolved from the depiction of specific historical events into an ever-repeated and only slightly varied type of abstract image of the Shah-Jahani power structure. Other subjects included court festivals, the spectacle of the elephant fight and the always popular imperial hunt. The Akbari theme of the important military action (largely abandoned in the *Jahangirnama*)[14] was also revived and reinterpreted according to the new principles of Shah-Jahani painting: it no longer depicted acts performed by the Emperor himself but by his agents, who were shown taking fortresses and cities in his name, and punishing rebels on his behalf. The pursuit of naturalism, always a characteristic of Mughal painting, was channeled into specific areas within the context of these themes.

THE JHAROKA IMAGE AND THE FORMATION OF THE MUGHAL GROUP PORTRAIT

Almost one third of the illustrations to the Windsor manuscript show the assembly of the *jharoka-i khass-u-'amm*. The Emperor's autocratic rule required confirmation through regular public appearances, and special emphasis was therefore placed on his public audiences, which came in two forms. The most open was that which took place in the *jharoka-i darshan*, the viewing window on the outer wall of the palace (Figs 157, 158). Here, the Emperor would appear every day at sunrise to comply with the ancient requirement of Persian and Indian kingship to be accessible – at least visually – to all of his subjects. From the *jharoka-i darshan* he proceeded to the *jharoka-i khass-u-'amm*, a balcony projecting from the back wall of his audience hall (the *Dawlat Khana-i Khass* or *Diwan-i 'Amm*), in the courtyard of the *khass-u-'amm*. This assembly or *darbar* of the grandees of the empire (*khass*), the lower ranks, and wider public (*'amm*) was Shah-Jahan's most public form of holding court and conducting state business within the palace. The *darbar* took place twice, in the morning and in the afternoon. The nobles and their retinues had to position themselves according to their rank below the *jharoka* in the audience hall and in the courtyard, in areas delineated by railings. The higher a noble's rank, the closer was his position to the Emperor. All participants in the *darbar* had to stand – although they were allowed to support themselves on ceremonial staffs – and nobody was to move or to speak without being requested to do so.[15] When everybody was in place, the Emperor appeared in the *jharoka*. The *bakhshi*s then presented the requests and petitions of the *mansabdar*s and the Emperor dealt with appointments, salaries, awards, and raises in rank. Viewing of the imperial horses and elephants, and the reception of foreign dignitaries and delegations, also took place during this assembly, and it provided a stage for the great court festivals, in particular *Nawroz* (the Persian New Year) and *julus* (the anniversary of the Emperor's accession). The *jharoka-i khass-u-'amm* was the administrative center of the Mughal empire and the focal point of court events, where the power and pomp of the 'Great Moghul' were enacted.[16] The frequency of its representations in the Windsor manuscript takes account of the fact that the *jharoka*, in reality as in painting, had become an image of the Mughal state. As the 'stage for enacting political rituals, the *jharokha* thus expressed themes central to Mughal political culture: the subordination of all state servants ..., the corporate solidarity of the ruling class, and the precise position of each member relative to others in the graded hierarchy of state service.'[17]

The *jharoka* paintings were executed by several artists but all based their *darbar* scenes on the same compositional formula. The Emperor appears in a central position in the *jharoka*, a raised seat projecting from a gallery in the back wall of the audience hall, sheltered by a cupola, in front of which is hung a canopy: below him his nobles are assembled in two groups facing each other. This basic scheme, the architecture of which was modeled on the actual architecture of the palaces in the capital cities of Agra and Lahore (see Fig.129) – could vary in its details to accommodate the peculiarities of a specific place, period, or event. Within these parameters there was also a margin for artistic variation, where the painter could express himself more freely.

The image – like the ceremony itself – originated in Akbar's time: we find among the illustrations of the *Akbarnama*, the history of Akbar's reign, a number of representations of the Emperor appearing before his nobles either in an open structure under a cupola on the roof of a palace building or at an upper-story window (Figs 12, 27). These structures form part of a larger architectural setting given in uncoordinated perspectival views. The architecture is quite recognizably Akbari, even though in no instance can it be verified against surviving buildings: these are free paraphrases of the architecture of the period. Neither the settings nor the form of Akbar's audience show any uniformity. All the participants in

Fig.27 Kesu Kalan and Madhu Kalan, *Akbar receiving the submission of the rebel brothers* ..., circa 1590.

these public *darbars* appear as small figures, in keeping with the schematic figure conventions of Persianate painting, although more realistically interpreted. They are usually shown not standing still but moving about in the architectural settings, gesticulating and apparently torn by different impulses. The small format of the figures and their involvement in their activities gives the painter little opportunity to endow anybody's face with personal traits beyond a very general likeness. This is also true of the Emperor, who appears usually in three-quarter profile.

At the end of Akbar's reign, his painter Manohar began to take a closer look at *darbars*, depicting them as small private meetings at which the Emperor appears not in a formal *jharoka* setting but on a canopied platform or throne surrounded by a few selected nobles (Fig.28).[18] The artist suppresses action and gives his quietly standing figures more substance, emphasizing in particular the heads in relation to the bodies, which allows him to define character more strongly and to convey inward psychic life. These tendencies were further developed in depictions of *darbars* under the succeeding imperial patron, Jahangir.

Jharoka scenes from Jahangir's time were rare – more frequent were representations of the Emperor surrounded by a number of close nobles, evolved from the type of private *darbar* created by Manohar (see Fig.14).[19] All *darbar* scenes placed a new emphasis on the realistic representation of the setting as well as of those taking part in the assembly (Figs 13, 29). The painting of the *jharoka* corresponds to its real form, for which we have evidence from Akbar's reign onwards, in such palaces as those at Fatehpur-Sikri and Mandu, or in the *Diwan-i 'Amm* of the Lahore fort (Fig.129). This tendency to show the architectural elements of the *jharoka* realistically was enhanced by the use of perspective, but these views are still, as in Akbari *darbar* scenes, not spatially coordinated according to a unified viewpoint. The more realistic portrayal of those who take part in the *darbar* corresponds to the more accurate representation of the individual architectural elements. The figures appear, however, as a mechanical collection of individual half- or full-length portraits, mostly in profile, all in a linear array on a single imaginary plane that contradicts the perspective of the architectural setting. They are crammed into the space below the Emperor, and one has to take a close look at the dense crowd of overlapping figures to be able to make out where it has been divided into two facing groups to give the profile portraits some inner logic. The main point of consistency is in the adherence to pure profiles for the depiction of Jahangir and most of his courtiers in all *jharoka* and *darbar* scenes. This was a radical departure from the three-quarter profile preferred in Akbari paintings.

The next step in the development of the *jharoka* image was aimed at systematizing and disciplining it, to complete its transformation from the record of a historical gathering into a standardized image of the imperial authority of Shah-Jahan. This process can be observed in a comparison of a *darbar* scene in Boston (Fig.29) with one of those painted by Murar (**32**), though almost any of the *jharoka* assemblies of Shah-Jahan would serve to illustrate our point. The Shah-Jahani *jharoka* image differs from the earlier Jahangiri representation in that it organizes the figures into a composition of strict bilateral symmetry within the architecture. The same is true of the real court session, whose architectural setting was recreated by Shah-Jahan soon after his accession in 1628. He replaced the tent erected in front of the wall supporting

Fig.28 Manohar, *Akbar receiving Mirza Aziz Koka* (detail), circa 1602–4.

Fig.29 Manohar, *Darbar of Jahangir* (detail), 1620s.

the *jharoka*, which had served until then as an audience hall, with a larger, permanent wooden hall 'in all palaces of the empire.'[20] The new halls, called *Dawlat Khana-i Khass-u-'Amm* or *Diwan-i 'Amm* were also referred to as *Chihil Sutun* ('forty-pillared') to show their inspiration from Persia, and later (before 1637) replaced by the stone versions which we still see today (Fig.104). From their appearance in *darbar* scenes and from poetic contemporary descriptions, we learn that the new wooden *Chihil Sutun*s had green pillars – like cypress trees – topped by red elements (**14, 32, 43** and Figs.144, 146 and 149), and that they supported painted ceilings (**44** and Fig.143). The outer intercolumniations of the halls were closed by a railing similar to that which had surrounded the tent hall, to allow access only to those with rank above 200.[21] The arrangement of the pillars within the wooden halls is uncertain but it may – as in the later stone version – have been based on a hypostyle mosque plan: the *jharoka* took the place of the *mihrab* in the wider central nave, and enhanced Shah-Jahan's appearance in the *darbar* with religious overtones, representing him as the *qibla* of his subjects.

Although Shah-Jahan's halls were hailed as a new achievement, highly charged with symbolic meaning, only so much is shown of their architecture in the *jharoka* paintings as was necessary to identify the setting and to structure the composition according to the principle of *qarina*, the Shah-Jahani ideal of bilateral symmetry. Works composed according to *qarina* repeat the elements of one half, reversed, on the other side of a central axis, the main emphasis being placed on the features in the center.[22] The *jharoka* and the pillars of the hall are rendered in a frontal view, replacing the earlier foreshortened views of these elements in Akbari and Jahangiri *jharoka* scenes. This change is clearly intended to translate the three-dimensional reality of the assembly into a two-dimensional abstraction, in which space is predominantly expressed by layering figures and objects and where architecture functions in the manner of a grid to organize the composition. This flattening is characteristic of Shah-Jahani adaptations of the scheme, but perspectival elements are not entirely eliminated; they are used in a calculated way to set symbolic accents, and to center the composition on the imperial axis where the Emperor presides in a pyramidal arrangement over his nobles. The courtiers, who are again arranged in two balanced groups facing each other, appear as a collection of self-contained, idealized individual portraits that are unified not psychologically – they look straight ahead and their gazes meet only because they are placed opposite each other – but solely through their subordination to the dominant figure of the Emperor. This subordination is brought about by the architectural framework which provides vertical hierarchy and centralizing symmetry.[23]

The Emperor, always identified by a golden sun nimbus, appears in profile, as do most of his dignitaries. This preference for side views in the portrayals of the ruling elite can be seen in Jahangiri paintings, but Shah-Jahani art demonstrates a rigid attachment to it. While Akbar's likeness was usually given in three-quarter profile (Figs 27, 28), and this view was sporadically employed for representations of Jahangir,[24] no portrait of Shah-Jahan is known in which he is represented other than in pure profile. Neither contemporary texts nor existing scholarship offer an explanation for this preference,[25] but it accords with the frontal representation of the architecture in Shah-Jahani painting in its tendency to eliminate three-dimensional space.

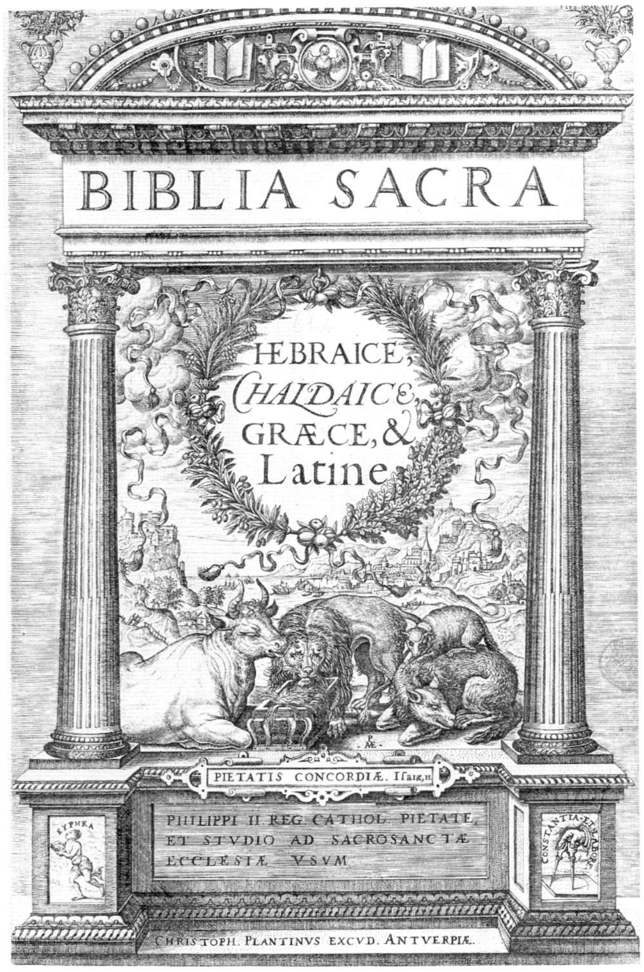

Fig.30 Title page of the *Royal Polyglot Bible*, 1569.

At first glance there seems to be a contradiction in the increasing interest in realistic portrayal and the preference for the side view. It appears that artists were deliberately limited to the profile for the purposes of Shah-Jahani group portraiture, in order to submit the individuality of the one portrayed to the corporate identity of the group, the ruling elite. The individual had to be recognizable, but only as a member of a superimposed system.[26] (The figures often have their names written on their clothes, as if to declare that the stimulus for portraiture is not to show an individual's character but merely to identify him.) The fact that the person portrayed does not face the beholder puts a visual distance between the figure and the audience and emphasizes his affiliation with the court. Even more importantly, the profile view was a means to protect Shah-Jahan and his elite courtiers from realistic portrayal: this kind of representation clearly aims to exclude all accidental factors, such as the distortions produced by foreshortening and oblique views, which, as Plato argued in condemning perspective as a deceiving reproduction of reality, do not produce a true image of the model, but only show it as it appears to the eye.[27] The standardized and idealized profile representation emphasized that the ruler and those in the state who mattered were above what is human, imperfect and subject to change.[28] Indeed, we could point out that, as a rule, official Shah-Jahani painting applies freer views to persons of no or low rank, almost amounting to a form of class distinction, as in the depiction of the nameless bearers of the imperial insignia (**5**, **6**, **13**, **19**, **32**, **39**, **43** and **44**), foreigners (**17**, **19**, and **20**), or those who had fallen out of favor or disgraced themselves (**16**, **36** and **40**).[29]

Since the architecture had to be adjusted to accommodate the plane of the figures and thus maintain the stability of the composition, realism in *darbar* images could manifest itself only in the detailed treatment of surface texture, and the painters refined their skills in this through the systematic study of the latest European techniques.[30] There were, however, artists who would not always conform to these conventions, among them the self-willed Abid (**37**) or the genial Payag (**39**). Payag in particular was able to fill the petrified *darbar* group portrait with new life by engaging closely with personality and striving for psychological unification.

In the *jharoka* scenes, symmetrical composition, types of figure arrangement and differing modes of representation are used to symbolize an abstract concept, namely the corporate identity of the ruling elite under Shah-Jahan. In addition, Shah-Jahan's position is strengthened with images symbolizing the righteousness and justice of his government. Besides the carefully orchestrated flower ornamentation, which illustrates – beyond Mughal naturalistic interests – the bloom of the empire under his benign rule,[31] pictorial symbols are placed below the *jharoka* to support, in the true sense of the word, Shah-Jahan's imperial authority. Such imagery had been introduced in Jahangir's time, when techniques of European allegory were put to the service of Mughal art to express, in a new fusion of literature and painting, concepts of rulership which had hitherto been formulated only in writing.[32] The decisive impulse toward generating Mughal political allegory was provided by the illustrations to the 'Royal Polyglot Bible,' sponsored by King Philip II of Spain (r. 1556–1598) and printed by Christophe Plantin in Antwerp between 1568 and 1572. The Bible – together with a host of Christian images – was brought to the Mughal court by the first Jesuit mission in 1580 and among their evangelistic endeavors the fathers acquainted the Mughal court

Fig.31 Grisaille wall-painting of lion and ox (detail of **39**).

with the content of its pictures. The Mughals were quick to realize the potential of such images for their own purposes and surrounded themselves with Christian symbols to enhance their aura as semi-divine rulers (**38**, **39**, and **44**). The first title page of the Polyglot Bible in particular, showing an ox, lion, wolf and sheep lying down under the rule of the Messiah (Fig.30), had a great impact on allegorical representations of Jahangir and Shah-Jahan. Animals modeled upon those in the Bible – of which the combination of lions with a sheep or with an ox, bull or cow proved most influential – became a leitmotif of the just rule of the 'Great Moghul' (**10**, **39** and Figs 31, 145 and 149). The success of this image in Mughal art was based on the fact that it related generally to the role that animal symbolism played in the Islamic pictorial tradition, for example, in illustrations of the moral tales known as the *Mirror for Princes*. More specifically it represented a visual expression of the ancient idea that peace among the animals was brought about by a just ruler – above all by the Koranic model of rulership, Solomon – which had been used extensively to eulogize Muslim princes, since at least the time of Firdawsi.[33]

In the *Padshahnama*, lions with cows or sheep, often appearing on globes (**10**, **39** and **43**), and other imagery, like the scales or chain of justice (**32** and Fig.145) or shaykhs (**44**), some holding globes or swords (**37**, **39** and Fig.149), symbolize the golden age brought about by the justice of Shah-Jahan's worldly and spiritual rule. Given Shah-Jahan's increasing orthodoxy, one might ask whether such images depict the decoration of the real *jharoka*s or are merely a convention of painted architecture. The image of *Orpheus playing to the beasts*, which was re-interpreted as a Solomonic symbol in the 1640s in the decoration of the *jharoka* at Delhi (Fig.111) – to illustrate Shah-Jahan's justice – indicates that such pictures were indeed seen in this setting. The concept of Solomon as the patron of figural arts excused the use of images, in particular in the context of throne decoration.[34]

The *jharoka* image was considered so successful that for a considerable length of time it was not adapted to the changes in the real architecture that occurred when Shah-Jahan had the *jharoka* at Agra remodeled in the form of a raised loggia at some time before 1637 (Fig.104) and had a new vaulted form created for the palace at Delhi (Fig.111) in the 1640s.[35] Later *jharoka* images merely adapted the architectural vocabulary to the latest decorative styles,[36] and only towards the end of Shah-Jahan's reign was the pictorial *jharoka* setting modeled more closely on the actual architecture (see Fig.154).

OUTDOOR SCENES AND LANDSCAPES

The outdoor scenes of the *Padshahnama* show *darbar*s in the imperial encampment, processions, pilgrimages, hunts, elephant fights, sieges and the surrender of fortresses and cities, and the punishment of rebels. The same hierarchic principles that determined the representation of figures and the architectural setting in *darbar*s and court festivals were applied to these narrative images of action in landscape.[37] Although related to historical events, the action is minimized, and, when it does occur, the main protagonists once again perform in a formalized, ceremonial manner. This is particularly true of the Emperor, his agents, or the princes. Their profile figures often contradict the logic of the action.

In scenes that involved spaces larger than the palace halls and courtyards, the painters had to solve an even greater conflict

Fig.32 Manohar, *Majnun mourns his father's death*, 1595.

between the formal impositions of linear, hierarchic court representation and the three-dimensional, naturalistic interests of the period. The principal source for all solutions generated by Shah-Jahan's studio was again Jahangiri painting, where a guarded treatment of space characterized the representation of landscapes. Wide vistas with continuous transition from the foreground into depth – Indianized adaptations and transformations of the internationally acclaimed Flemish 'world landscapes'[38] that had already reached the Mughal court before the Polyglot Bible (Fig.30), and which formed part of the experiments with landscape in Akbar's painting studio (Fig.32)[39] – were not entirely abandoned in Jahangiri painting, but were confined to the far distance. Such backgrounds assume two forms. The first is as an atmospheric backdrop to a formal, full-length portrait, in which a panoramic landscape painted in the thinnest colored washes appears, without a middleground, behind the central figure, who is painted in opaque colors and, of course, in a linear side view (Fig.33).[40] In its other form, the illusionistic vista appears, marginalized and reduced, at the top of the plane of the figure ground – where the top denotes distance (Fig.34).[41] Both versions provided a relatively undistracting environment for the figures, animals and plants seen at close range that constituted the main artistic interest of Jahangir's atelier. The artists of the *Padshahnama* were to perfect this exploration of antithetical stylistic modes to further their aims of hierarchically correct representation.

As in the architectural settings the principal figures had to appear in flattened, two-dimensional arrangements, and the painters were confronted with the task of creating a corresponding environment in landscape. They treated landscape in the first instance as a means to organize the figure composition and to support its content. They had to eliminate three-dimensional space as far as possible in the fore- and middle grounds, the areas reserved for the main protagonists. On the other hand, to make the event appear real, the two-dimensional landscapes had to be rendered as naturalistically as possible. Shah-Jahan's artists found different solutions to this common problem, depending on their ingenuity and talent for exploring the limited artistic freedom allowed them. They perfected the established conventions of Mughal painting – a structured, flat figure ground that extended into illusionistic vistas – and at the same time creatively synthesized these compositional devices with fresh sources. They also exploited the two antagonistic idioms of representation, the linear and the three-dimensional, in analogy to the content of the painting.

The simplest means of producing a two-dimensional landscape environment tinged with a touch of realism was based on the Jahangiri formula, using a flat, minimally structured terrain as a neutral backdrop for the figure arrangement, and expanding it at the top into a distant topographical vista (Fig.34). This formula was used by various painters when they were commissioned to illustrate ceremonial events, such as the solemn wedding processions along the Jumna River at Agra (**21–22**). Murar and Bhola interpreted it in architecture for their processions of the princely bridegrooms in the courtyard of the *Diwan-i 'Amm*, an outdoor space in the interior of the palace (**23–24, 27–28**). Even what appears to be the most realistic landscape of the Windsor manuscript, the setting for the elephant fight (**29**) is based on this formula (while the fight was not a formal event, it did involve the Emperor and the Princes, and a neutral background was needed to set off the hierarchically correct representation of the figures.) At the top, the masterful illusionistic rendering of the riverfront at

Fig.33 Bichitr, *Asaf Khan*, late 1620s.

Fig.34 Govardhan, *Music at an encampment* (detail), 1620s.

Agra (Fig.61), in Europeanizing aerial perspective,[42] draws our attention away from the formula on to which this observation from nature is grafted.

A specific variant of this landscape scheme was employed for scenes of the ceremonial surrender of fortresses or cities: here, Tezdast (*The capture of Orcha*, **35**) and Dhola (*The surrender of Udgir*, **40**) present topographical features in detailed but largely imaginary bird's-eye views of the conquered buildings, embedded in illusionistic landscape vistas, and such concise views of fortresses and cities are integrated into other types of landscape in the Windsor manuscript.[43] In his painting of the siege of Daulatabad, Murar modifies the scheme to the extent that the main focus is on the topographical feature itself (**31**; Fig.26); he compresses the figure arrangement of the conquerors and conquered in the lower foreground to give maximum space to his stunning depiction of the fortress, which, in its realistic approach, has no equal in Mughal painting with the possible exception of *The surrender of a fortress*, thought to be Qandahar (Fig.147). The closest parallels are bird's-eye city views in European geographical works where such depictions had became a recognized and well-developed form of topographical art since De' Barbari's magnificent view of Venice of 1500. Georg Braun and Frans Hogenberg published six volumes of engraved views of towns throughout the known world in their *Civitates orbis terrarum* (1572–1618), and it is likely that either these or related works reached the Mughal court (see Fig.110). City views also appear in the most popular geographical work of the period, Sebastian Münster's *Cosmographia*, which was published in several languages and editions from 1550 to the 17th century (Fig.35, 37).[44] We know that such works found their way into the imperial library of the Mughals at an early stage; in 1580 the first Jesuit mission presented Akbar with an atlas sent by the archbishop of Goa,[45] and Sir Thomas Roe learned at the beginning of his embassy in 1615 that 'the Great Mogull' made 'more estimation' of maps of the world 'then off all other presents.'[46] European geographical works have, as yet, not been considered as a source for Mughal painting, despite the fact that the Mughals showed great interest in European forms of cartographic representation and assimilated them for their own artistic purposes. In the allegorical depictions of Jahangir and Shah-Jahan discussed above (pages 136–7), for example, miniature adaptations of European maps and globes were intended to give a realistic and contemporary touch to the 'Great Moghul's' claim to world rule (**10**, **38**). For views of fortresses and cities, however, the artists of the *Padshahnama* consulted European models only with regard to representational techniques – the aerial view of an agglomeration of buildings and a characteristic multiple perspective – as we find in the depiction of Daulatabad, where the fortified hill is rendered frontally while its three rings of fortifications are seen from above (compare Figs 109 and 110): the architectural elements were usually substituted by Indian forms, although these, with the exception of Daulatabad, were largely imaginary (**20**, **34**, **35**, **40** and **41**).

The second means of introducing three-dimensional elements into a landscape which had to be structured two-dimensionally is found in the works of the 'Kashmiri Painter.' This artist, whose identity is not known, also exploited European cartographic sources. His trademark is a very specific method of structuring the tilted plane that forms the basis for all the landscapes in the Windsor manuscript. The Kashmiri Painter introduces into his landscapes, such as that passed by the royal procession (**34**), a multitude of rather mechnically layered hillocks topped with ever

Fig.35 Sebastian Münster, Map of Ostia, from the *Cosmographia*, 1628.

more minute trees, bushes and buildings, and populated by minuscule figures, all suggesting a bird's-eye view of the landscape (Fig.36). In European pictorial maps of the sixteenth century, hillocks were a standard device to indicate mountainous areas, tiny trees denoted forests, and buildings inhabited places (compare Figs 35 and 37 with **34**, **41** and **42**). The Kashmiri Painter seems to have become interested in the hillock formula because it provided a means to give depth to two-dimensional surfaces, as well as a context for minute detail, and the close layering of the hills conveniently echoed the layering of groups of figures. Moreover, just as the formula of the flat surface surmounted by a strip of distant landscape was transformed by a highly talented naturalistic artist in the elephant fight (**29**), so the hillock scheme found an inspired illusionistic interpretation in the splendid landscape of Shah-Jahan's antelope hunt (**33**), where its skillfull handling – suggesting continuous and infinite space – is a far cry from the humble woodcuts of its cartographic sources (compare Fig.62 with Fig.37). It throws a significant light on the selective, purposeful approach of the Shah-Jahani artists to their sources that European art was explored not only for its realism, but also for schematic forms of representation that might serve the artistic intentions of Mughal painting. Bird's-eye views of areas populated by tiny figures corresponded to the way imperial patrons liked to see their landscapes.

One of the attractions of the hillock scheme was that it blended well with formally related and well-established compositional devices from the Persianate painting tradition – schematic rock formations – which were the third solution used by the artists of the *Padshahnama* to structure their landscapes according to the requirements of the official court style. This time-honored motif appeared early, in the 'scale patterns' of Assyrian reliefs,[47] and had been employed in Europe and in Asia,[48] where such rock formations appeared prominently in the Buddhist cave paintings of Ajanta and Ellora.[49] Through Buddhist channels the motif made its way from India into China and from there – in particular during the Mongol conquest of the thirteenth century – into Persian painting, where its fantastic interpretations became a characteristic element of sixteenth-century Safavid landscapes (Fig.80).[50] In this form schematized rocks were introduced into Akbari painting by Abd al-Samad.[51] They were amalgamated with other landscape schemes and gradually given a more naturalistic interpretation,[52] of which there are several instances among the illustrations to the *Padshahnama*.

The painters of the *Padshahnama* used the rock formation in various ways. The Kashmiri Painter integrated it into the schematic range of hills in his representation of Shah-Jahan's pilgrimage to Ajmer, to reinforce the idea of a mountainous region (**41–42**). If he is also the painter responsible for the splendid illusionistic backdrop to the submission of Rana Amar Singh (**7** and Fig.79), he surpassed himself there by merging the two schemes. Balchand used the Safavid rock formation as a deliberate retrospective Persian allusion, but couched his reference in naturalistic pictorial language (**30**). Abid also introduced the formula, again naturalistically re-interpreted, to his illustration of the death of Khan Jahan Lodi, apparently because the motif was traditionally connected with a row of spectators, here used to highlight the didactic message of the painting (**16**).

Another device used by the artists of the *Padshahnama* to structure terrain two-dimensionally are clearly defined, often differently colored, silhouettes forming ridges that traverse the landscape horizontally. The formula was much used in Timurid and Bukhara painting (Fig.98). In the hands of Balchand and, more especially, of his brother Payag it underwent a similar naturalistic transformation to the rock scheme (**15**, **18**, and **36**). The silhouetted ridges recommended themselves in particular for the organization of simultaneous action in a composite representation, and for the depiction of trenches in sieges, and also allowed the painters to combine a neutral flat background with carefully controlled recession into space.

Despite this reserve toward open space, creating surroundings analogous to the linear figure compositions, realism could mani-

Fig.36 Hilly landscape (detail of **34**).

Fig.37 Sebastian Münster, Map of Istria (detail), from the *Cosmographia*, 1628.

fest itself in the formal areas of a painting when it supported the content. In a particularly telling example, Payag's depiction of Khan Dawran receiving the heads of the Bundelas (**36**), the flat surface on which the Mughal commanders advance changes into a more deeply modeled and densely forested terrain in the area occupied by the rebels, mirroring their three-dimensional rendering. But the most inclusive and subtle means of introducing realism into the two-dimensional structure of the landscape was – as in palace settings – by way of a microscopic treatment of every surface texture, and of vegetation and animal life. In this context the best artists developed their watercolor technique so skillfully that they could almost reproduce the effects of oil painting. Abid (**16**), Dawlat (**46**), the 'Kashmiri Painter' (**7**, **34** and **41**) and Payag (**18** and **36**) would apply layers of color and then work them – employing their brushes like a pen or an engraver's burin[53] – with such minute stippling, parallel lines and cross-hatching that one can fully appreciate their painstaking efforts only with the help of a microscope or a strong magnifying glass. The effect was heightened by thinly applied gold washes. Among the most amazing expressions of this miniature realism are the landscapes of royal processions or the royal hunt, which reflect the imperial patron's own personal experience of nature.

Unrestrained recession into space could occur in the backgrounds where it did not disturb the main formal figure composition. To explore these margins for free representation to the fullest advantage, some painters developed a stunning illusionism, which, in this miniature format and in its best and freest expressions – such as the stampede on the riverfront at Agra or the troops storming through the jungles of Bundela country (**29**, Fig.61 and **35**, Fig.116) – has hardly an equal in the history of painting.

CONCLUSION

The miniatures of Shah-Jahan's *Padshahnama* are highly complex creations, which, in their artistic ambition and dialectic of form and content reach far beyond their apparent function as illustrations of a historical narrative in the tradition of Islamic miniature painting. Each episode contains ahistorical references, opposing the accidental and changing forces of history with that which was not supposed to change, namely the hierarchy and order of Shah-Jahan's rule. Programmatic statements were expressed through aesthetic means; artistic style could serve as an interpretational key. Through the manipulation of antithetical modes, Mughal painting developed its own pictorial code, to supplement, comment on and paraphrase historiography. The formal linear idiom stood for the power structure of Shah-Jahani rule, for the forces that regulated the system. The use of naturalism was much more complex: besides expressing genuine aesthetic interests, naturalism had to grade strata within the power structure and identify that which was outside it; at the same time, it had also to support the system, permeating it subtly but thoroughly to give Shah-Jahan's ordered world the utmost appearance of reality. Thus the *Padshahnama* paintings provide, in miniature format, the most detailed pictorial records of Mughal court culture ever produced, a congenial visual equivalent to the detailed literary account in Abu'l-Fazl's earlier *A'in-i Akbari*.

The formality of the Shah-Jahani court style, coupled with its concern for painstaking detail, has led scholars to misjudge its intentions as an empty aesthetic formalism.[54] However, no other instance in the history of art comes easily to mind where the artistic form was so methodically manipulated for non-artistic aims, namely imperial ideology. If we are allowed to make connections over large expanses of space and time, a surprisingly close parallel may be found in Egyptian art: Shah-Jahani art, with its hierarchical impositions (frontally presented settings and figures in pure profile) against which work genuine naturalistic interests (expressed in particular in plant forms and 'flowermania'), appears almost as a transposition of Egyptian art into another culture.[55]

In the realm of more tangible historical connections, the occurrence of 'at least two separate and very different visual idioms' in Timurid painting of the fifteenth century has been identified recently by Thomas Lentz.[56] The formal idiom of manuscript illustration, conceived as a Timurid link to the Persian past, was juxtaposed – in the context of wall-painting – with freer modes of representation derived from Chinese art, reflecting a renewed interest in expanding cultural and aesthetic aspirations. It is noteworthy, I think, that the Mughals, in their comparable quest, turned not to China but toward Europe for inspiration; from the time of Greco-Buddhist Gandhara art this led to the first systematic exploitation of Western forms for the aims of Indian art. European realism best served the Mughals' interest in nature and their close and rational observation of the visual world. At the same time it provided fresh ideas and representational techniques allowing them to use the visual arts as a vehicle for the expression of abstract concepts, as Abu'l-Fazl acknowledged: 'Although in general a picture represents a material form and [here in particular] the painters of *firang* quite often express, by using rare forms, our mental states and [thus] they lead the ones who consider only the outside of things to the place of inner meaning.'[57]

This approach agreed in principle with the traditional aims of Indian art, and the Hindu masters who dominated Shah-Jahan's imperial studio would have been the most likely to explore naturalism for non-naturalistic purposes. For all its dependence on Persian traditions and European realism, and despite its programmatic support of Mughal rule, Shah-Jahani painting is here truly Indian in its intention.

Its syncretistic and programmatic approach and high standards of artistry set Shah-Jahani painting apart from contemporary Safavid and Ottoman painting. In the context of book illustration, only the Mughals achieved such ambitious history painting. The Ottomans did not match their written descriptions of the sultans' activities with paintings of a similarly detailed realism, and the Safavids, lacking a particular urge towards self-representation, had no systematic historiography and hardly any history painting. Confronted with the vastness and complexity of their empire and its heterogeneous traditions, the Mughals felt a greater need to measure and support their imperial authority with a carefully constructed image of rulership. Each emperor found his own means of translating this image into the arts, while building on the tradition of those who preceded him. For Shah-Jahan, these means were the absolute control of the pictorial world and its laws of representation.

1. See page 162.
2. The use of this argument to enforce absolute rule was also popular with Shah-Jahan's contemporaries in Europe. See, for instance, Warnke 1993, pp.224ff.
3. *'Amal-i Salih*, III, p.18. For a fuller translation of this passage, see Koch 1982b, p.259; for Kanbo's subsequent remarks adducing Solomon to legitimize the patronage of the figural arts, see Koch 1988, p.24.
4. *Padshahnama*, I/I, p.149; Nur Bakhsh 1903–4, p.190–91; *cf. 'Amal-i Salih*, I, p.206. See also Koch 1982b, p.259; Koch 1991b, p.96.
5. Qazwini, *Padshahnama*, fol.139a (re-foliated 140); translated by Ebba Koch.
6. Particularly explicit statements on this point are made by Akbar's historian Qandahari: *Ta'rikh-i Akbari*, pp.144f., 147; translated in Brand & Lowry 1985, pp.290–91, 294. Shah-Jahan's authors repeatedly refer in a similar way to the Emperor's patronage of architecture; see note 4, above, and *Shah Jahan Nama*, p.570.
7. *Padshahnama*, as quoted in note 4, above; *cf. 'Amal-i Salih*, I, p.206.
8. Shah-Jahan's return to the Sunni fold began to manifest itself openly in his tenth regnal year, when he replaced the solar calendar, used since Akbar, with the Islamic lunar calendar. See *Shah Jahan Nama*, p.xix. The effect of Shah-Jahan's new orthodoxy on his artistic patronage is discussed in Koch 1988, pp.28, 35, with further literature.
9. Brown 1975, p.87; Beach 1978, p.29; Leach 1995, I, p.353.
10. A reference to the Shah Burj pavilion, which was the location of the most private and personal of Shah-Jahani court assemblies, reserved for the imperial princes and highest dignitaries.
11. A measure of time equivalent to 24 minutes.
12. Qazwini, *Padshahnama*, fols 140a–140b (re-foliated 141–42); translated by Wheeler Thackston. Lahawri's text underwent a similar editing process; it was first checked by Allami Sa'dullah Khan, then read out to Shah-Jahan, who freqently ordered emendations. See *Shah Jahan Nama*, p.xxi.
13. For the most recent discussion of the illustrations of the *Jahangirnama* see Beach 1992b, with further literature.
14. For the effect on artistic themes of the different personalities of these Emperors, see the discussion by Milo Beach above, page 118.
15. See the description in *Chahar Chaman*, quoted above, pages 113–14.
16. Koch 1994, in particular pp.143–46.
17. Eaton 1993, p.160. See also Asher 1993 and Necipoğlu 1993.
18. McInerney 1991, in particular pp.59ff.
19. See also McInerney 1991, figs 12, 13, 15.
20. For this and the following see Koch 1994, in particular n.2.
21. See the definition of the ranking system on page 82, note 2.
22. Koch 1991a, pp.93, 99, 109 and 141. The term *qarina*, which can be translated as 'counter-image,' was used consistently by Shah-Jahan's authors in their descriptions of imperial architecture; see, for example, the reference to the Taj Mahal and its flanking buildings in *Padshahnama*, 2, p.327.
23. The point is discussed in more detail on pages 167–68, below.
24. See the portrait, perhaps by Hashim, *circa* 1620, in the British Museum, London: illustrated in Beach 1992a, fig.81.
25. Losty 1990.
26. 'It is a commonplace that a recognisable transcription of the features can be achieved more easily in profile than in any other way.' Pope-Henessy 1989, p.35.
27. Plato 1987, pp.363–64 (Book 10, 598). For Plato's views on this point, see Schäfer 1986, in particular p.87. As Aflatun, Plato was well known at the Mughal court: see Koch 1988, pp.29–30, 34–35, pl.50.
28. *Cf.* the discussion on page 203, below, with further literature.
29. At times, however, the three-quarter face was used when no other model was available; in the depiction of Timur (3), the three-quarter view has a historicizing connotation, conforming to the conventions of Timurid miniature painting.
30. See below, page 141.
31. See also page 200, below.
32. For this and the following see Koch 1982a; Koch 1982b.
33. Koch 1988, in particular pp.29–35.
34. Koch 1988, pp.28, 34.
35. See Koch 1988, pp.13–15.
36. See **32** and **43**.
37. There is as yet no study of Mughal landscapes, with the exception of a brief treatment by Robert Skelton (1980).
38. Gibson 1989.
39. Early examples of these wide vistas appear in the astrological manuscript at Rampur dated to the late 1560s; illustrated in Beach 1987a, pl.49.
40. Early versions of this type of landscape appear in the œuvre of Kesu Das, for example, in his painting of a Jesuit of the late 1590s, in an album probably made for Sultan Salim (Jahangir); illustrated in Leach 1995, I, cat. no.2.165, pl.44. The image was refined by the great masters of the imperial studio: see Bichitr's full-length portraits from the 1620s of I'tibar Khan (Musée Guimet, Paris) and of Asaf Khan (Victoria & Albert Museum, London). Abu'l-Hasan's *Jahangir holding a globe and viewing his troops* seems to survive only in mid 17th-century (Freer Gallery of Art, Washington, DC) and later versions (see Beach 1981, no.18b). For color illustrations see Okada 1992, pls 55, 196, 197.
41. Govardhan's *Rustic concert* and *Jahangir receiving the captured Mirza Hasan at Sikandra* (1620s) are both in Dublin, Chester Beatty Library; illustrated in Leach 1995, I, cat. nos 3.1, 3.21, pls 52, 57.
42. In Akbar's period aerial perspective, which suggests distance by a succession of increasingly light color areas, became a confidently handled feature of the illusionistic repertory of Mughal artists after its adoption by masters such as Kesu Das from Flemish sources, most likely independent miniatures or small oil paintings on copper.
43. See also Fig.148.
44. See Harvey 1980.
45. Monserrate 1992, p.28.
46. Roe 1990, p.45.
47. For illustration, see Curtis & Reade 1995, no.24.
48. For the migration of this motif into Byzantine and Western art, see Clark 1973, pp.10–12; Jahn 1975.
49. Strzygowski 1922, pp.4ff; Jahn 1975, pp.21–22, n.6; Gombrich 1976, pp.11–13.
50. For the various points touched on in this outline see William Watson 1980; Soucek 1980; and Brend 1980.
51. See for example Okada 1992, pls 60–62, 65.
52. See in particular the landscapes in the *Khamsa* of Nizami of 1595, British Library, London, Oriental and India Office Collection, MS.Or.12208; illustrated in Brend 1995, figs 1, 2, 6, 8, 10, 12, 14, 17 , 18–21, 24, 26, 29, and 35–39. For further adaptations of the motif see, for instance, *Jahangir watching a spider fighting a snake*, in the Raza Library, Rampur; illustrated in Brown 1975, pl.19.
53. This technique, especially when hatching is used instead of shading, resembles the translation of Western prints into miniature paintings.
54. See, most recently, Leach 1995, I, p.354: 'He [Shah-Jahan] seems to have viewed painting purely as a decorative art and therefore did not require the same expressiveness from his miniaturists as did Jahangir.'
55. In my analysis of the principles of Shah-Jahani painting I have been greatly helped by Schäfer 1986.
56. Lentz 1993, in particular p.256.
57. *A'in-i Akbari*, I, p.11; trans. I, pp.102–103; here as translated in Koch 1983, p.193, n.73.

Fig.38 Raja Bikramajit and Afzal Khan offering tribute (detail of **9**).

Fig.39 A laughing shaykh (detail of **37**).

Fig.40 Holy men awaiting alms after the weighing of Shah-Jahan (detail of **13**).

Fig.41 Courtiers at a *darbar* (detail of **37**).

Fig.42 Musicians at the wedding of Dara-Shikoh (detail of **26**).

Fig.43 Female musicians celebrate the wedding of Awrangzeb (detail of **45**).

Fig.44 Musician with a kettledrum (detail of **24**).

Fig.45 The *nawbat* ensemble (detail of **12**).

opposite Fig.46 Musicians in a wedding procession (detail of **22**).

Fig.49 Europeans escaping from a mined building at Hoogly (detail of **20**).

Fig.50 An explosion at the siege of Qandahar (detail of **18**).

opposite above Fig.47 The imperial encampment outside Hoogly (detail of **20**).
opposite below Fig.48 The fortress at Orchha (detail of **35**).

Fig.51 The death of the rebellious Bundelas (detail of **36**).

opposite above Fig.52 Imperial soldiers scale the walls of Fort Dharur (detail of **15**).
opposite below Fig.53 The elephant Surat Sundar controlled by men with *charkhi*s (detail of

Fig.55 A pair of sarus cranes (detail of **30**).

Fig.56 Ducks in a marshy creek (detail of **30**).

Fig.57 Lions defending their cubs (detail of **46**).

Fig.58 Rabbits by a pool on the battlefield (detail of **18**).

opposite Fig.54 Black buck antelope by a stream (detail of **33**).

Fig.59 The countryside beyond Fort Dharur (detail of **15**).

Fig.60 Fireworks above Agra at the wedding of Awrangzeb (detail of **45**).

Fig.61 The riverfront at Agra (detail of **29**).

Fig.62 Farmers fetching water (detail of **33**).

Fig.63 The Jumna River at Agra (detail of **21**).

overleaf Fig.64 Dancers celebrate the weighing of Shah-Jahan (detail of **12**).

Catalogue

Padshahnama of Abdul-Hamid Lahawri

Copied by Muhammad-Amin of Mashhad in AH 1067 (AD 1656–7)
239 folios, with 10 double- and 24 single-page illustrations, and one double-page illumination; text written in *nastaliq* script;
18th-century lacquer covers
Full page 58.5 × 37 cm
Text panel 34.6 × 19 cm

Colophon (folio 239a): *Katabahu'l-'abdu'l-faqiru'l-mudhnibu Muhammad-Amin al-Mashhadi 1067* ('Written by the poor, sinful slave Muhammad-Amin al-Mashhadi 1067 [AD 1656–7]')

Exhibited: British Library, 1982 (see Losty 1982, no.82); British Museum, 1976 (see British Museum 1976, no.167); British Museum, 1960; Royal Academy of Arts, 1947–8
Published: Komala 1984; Beach 1981, p.177; Leach 1981, pp.148–50; Beach 1978, pp.78–85; Barrett & Gray 1963, pp.112–14

Each of the original text pages of the manuscript has been set within borders of a later date that are edged with new marginal rulings and a repeated decorative design in gold. With one exception, each page is stamped with an owner's seal (Fig.67), inscribed: *Wazîru'l-mamâlik Âsafu'd-dawla Âsafjâh Yahyâ Khân Bahâdur Hizibr-Jang 1190* ('Waziru'l-mamalik ['Vizier of the Realm'] Asafuddawla ['Asaph of the State'] Asafjah ['like Asaph in status'] Yahya Khan Bahadur Hizibr-Jang ['Lion in Battle'] 1190 [AD 1776–7]').

Paintings made separately have been glued to the reverse of text folios where the scribe left space for illustrations, and on these folios the border paper has been laminated to increase its thickness. All the paintings were once given separate cover sheets which provided a summary in Persian of the text passage accompanying each illustration. Several of these sheets, which seem to be of eighteenth-century date, are now missing and – as described elsewhere in this study[1] – many of the episodes related do not correspond to the actual subject of the accompanying illustration.

SIR JOHN SHORE, who became Lord Teignmouth in 1798, was Governor-General of India between 1793 and 1798. His refusal to accept for himself the gift of the *Padshahnama* manuscript from Asafuddawla, Nawab of Oudh, has been credited to his personal integrity (see page 13, above). A poet and scholar interested in Indian languages, he was described by a contemporary as 'a good man, but as cold as a greyhound's nose.'[2] Asafuddawla, who ruled as Nawab of Oudh from 1774 to 1798, on the other hand, was quite a different sort of personality, famed for his dissipation and financial recklessness. In a letter to the Court of Directors of the East India Company dated 19 February 1797, Lord Teignmouth described his visit to Lucknow in 1797, at which time the gift of the *Padshahnama* may have been offered: 'The Nawab and myself visit daily, and are in the best humour imaginable with each other. His disposition is naturally good, but irritated by bad advisers, mean associates, and absolute power; which, however, he does not exercise cruelly. He promotes rather than performs

Fig.65 Manuscript of the Windsor *Padshahnama* of Abdul-Hamid Lahawri, 1656–7.

Fig.66 Lacquer binding of the Windsor *Padshahnama*.

Fig.67 Folio 50a, detail showing seal of Asafuddawla.

bad actions. A few years ago, an Englishman, for His Excellency's amusement, introduced the elegant European diversion of a race in sacks by old women: the Nawab was delighted beyond measure, and declared that although he had spent a crore of rupees, or a million sterling, on procuring entertainment, he had never found one so pleasing to him.'[3]

Asafuddawla, whose eccentricities have dominated discussions of his character, was also quite genuinely interested in creating at his court the splendor formerly associated with Mughal rule, and this may have been one incentive for his acquisition of the *Padshahnama*. We lack, however, the crucial information about the source from which it was acquired. MCB

1. See pages 15 and the remarks on the condition of individual folios in the descriptions that follow.
2. Bence-Jones 1982, p.5.
3. Quoted in Edwards 1960, p.15.

1–2
Double shamsa

Folios 1b–2a, *circa* 1657
Original page (within gold borders) 46.2 × 28.5 cm (folio 1b); 45.8 × 28.6 cm (folio 2a)

Prior to the execution of the illumination, a geometric grid was impressed into the paper and is still visible beneath the central gold areas. Seals and a scribal notation written on the original paper at the edge of the *shamsa* have been erased, but remain visible (although illegible) under ultraviolet light.

Fig.68 Husayn Naqqash, *Shamsa*, from a manuscript of the *Diwan* of Amir Khusraw Dihlavi, dated 1597–8.

IMAGES OF A *shamsa* (sunburst) often appear as the frontispiece to important Islamic manuscripts and albums. As described by Anthony Welch, the *shamsa* 'functions not only as a symbol of God but also of royalty, supported by God: its solar centre is a symbol of the *axis mundi* and reveals the passion for geometry, as a visual metaphor of divine order, that flows throughout Islamic culture. It is also a symbol of both divine *tawhid* (unity) and divine *nur* (light), centripetally pulling the outer world toward it at the same time that it centrifugally expands into it.'[1]

Two pages now in the Kevorkian Album in the Metropolitan Museum, New York, are the finest and most complex examples known from the Shah-Jahan period.[2] Together with this double-page frontispiece in the *Padshahnama*, and a fifth page in a private collection, they have been attributed to a single artist by S.C. Welch.[3] The consistency of Shah-Jahan period examples can be seen by comparing these designs with earlier, more experimental and rugged *shamsas* found in volumes associated with the previous three Mughal emperors: one from an anthology of literary works, datable to *circa* 1550;[4] the second executed by the painter Husayn Naqqash for a manuscript of the *Diwan* of Amir Khusraw Dihlavi, dated 1597–8, now in the Walters Art Gallery, Baltimore (Fig.68); and the third in an unidentified album of *circa* 1625.[5]

During this period, from the mid-sixteenth to the mid-seventeenth century, these designs were executed with increasing precision and refinement – a development parallel to that of portraiture or historical illustration. The late sixteenth-century example reproduced here, drawn by Husayn Naqqash,[6] shows the spontaneity of a painter rather than the precision of an illuminator. The Jahangiri painter Mansur and his colleague Dawlat, who worked on the *Padshahnama*, were also among those few artists who mastered both figural illustration and illumination.[7] ❧ MCB

1. A. Welch 1995, p.460. See also the discussion by Ebba Koch below, page 160.
2. S.C. Welch *et al.* 1987, nos 1 and 5. Since they are now placed in an album formed in the nineteenth century, it is possible that these pages were not intended originally for the same volume.
3. As given in A. Welch 1995, p.463 and no.3.
4. Reproduced in color and discussed in Soudavar 1992, no.133. An inscription in the center reads, 'For the library of Prince Muhammad Humayun, renowned *shah*.'
5. Soudavar 1992, no.130. Inscriptions inform us that the book entered Shah-Jahan's library on the day of his accession (the usual formula to proclaim ownership of a book belonging to the previous regime), and that later, in 1630 at Burhanpur, he gave the volume to his 'true disciple of perfect understanding, Ja'far Khan.'
6. For a list of works by Husayn Naqqash, see Verma 1994, p.181.
7. An unpublished illumination by Mansur is found on fol.1v of the *Diwan* of Amir Khusraw Dihlavi, Walters Art Gallery, Baltimore, acc.no.w.624.

3–4
Timur and Shah-Jahan

Folios 2b–3a, *circa* 1657
Original page (within gold borders) 46.7 × 26.1 cm (folio 2b); 46.6 × 26.1 cm (folio 3a)

Inscription: [folio 3a, in cartouche] *hazrat-i Shâhjahân bâdishâh* ('His Majesty the Emperor Shah-Jahan')

The original paper on which each of these portraits and illuminations was placed extends beyond the gold marginal ruling on the inner edge of the present pages and forms a clear contrast with the more recent borders with which the manuscript has been refurbished (Fig.3). Since the obverse and reverse sides of folio 2, respectively termed folios 2a and 2b, were painted on a single sheet, these extensions were inevitable if the full design of the *shamsa* on folio 2a was to be maintained. The opening text on folio 3b, however, is on a separate sheet glued to the back of the illuminated portrait of Shah-Jahan, indicating that the initial illuminated pages were not necessarily created at the same time as, or specifically for, this copy of the *Padshahnama* text.

THE MANUSCRIPT could not open with a clearer statement of the importance of the *Padshahnama* as an official chronicle, for these two images are included specifically to proclaim the dynastic legitimacy of the Mughal Emperor. In the left-hand portrait, Shah-Jahan faces his ancestor Timur, from whom he symbolically receives his crown. Better known in the west as Tamerlane, the Emperor Timur was a Central Asian warlord of Turko-Mongol lineage whose family ruled much of the Persian world in the fifteenth century, and whose defeat of Delhi in 1398 provided a precedent for the later Mughal conquests in north India. Because of Timur's power and fame, the Mughals promoted themselves as Timurids (descendants of Timur) before any other identification. One of the titles regularly used by Shah-Jahan was *Sâhib-Qirân-i Sânî* ('Second Lord of the Auspicious Planetary Conjunction'), which refers specifically to Timur, who had been known as *Sâhib-Qirân* ('Lord of the Auspicious Planetary Conjunction'). This Timurid reference is nothing new within Mughal imagery. Among the opening illustrations of the imperial copy of the *Akbarnama*, the official biography of Shah-Jahan's grandfather Akbar presented to that emperor in 1596, a scene of festivities at the birth of Timur is carefully placed to be seen in relation to a depiction of Akbar's own birth.[1] And in the description of the great Peacock Throne which Shah-Jahan commissioned as an emblem of his wealth and power, the *Shahjahannama* states that 'among the precious stones used upon it was a ruby worth one lakh of rupees that Shah 'Abbas Safawi had sent to the late Emperor Jahangir. And on this ruby were inscribed the names of the great Emperor Timur Sahib-i-Qiran, Mirza Shahrukh, Mirza Ulugh Beg, and Shah 'Abbas – as well as the names of the Emperors Akbar, Jahangir, and that of His Majesty himself.'[2] Such emphasis on the routes by which sovereignty was transmitted is found frequently, and the *Padshahnama* is no exception. Timur was considered the progenitor of the Mughal dynasty, and by juxtaposing his image with that of Shah-Jahan, the double-page asserts that Shah-Jahan was Timur's worthy heir.

Within the *Padshahnama* illustrations, the artists alternate between painting familiar figures by the repetition of established prototype portraits and creating uniquely probing depictions of individuals, often less important men who either captured the artist's attention or for whom there

Fig.69 Govardhan, *Timur presenting the imperial crown to Babur*, circa 1635.

Fig.70 Timur enthroned (detail of 3).

Fig.71 Timur enthroned (detail of Fig.69).

was no existing portrait to use as a model. Here, the composition used for the figure of Timur is one found frequently in works of the period (see, for example, Figs 69–71), a thoroughly appropriate solution for the portrayal of a long-dead figure who had assumed iconic status.[3] Only the most inventive and incisive artists ignored established portrait formulas to invent compositions or characterizations afresh. ❧ MCB

Shah-Jahan sits on a golden bejeweled throne, raising his right hand to receive the Timurid crown from his ancestor, Timur, who sits opposite him on a throne of a similar square shape. Both thrones have ornamental backrests furnished with large cylindrical cushions, which are a characteristic component of Mughal thrones.[4] The Emperor Shah-Jahan is shown in old age but the representation evokes his accession when, on 14 February 1628, 'donning the robe of kings (*qaba-yi Khusrawi*) ... he came to the throne of the empire of Akbar with the fortune of Timur and the success of Babur ... and sat with royal ease ... under a golden parasol (*chatr*)[5] ... the fringes of which – looking like the rays of the sun – ... encircled the head of the [imperial] sun.'[6] Inspired by the fire- and sun-worship of Akbar[7] and ideas of 'light-rulership' explored by Humayun, Akbar, and Jahangir,[8] Shah-Jahan based his absolute rule *inter alia* on the claim to be a 'sun king,' 'the sun of the sky of empire and caliphate ... whose shining forehead illuminates the eyes of the world and worldlings.'[9] Integrated in this concept was the notion of the kingly quality *farr-i shahinshahi*, believed to be the ruler's share of the *farr-i izadi* ('divine effulgence'), the ancient Persian concept of the manifestation of the sacred element of fire or light in the person of the rightful ruler. Since, according to *hadith* (the sayings and traditions of the Prophet), the sultan is 'God's shadow on earth'[10] the divine shadow was assimilated in Islamic thinking with the divine effulgence.[11] The attraction of the concept of the king as the embodiment of light was heightened for the Mughals as rulers of India by the fact that it related to ancient Indian ideas of cosmic rulership, according to which the king 'surpasses all created beings in lustre ... burning eyes and hearts like the sun ... being fire and wind, sun and moon ... a great deity in human form.'[12]

This concept of the divine right to rule was expressed in Shah-Jahani art with various solar symbols, such as the image of the *shamsa*,[13] or, in imperial portraits, with the halo. The use of this ancient oriental symbol[14] had been revitalized at the Mughal court in response to the Christian images brought to India by the Jesuits.[15] Jahangir adopted the sun halo as an imperial emblem and was followed in this by Shah-Jahan. At times, portraits of earlier rulers were also haloed 'in retrospect' (see, for instance, the portrayal of Akbar in **5**). In this juxtaposition with the portrait of Shah-Jahan, however, Timur's head is more informally encircled by an extension of the backrest of his throne, so as not to undermine the ruling emperor's status and, also, to indicate that Timur was an ancestor from another, distant era. ❧ EK

1. See the British Library *Akbarnama*, MS.Or.12988, fol.20b (*The infant Akbar placed in the care of his nurses*) and fol.34b (*Astrologers casting horoscopes during the celebrations for the birth of Timur*).
2. *Shah Jahan Nama*, p.147.
3. Other identically posed examples include a depiction of Timur in the Ardeshir Album (Sotheby & Co. 1973, lot 22). This image, like the small portrait of Shah-Jahan by Abu'l-Hasan in the Walters Art Gallery, Baltimore (MS.W.668, fol.45; Beach 1978, no.29), may have been intended for the opening double-pages from a now unknown volume.
4. There is as yet no comprehensive study of Mughal thrones. Shah-Jahan used various types, about which Aziz (n.d., pp.23ff.) gives some information; a throne similar to the one in the Windsor painting is described on p.27.
5. The parasol, an ancient and widely used symbol of rulership, is discussed as such by Abu'l-Fazl (*A'in-i Akbari*, I, p.52.)
6. These phrases are extracted from Kanbo's highly poetic description of Shah-Jahan's accession, *'Amal-i Salih*, I, p.188.
7. *Muntakhab-ut-Tawarikh*, II, p.332; *A'in-i Akbari*, I, p.50.
8. Skelton 1988, pp.180–81.
9. *Chahar Chaman*, fol.20a; see also Koch 1994, p.155, with further literature, on the 'sun-rulership of Shah-Jahan,' who preceded Louis XIV of France (d. 1715) in the systematic exploitation of the concept of the 'sun king' to further the aims of seventeenth-century absolutism.
10. *Cf.* the inscription on **10**, given on page 167.
11. *Nasihat al-Muluk*, pp.xli, 45, 73–74, 85, 110; for the Mughal context see Hardy 1985, p.117. See also Soudavar 1992, pp.410–16.
12. *Manu-smriti*, p.217.
13. See the discussion by Milo Beach above, page 159. For a Mughal definition of the *shamsa*, see *A'in-i Akbari*, I, p.52.
14. Ettinghausen 1961, p.100.
15. Koch 1982a. See also page 200.

5

Jahangir receives Prince Khurram on his return from the Mewar campaign

Ajmer, *Diwan-i 'Amm*, 20 February 1615
Folio 43b painted by Balchand, *circa* 1635
Image area 30.4 × 20.1 cm

Inscriptions: [on wall below throne] *Sanâ-yi shâh-i jahân zikr-i shabnashînân bâd / du'â u dawlat-i û wird-i subhkhezân bâd / Ba har taraf ki kashad tegh yaksuwâra chu mihr / sipâh-i khasm chu anjum zi ham gurezân bâd* ('May praise of the king of the world be the remembrance of those who keep vigil by night; may prayers for his fortune be the litany of those who rise with the dawn. In whatever direction the lone rider wields his sword like the sun, may the army of his enemy be as dispersed as the stars')
[below throne] *banda-i dargâh / raqam-i Bâlchand* ('slave of the court, drawn by Balchand')
[on book held by courtier at lower left] *shabîh-i Bâlchand* ('likeness of Balchand')

References: *Padshahnama*, I/I, pp.174ff; *Shah Jahan Nama*, p.6; *Jahangirnama*, folios 107b–108a; *Tuzuk-i-Jahangiri*, I, pp.276–78
Published: Smart 1991, p.136 and figs 1–2; Losty 1982, no.82 and pl.XXXII

THROUGHOUT THE PERIOD of Akbar's rule, the chiefs of Mewar, whose capital was at Udaipur, had steadfastly refused to follow the path of the Amber rulers, among others, and acknowledge Mughal overlordship. Since the Mewar rana was hierarchically the most important of all the Rajput leaders, his defeat was essential to Mughal policy. To help support the imperial troops in this effort, Jahangir moved his entire court to Ajmer in Rajasthan and remained there from 18 November 1613 to 10 November 1616. Prince Khurram led the successful campaign against the Hindu ruler and is here shown on his return, greeting the Emperor, who sits directly under a mural portrait of his father and predecessor, the Emperor Akbar (see the frontispiece). Like the opening illustrations of Shah-Jahan and Timur (**3–4**), this too is a symbolic statement of the Prince's political lineage. The painter tells us that Prince Khurram, soon to be given the title *Shah-Jahan* ('King of the World'), is a worthy successor to his grandfather.

The three men who attend the Emperor and Prince Khurram at the level of the *jharoka* – the raised seat or window used by the emperors for official appearances – are I'tibar Khan (Fig.73), a family servant, Firoz Khan, the chief eunuch, and a second eunuch, Khidmat Khan. The identification of the event is confirmed by the presence among the assembled courtiers of Karan Singh Sisodia of Mewar, who was sent to court as the representative of his defeated father the Rana. The Rajput prince, who is being greeted by Sadiq Khan, an imperial *bakhshi* (political or military official), is otherwise surrounded by men in service to Prince Khurram. These are Afzal Khan Allami, Abdullah Khan *Firoz Jang*, Wazir Khan, and Raja Bikramajit (Fig.78). Facing this group, on the Emperor's right, are the most important ministers and nobles at Jahangir's court: I'timaduddawla (the *Wakil*, or Prime Minister), Asaf Khan (Fig.120), Mahabat Khan *Khankhanan*, Khwaja Abu'l-Hasan *Ruknussaltana* (the *Mir Bakhshi*, chief *bakhshi* and fiscal administrator of the Empire), A'zam Khan (the *Mir Saman*, or Imperial Steward), and Mirza Rustam Qandahari.

While conquest was one way to assure the subservience of the great Rajput chiefs, the emperors also extended their control by marriage alliances. The most prominent of the courtiers, the noble closest to the Emperor at the right-hand side of this scene, for example, is the Rajput Suraj Singh Rathor of Jodhpur, whose sister Jagat Gosaini (or Jodhi Bai) was married to Jahangir and was the mother of Shah-Jahan. Suraj Singh's importance at court is indicated by his inevitable attendance at major ceremonial events, including several recorded in additional *Padshahnama* illustrations, and by a series of independent portraits painted by the chief imperial artists, such as Bishandas (Fig.72).

Included among the courtiers outside the gold railing to the inner enclosure is the artist, Balchand (Fig.135). This painter was active at the time of the

1. Abdullah Khan
2. Khwaja Abu'l-Hasan (*Mir Bakhshi*)
3. Afzal Khan
9. Asaf Khan
13. A'zam Khan (*Mir Saman*)
15. Balchand
16. Mirza Bhao Singh
18. Raja Bikramajit
21. Dawlat Khan
23. Firoz Khan
27. Habash Khan
31. I'tibar Khan
34. I'timaduddawla (*Wakil* and *Diwan*)
38. (The Emperor Jahangir)
42. Karan Singh Sisodia
43. Khan Alam
48. Khidmat Khan
49. Kishandas
51. Mahabat Khan
59. Muqarrab Khan
63. Murtaza Khan
64. Musavi Khan
73. Rai Khan
74. Riza Bahadur
75. Mirza Rustam Safavi
76. Sadiq Khan
81. Sarbarah Khan
85. (Shah-Jahan [Prince Khurram])
89. Raja Suraj Singh Rathor
90. Wazir Khan

Note: unless given within parentheses in this and following keys, names are inscribed within the illustration. Numbers cross-reference those named to the *Biographical listing* on pages 230–34.

Fig.72 Bishandas, *Suraj Singh Rathor* (detail), dated 1608.

Fig.73 Bichitr, *I'tibar Khan* (detail), *circa* 1623.

Fig.74 Marble relief, dado of the Shah Burj Pavilion. Agra fort, 1630s.

Fig.75 Floral wall decoration (detail of **5**).

Fig.76 *Jali* screens in the courtyard of the Shah Burj Pavilion. Agra fort, 1630s.

Fig.77 *Jali* screen (detail of **5**).

ceremony in Ajmer and possibly even present at it,[1] for artistic workshops regularly traveled with the imperial party. Artists' self-portraits, virtually unknown during the period of Akbar's rule, became increasingly prominent under Jahangir and Shah-Jahan.[2] MCB

It is no coincidence that the historical illustrations of the manuscript begin with Khurram's victory over the Rana of Mewar in 1615, the first and – in Mughal eyes – most glorious of a series of military achievements which firmly established the position of the Prince as designated successor to Jahangir.[3] Succession to the Mughal throne was not regulated by primogeniture and depended largely on the position which the sons of the emperor could carve out for themselves during their career as princes, and at the crucial moment of the death of their father.[4] Such pictures were intended to put Prince Khurram's accomplishments on record and thus legitimize in retrospect his coming to the throne as Shah-Jahan *padshah* ('emperor').

Jahangir receives Prince Khurram on his return from the Mewar campaign is the first of the eleven *darbar* or court assembly scenes in the Windsor manuscript showing an emperor in the *jharoka-i khass-u-'amm*, that is, the architectural frame for his appearances before the court in the assembly of *khass-u-'amm*, his most public form of conducting state business.

Although the fort of Ajmer, where Jahangir presumably received Khurram,[5] has no *jharoka* of the type shown in the painting, Balchand takes care to be historically accurate in the rendering of the hall erected in front of the wall that supported the Emperor's seat. In Jahangir's time this was a temporary construction in the form of an *iwan az parcha* (a tent-like structure made of cloth)[6] partly obscuring the *jharoka*. Balchand also decorates the wall above the dado in typical Jahangiri fashion, with figural elements (Akbar's portrait) and with *chinikhana* motifs, small decorative niches filled with vessels, in this case precious china ware, glasses and bottles.[7] The dado itself is not historicizing but shows meticulously rendered naturalistic flowering plants,[8] the favorite decorative motif of Shah-Jahan's own architecture (Figs 74, 75).[9] The *jalis* or pierced stone screens that constitute the panels of the railing of the *jharoka* have geometrical patterns common to both periods (Figs 76, 77). The setting thus amagalmates retrospective and contemporary elements, all rendered with the same elegant realism and detail as the portraits of those who take part in the *darbar*. This endows the painting with an intricate quality that counterbalances the formality of the composition. EK

1. For further discussion of Balchand, see page 213, below.
2. The depiction of painters included at full court audiences seems to begin in earnest soon after 1615. The earliest examples include self-portraits by Bichitr (Beach 1981, no.17a) and Nanha (see Fig.17 and page 129, note 31, above).
3. *Tuzuk-i-Jahangiri*, pp.272ff.; Prasad 1922, pp.237–44; Saksena 1976, pp.15–17.
4. Khosla 1976, pp.82–117.
5. In his autobiography, Jahangir tells us remarkably little about his residence at Ajmer. From Shah-Jahan's chronicler Waris we learn that Jahangir used the fortified palace in the city (built by Akbar in 1570) until he had constructed marble pavilions on the shore of the Ana Sagar lake (*Padshahnama*, fol.472a). The pavilions apparently did not include a *jharoka-i dawlat khana-i khass-u-'amm* because this was added by Shah-Jahan at some time before 1637. On the problem of dating the pavilions, see Koch 1982 c, p.337, n.18.
6. *Cf.* **9, 37–39**; Koch 1994, pp.144, 158, n.3. See also Andrews 1987, p.155; and pages 134–35, above.
7. Koch 1991a, pp.70, 89.
8. Skelton 1972.
9. Koch 1991a, pp.95, 100–101. See also page 200, below.

6–7
The submission of Rana Amar Singh of Mewar to Prince Khurram

Gogunda (Mewar State), encampment,
5 February 1615[1]
Folio 46b painted by La'lchand and
folio 47a attributed to the 'Kashmiri Painter',
both *circa* 1640
Image area 32.4 × 21.6 cm (folio 46b);
32.5 × 22.1 cm (folio 47a)

Inscriptions: [folio 46b, bottom] *'amal-i La'lchand* ('work of La'lchand')
[above painting] *Taswir-i hazrat-i Shâhjahân dar zamân-i shâhzâdagi* ('depiction of His Majesty Shah-Jahan during the time he was prince')

References: *Padshahnama*, I/1, pp.171–72; *Shah Jahan Nama*, p.6; *Tuzuk-i-Jahangiri*, I, pp.273–74

A gold ruling is visible along the top and right inner edges of folio 46b, on the same paper as the illustration itself. This is distinct from the wider gold ruling placed around the scene when it was remounted within the present borders, and is probably what remains of untrimmed margins from an earlier use of the illustration in a different volume. The reverse of folio 47b (the left-hand illustration) bears eleven rather than the usual twenty-one lines of text, necessitated by the close sequence of illustrations in this section and the need to maintain coordination of text and image.

1. (Abdullah Khan)
3. (Afzal Khan)
7. (Amar Singh Sisodia)
18. (Raja Bikramajit)
54. (Mirmiran)
57. (Muhammad Taqi)
68. (Shaykh Nazir)
85. (Shah-Jahan [Prince Khurram])
89. (Raja Suraj Singh Rathor)
91. (Zulfaqar Khan)

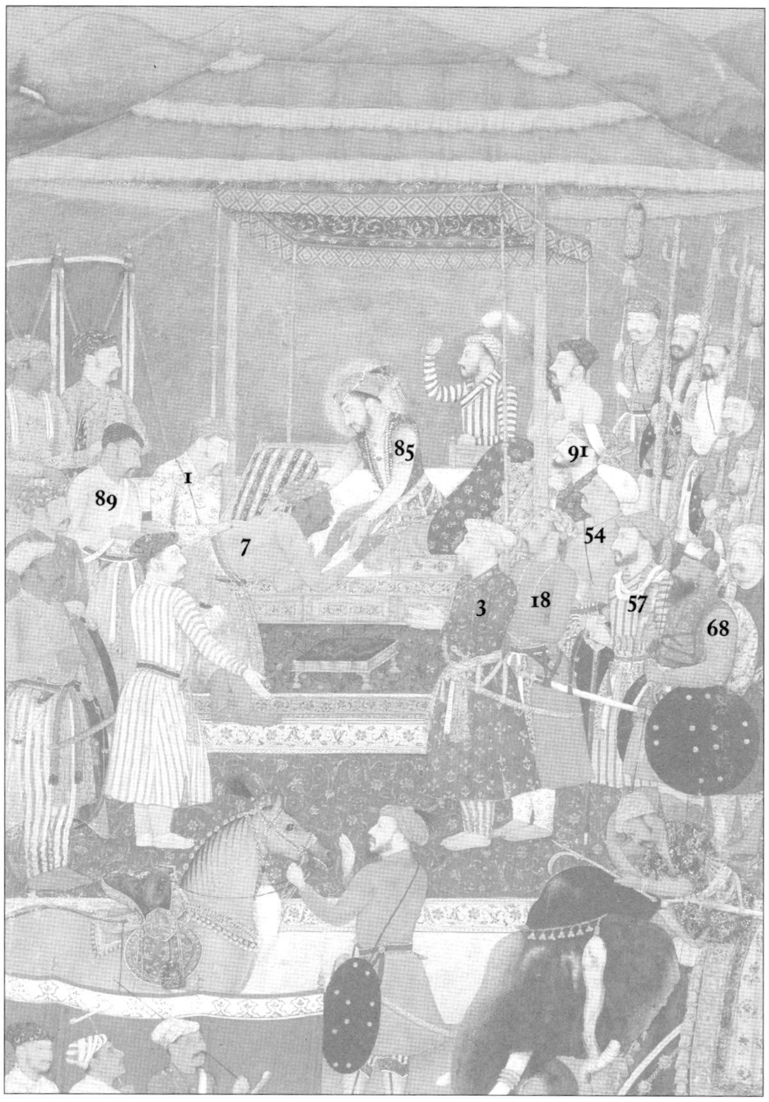

Fig.78 La'lchand, *Raja Bikramajit*, circa 1630.

THE COURTIER WHOSE hand rests so firmly on Amar Singh's back is Abdullah Khan *Firoz Jang*, one of the most brutal of Khurram's followers. The English traveler Peter Mundy had contact with him in India, and wrote that he was '... of a Cruell and Covetous nature, not sparinge any, where anythinge is to be had, litle regardinge lawes, trade or humanitie ...'[2]

While not inscribed, several additional figures can be identified from other portraits. All are men associated with the Udaipur campaign: Suraj Singh Rathor (Fig.72); Mulla Shukrullah, who was given the title *Afzal Khan* following the Udaipur campaign; Raja Bikramajit, known also as Sundardas, and titled *Rai Raian* (Fig.78); and Muhammad Beg, who bore the title *Zulfaqar Khan*. An almost identical depiction of the same event, including many of the same courtiers, but made slightly earlier and by a different painter for an imperial *Jahangirnama* manuscript, is discussed elsewhere (see pages 121–22, above).[3]

Relatively few works are known by La'lchand, the artist of the right-hand page. He excelled in visually quiet but intense portraits (for example, Fig.78), court scenes, and studies of holy men.[4] In only five of the ten double-page compositions in the *Padshahnama* are both halves arguably by the same artist. In several of the examples where two artists were at work, such as the following scene (**8–9**), artistic differences between the two halves are subtle and there is no visual incongruity. Here, on the other hand, the left half is so idiosyncratic and distinct that it has long been thought to be a later replacement rather than an illustration of the Shah-Jahan period.[5] The indifference to convincingly individualized portrait studies or to differentiation of facial expressions in the left-hand page, the heavy shading on the faces, and the figures' awkward gestures and proportions all indicate, at the very least, a difference of hand in the two illustrations, as well as a different level of competence. Since the left page uses identical pigments to continue the specific composition of the right, however, it was almost certainly made contemporaneously to complete that composition. A similar combination of high imperial style with this more puzzling variant recurs within the single-page depiction *Shah-Jahan receives the Persian ambassador, Muhammad-Ali Beg* (**17**). The painter of both these anomalous pages is an artist provisionally referred to as the 'Kashmiri Painter.'[6] MCB

The formal surrender of the Rana of Mewar in February 1615 is shown taking place in Prince Khurram's encampment, set up on the shores of a lake surrounded by hills, a landscape characteristic of the region of Udaipur.[7]

The submission is staged as an outdoor *darbar* in a tent complex, the components of which were the functional counterparts of the ceremonial building types used in the palace. Khurram sits on a canopied throne above which rises a *bargah*

Fig.79 Distant rock formation (detail of **7**).

Fig.80 Sultan Muhammad, *Bahram hunting the lion* (detail), dated 1539–43.

(audience tent), whose characteristic two-tiered form has been analyzed by Peter Andrews.[8] Throne and tent are placed on a stepped dais covered with carpets; this is the area reserved for the inner circle of the Prince and for the bearers of the imperial insignia on the right.[9] The left side of the dais is surrounded by an only partly visible *saraparda* (a wall of canvas supported by stakes; *cf.* **46**), forming an enclosure into which a splendid grey horse from the royal stables is being led as a present for the Rana. An elephant, also intended for the Rana, appears on the right. The depiction of the *saraparda* continues on the left-hand page, where the tribute of the Rana is being brought into the imperial enclosure. The setting is completed by the tented version of a *naqarakhana* (a structure sheltering the court orchestra; *cf.* **8–9**) in the background of the same page. All the tents are red: this was the imperial color and its use was a royal privilege, granted at times as a special favor to the imperial princes.[10]

The main composition was not invented by La'lchand, the painter who signed it. He based it on an almost identical version of the same scene made earlier for Jahangir's history, the *Jahangirnama* (Fig.17). The extent to which the illustrations of the *Padshahnama* depend on those of the *Jahangirnama* is still little studied, but a comparison of the two paintings clearly shows the new direction in which Mughal painting under Shah-Jahan was heading. La'lchand flattens the setting, particularly in his rendering of the tent, the dais and the carpets, from which he eliminates all perspectival elements. The artist focuses instead on portraiture: the group has actually become an assembly of individual full-length portraits. Thus, while on the one hand the composition has become more two-dimensional and formal, on the other hand the figures have acquired – within their linear contours – a stronger physical and psychical presence.

Since this is an outdoor scene it also involves landscape and each of the two painters, La'lchand and the Kashmiri Painter, to whom the left-hand page is ascribed, supplied his own version. La'lchand adopted the landscape of his model, with its vertical rendering of the surface of the lake, as a background for the figure composition although, in accordance with the new formalizing tendencies, he ordered the hills in a more regular formation. The barren, light-brown slopes appear to be based on actual observation, which is borne out by the *Jahangirnama* painting itself, where the artist Nanha showed himself making a drawing on the spot.

The landscape of the Windsor version is extended on the left-hand page with an entirely different terrain. We recognize in the rocky structures – heavily disguised by their naturalistic Mughal interpretation – an artistic stereotype of considerable antiquity. This is the motif of schematic rock formations which the Mughals knew best through its expression in Persian painting where, in the sixteenth century, the Safavids had invented particularly fantastic versions (Fig.80).[11] The Kashmiri Painter exploited the old formula with great originality, showing off his skills in the microscopic illusionistic background mode, another typical feature of the *Padshahnama* illustrations which existed side-by-side with the two-dimensional, formal idiom.[12] The foothills of his rocks are structured as a succession of small mounds, which the artist uses in a more schematic form in several other landscapes in the Windsor manuscript (**34**, **41–42**). Closest to the confident illusionistic handling of the landscape shown here is the hilly region in the background of *A royal procession* (**34**), where we also find similar lines of trees. Usually the multitudes of minute figures that populate the Kashmiri Painter's landscapes do not form large, dense groups. In rendering the troops witnessing the submission, however, he drew his figures together into a masterfully painted illusionistic crowd gathering on the slopes behind the encampment, and – as we would perceive them from a distance in real life – gradually losing their local color to tone in with their surroundings. ❧ EK

1. R. Sharma 1962, p.120.
2. Mundy 1907–24, II, p.152.
3. A third, later version is published in Enderlein & Hickman 1979, no.27.
4. For further discussion of La'lchand, see page 216 below.
5. See S.C. Welch 1963, p.103, where the author remarks that 'history painting was continued on a grand scale in the *Shah Jahan-nama* [*Padshahnama*] of the Windsor Castle Library .... It contains 43 [*sic*] miniatures, some of which were added at Lucknow or Faizabad.'
6. For further discussion of this unusual painter, see page 216 below.
7. The Mughal interest in the Mewar region is borne out by the detailed descriptions of the campaign by the court historians. The *'Amal-i Salih* in particular presents us with an accurate record of the topography of Udaipur and its environs (I, pp.60–61). Usually, Shah-Jahan's authors gave such detailed attention only to the landscape of Kashmir.
8. Andrews 1987, in particular p.150. For a comprehensive treatment of the Mughal camp, see Andrews, forthcoming.
9. The insignia are discussed below, pages 194–95.
10. See, for instance, *'Amal-i Salih*, I, p.36; II, p.58; III, p.113.
11. See also the discussion on page 140, above.
12. See page 141, above.

8–9
Jahangir receives Prince Khurram on his return from the Deccan

Mandu, *Diwan-i 'Amm*, 10 October 1617
Folios 48b painted by Ramdas and folio 49a painted by Murar, both *circa* 1640
Image area 31.1 × 20.8 cm (folio 48b); 30.8 × 20.9 cm (folio 49a)

Inscriptions: [folio 48b, on firework held by attendant at bottom left, written in *devanagari* script] *amal[-i] Râmdâs* ('work of Ramdas') [folio 49a, on steps to throne]: *'amal-i faqîr-i bekas Murâr* ('work of the poor, destitute Murar')[1] [on book held by figure at lower right]: *Murâr* [below painting]: *Murâr musawwir* ('Murar [the] depicter')

References: *Shah Jahan Nama*, p.7; *Jahangirnama*, folios 155b–156a; *Tuzuk-i-Jahangiri*, 1, pp.393–95
Published: S.C. Welch 1995, p.322 and fig.3[2]

The reverse of folio 49b (the left-hand illustration) bears eleven rather than the usual twenty-one lines of text, necessitated by the close sequence of illustrations in this section, and the need to maintain the coordination of text and images.

1. (Abdulcclah Khan)
2. (Khwaja Abu'l-Hasan) (*Mir Bakhshi*)
3. (Afzal Khan)
9. (Asaf Khan)
17. (Bhim Singh Sisodia)
18. (Raja Bikramajit)
30. (Islam Khan Mashhadi)
34. (I'timaduddawla) (*Wakil* and *Diwan*)
38. (The Emperor Jahangir)
42. (Karan Singh Sisodia)
51. (Mahabat Khan)
54. (Mirmiran)
61. Murar
64. (Musavi Khan)
70. (Qilich Khan Turani)
74. (Riza Bahadur)
75. (Mirza Rustam Safavi)
76. (Sadiq Khan) (*Bakhshi*)
85. (Shah-Jahan [Prince Khurram])
89. (Raja Suraj Singh Rathor)

Fig.81 Hunhar, *Maharaja Bhim of Udaipur*, circa 1650.

REFERENCES IN BOTH the *Tuzuk-i-Jahangiri* and the *Shahjahannama* cite the unusual action of the Emperor Jahangir, who rose to embrace the victorious Prince Khurram on his return in 1617 from the Deccan campaigns. This identification of the scene is confirmed by the two Deccani nobles – recognizable by their distinctive, tall turbans – who stand at the lower left on folio 49a. Moreover, the foremost trays of tribute being offered are held by Afzal Khan Allami and Raja Bikramajit (Fig.78), Mughal officers who negotiated the tribute sent to Jahangir from the Deccani kingdom of Bijapur by Sultan Ibrahim Adil Shah II.[3]

In comparison to the historical chronicles we know, however, there are inconsistencies. The central figure in dark-green on the left is Karan Singh Sisodia of Mewar. He first attended court between 15 February and 5 June 1615, as shown in **5**. He returned only once, on 19 March 1616, when he stayed one month and 23 days,[4] and he was therefore not at Mandu for the ceremony shown. His brother Bhim Singh Sisodia (Fig.81) did attend, and is also included in the scene. It is possible that Karan was depicted as a symbolic reference to Khurram's earlier battle, and to universalize the picture into an even more splendid statement of Mughal power.[5] But it is also possible that this represents historical carelessness on Murar's part.

Murar's three-quarter profile portrait of Karan Singh is distinct from the full-profile depiction by Balchand in **5**,[6] but it is a virtual copy of a portrayal by the artist Abu'l-Hasan.[7] An inscription on **38**, which includes an identical image of Karan Singh, states that Abu'l-Hasan, Jahangir's favorite painter, was Murar's teacher, although this alone cannot explain the similarity. Murar, who included a self-portrait in the scene (Fig.140), is a skilled artist, his works sharply focused and exceedingly neat, but with none of the humanistic sensitivity of Balchand (**5**), the love of individualistic appearances shown by Payag (**39**), or the brilliance of pattern of Bichitr (**10**). Moreover, he is seldom an innovator. This work is only one indication that he preferred the refining of established formulas to seeking new solutions or facing innovative challenges. Despite his debt to Abu'l-Hasan, Murar remains a distinct, and less inventive, artistic personality.[8]

Two drawings exist that are closely related to this composition by Murar (Figs 82, 83). While all three illustrations use an identical overall composition, each differs in the position of specific courtiers. There is no evidence to determine whether either drawing is a preliminary study for this *Padshahnama* illustration or a copy for use elsewhere – the two usual purposes for such drawings – and it is uncertain that either is by Murar. Similar versions of other episodes indicate that artists were constantly responding to and adapting each other's work.[9] More importantly, the three versions of this scene each show that, while apparently the documentation of rigid protocol, compositions were also determined by artistic decisions.

Fig.82 Unknown artist, *Jahangir receives Prince Khurram on his return from the Deccan*, circa 1640.

Fig.83 Unknown artist, *Jahangir receives Prince Khurram on his return from the Deccan*, circa 1640.

Fig.84 Wall-painting (detail of **8**).

Many of the remaining dispersed pages made for Shah-Jahan's historical manuscripts are the less important halves of double-page compositions showing attendants assembled in a courtyard observing a ceremony (for example, see Figs 161–166). Because the paintings put into the Windsor *Padshahnama* seem to have been drawn from available stock and therefore were not always specific to the episodes described in the text, it is probable that some double-page compositions were of necessity constructed from illustrations never meant to appear together. The right half of **8–9** and the left half of **10–11**, for example, are generic images of attendant figures by different artists than those who painted the illustrations they face. There is therefore no assurance that these pairs were originally intended to be seen together, though Ramdas, who painted and signed the right-hand illustration, is a particularly appropriate partner for Murar. While he is more attentive to individualistic facial expressions and gestures than Murar, his style and compositions never challenge the status quo.[10] It seems likely that he also painted the attendant figures in **10–11**, for the character types are identical.

The scenes of attendants inevitably include depictions of the musicians of the *nawbat*, the official orchestra that marked major stages of court ceremonial. Recent research by Professor Bonnie Wade[11] argues that the *nawbat* ensemble provided a constant aural evocation of the emperor's presence, and she has identified the instruments: kettledrums played in pairs (*naqara*); a long, straight trumpet (*karna*, literally 'ear'); and a short double-reed aerophone (*surna*) that is predecessor to the modern *shanay*, a large reed instrument. She further argues that this combination of instruments was known by the end of the tenth century throughout the Islamic world as an emblem of sovereignty. François Bernier, a French traveler and physician, wrote of these musicians in a letter dated 1663: 'Over the grand gate, situated in the middle of one side of this court, is a capacious divan ... called *Nagar-Kanay*. In this place ... are kept the trumpets, or rather the hautboys and cymbals, which play in concert at certain times of the day and night. To the ears of a *European* recently arrived, this music sounds very strangely, for there are ten or twelve hautboys, and as many cymbals, which play together .... You may judge, therefore, of the roaring sound which issues from the *Nagar-Kanay*. On my first arrival it stunned me so as to be insupportable: but such is the power of habit that the same noise is now heard by me with pleasure; in the night, particularly, when in bed and afar, on my terrace this music sounds in my ears as solemn, grand, and melodious.'[12] MCB

Jahangir receives his victorious son – about to be given the title *Shah-Jahan* – in the palaces of the Malwa sultans at Mandu, which had been adapted for Jahangir's stay there in 1617, during the Prince's first Deccan campaign.[13] The Jahangiri adaptions included a *jharoka*, and what survives of its architecture is so close to Murar's rendering that he must have modeled his representation on the original.[14] In addition to this accurate depiction of the *jharoka*, Murar introduces the tented ceiling to characterize the setting as Jahangiri.[15] He also furnishes the architecture with a narrative element that draws attention to a significant aspect of the event. The special favor granted to the Prince – to come up to the Emperor in the *jharoka* – is highlighted by the partly visible red ladder leaning against the low marble railing of the gallery near the left margin.[16]

The historical references do not end here. Embedded in the architectural decoration is a veiled allusion to the fact that things were not always as rosy between Jahangir and his son as in this moment of triumph; specifically, that a few years later their relationship was to deteriorate to the extent that Shah-Jahan was in open rebellion. The central area below the *jharoka* – where we usually find symbolic images idealizing Mughal rule[17] – appears at first glance to contain merely a red *jali* panel. If we have a closer look, however, we can make out a wall-painting of a fat *mulla* being offered a bag of coins by a self-abasing Mughal official, and below this scene, another official driving away starving supplicants with a staff (Fig.84). This is clearly an allegory of bad government,[18] illustrating the conditions in the country as represented by Shah-Jahan's propagandists, who claimed that his accession as a *mujaddid* ('renewer') ended an era of corruption and religious decline.[19] ❧ EK

1. This artist's name has often been misread as Murad, but the signature and ascription to him on numerous folios of this manuscript make the reading Murar unmistakeable. The name Murar, or Murari, is an epithet of Krishna. (WT)
2. Fols 48b and 51a, which seem to be by the same hand, have been attributed by S.C. Welch to Payag, despite the signature on fol.48b of Ramdas.
3. *Tuzuk-i-Jahangiri*, I, p.387.
4. *Tuzuk-i-Jahangiri*, I, p.317. See also Somani 1976, p.258 (who miscalculates the year as 1617).
5. One precedent for this symbolic use of portraiture is the double-page *Jahangir greets the poet Sa'di*, by Abu'l-Hasan, where the 17th-century Emperor Jahangir, Karan Singh, and members of the court are shown with the great 13th-century mystic poet (Beach 1981, no.31). An inscription stating that the painting was made at Ajmer dates it to between February 1615, when Karan Singh arrived, and November 1616, when Jahangir left the city. With the exception of Sa'di's presence, the image is a perfect document of the hierarchies and appearance of the major courtiers at that time. For an explicitly allegorical scene in the Windsor *Padshahnama* manuscript, see **41–42**.
6. See also page 198, below, and Fig.120.
7. See Beach 1981, no.31.
8. For further discussion of Murar, see page 216.
9. See, for example, the two distinct versions of *The submission of Rana Amar Singh of Mewar to Prince Khurram* (**6**, and Figs 16 and 17).
10. For further discussion of Ramdas, see page 217.
11. Wade 1995.
12. Bernier 1891, pp.259–60. For comments by Peter Mundy, see Mundy 1907–24, II, p.210.
13. *Cf.* **37**. For the campaign, see *Tuzuk-i-Jahangiri*, I, pp.337ff.; *Shah Jahan Nama*, p.7; Prasad 1922, pp.270ff.; Saksena 1976, pp.20–23.
14. The arrangement of the three doors leading from the interior of the palace into the *jharoka* and its flanking galleries is similar in the painting. The details of the *jharoka* construction, including the heavy brackets which support it, and the leaf molding running along the edges of the floor and gallery, are almost identical, are as are the brackets and eaves which delineate the gallery from above. That Murar had an eye for architecture is also clear in his fine depictions of the fortress of Daulatabad (**31**) and the audience hall at Lahore (**44**). For the *jharoka* of Mandu see Brand 1991, esp. p.13.
15. See also **5**, **37**, **38**, **39** and page 135, above.
16. I owe this observation to Milo Beach.
17. *Cf.* in particular **10–11** and **39**.
18. These scenes were first noted by Robert Skelton (1988, p.185), who offered a different explanation from Jahangir's point of view, that he considered it inauspicious for the king to see misery. It must be remembered, however, that the painting was commissioned by Shah-Jahan.
19. Surprisingly unrestrained condemnatory statements on this are made by Shah-Jahan's first historian, Qazwini; *Padshahnama*, fols 10a–b (re-foliated 11a–b). See also *'Amal-i Salih*, I, pp.1–3, 186–87.

10–11

Shah-Jahan receives his three eldest sons and Asaf Khan during his accession ceremonies

Agra, *Diwan-i 'Amm*, 8 March 1628
Folio 50b painted by Bichitr, *circa* 1630 and folio 51a attributed to Ramdas, *circa* 1640
Image area 30.8 × 21.1 cm (folio 50b); 30.9 × 21.2 cm (folio 51a)

Inscriptions: [folio 50b, on wall below throne] *raqam-i banda-i kamtarîn Bichitr* ('drawn by the most insignificant slave, Bichitr') [folio 50b, on gold band above throne] *majlis-i julûs-i aqdas-i wâlâhazrat-i shâhjahânî / khilâfatpanâhî sâya-i âftâb zill-i ilâhî u sa'âdat / andûkhtan-i shâhzâdahâ-yi kâmkâr-i rafî' qadr-i 'âlî- / miqdâr az mulâzamat-i Sâhibqirân-i Sulaymân-iqtidâr / u sarafrâzî yâftan-i Khânkhânân-i sipahsâlâr az / zamînbos-i în âsitân-i zamîn-âsimân-i 'arsh-istiqrâr* ('assembly of the most blessed accession of His Most Exalted Majesty Shah-Jahan, refuge of the caliphate, shadow of the sun, shadow of God, and the attainment of felicity by the exalted princes through paying homage to the Sahib-Qiran as mighty as Solomon, and the ennoblement of the Khankhanan, master of the army, by kissing the ground before the heavenly threshold as stable as the divine throne') [folio 50b, below painting] *dîvân-i Shâhjahân Pâdishâh u rukhsat-numûdan-i si pisarânrâ* ('divan of the Emperor Shah-Jahan, and his giving leave to his three sons')

References: *Padshahnama*, I/I, pp.177–82; *Shah Jahan Nama*, pp.20–21 and pl.2
Published: S.C. Welch 1995, p.322 and fig.4

SHAH-JAHAN'S SUCCESSION to the throne was by no means certain when his father Jahangir died in November 1627 on his way from Kashmir to Lahore. The future emperor was then in exile far from the court in Nizamshahi territory in the Deccan, where he had been given asylum after his rebellion in 1622. His three eldest sons were kept as hostages at the court of Jahangir. However, through the ruthless machinations of his father-in-law, Asaf Khan, Shah-Jahan was able to come to the throne in Agra on 14 February 1628. Almost a month later Asaf Khan brought the Emperor's sons from Lahore, and extensive festivities consigned the dark shadows surrounding the accession to oblivion.[1] So began Shah-Jahan's rule, during which the Mughal empire experienced its phase of greatest prosperity and stability.

Bichitr's grand accession picture focuses on the ceremonial reception of the princes in the *jharoka* of the *Diwan-i 'Amm* or Hall of Public Audiences in the Agra fort. The gesture in which Shah-Jahan embraces his eldest son Dara-Shikoh suggests dynastic continuity; it recalls similar scenes with his own father Jahangir.[2] As in the ten other *jharoka* scenes in the Windsor manuscript, the composition follows an established pattern evolved from earlier representations in Akbari and Jahangiri painting.[3] Yet Bichitr's version expresses the new artistic ideals of the period in an almost canonical manner, and it was perhaps here that the scheme was laid down for other painters to follow.

Bichitr achieves an unparalleled harmony between architecture and figures by means of a balanced symmetrical composition. Most of the elements of one half of the painting on folio 50b are repeated in reverse on the other side of a central, vertical axis; one notices in particular the splendid symmetry of the flowers flanking the *jharoka*.[4] No other Windsor *darbar* scene adheres so closely and yet so elegantly to *qarina* or 'counter image,' the Shah-Jahani ideal of bilateral symmetry in composition.[5] A typical *qarina* scheme emphasizes the features of the central axis: here, it is the axis of the Emperor. He appears in strict imperial profile in the *jharoka*, presiding in a pyramidal arrangement over his nobles who are arranged below him in two balanced groups and who face each other, also in profile. (The only persons shown obliquely are the nameless bearers of the imperial insignia.) Each group is in turn subdivided horizontally by a railing, to express degrees in rank. At the bottom, the participants in the *darbar* come together to form a half circle, which again directs our view to the imperial axis. These centralizing tendencies in a two-dimensional system are intensified by carefully placed perspectival elements. The awnings of the tent ceiling,[6] the brackets supporting the *jharoka* and its flanking gallery and the eaves above, converge toward the center where the perspectival elements are grouped most densely. On the imperial level a foreshortened canopy is hung in front of the *jharoka*;[7] below, on the level of the nobles, a foreshortened dais and two stools guide our view toward an illusionistic opening, a painting within the painting. All these elements converge not in a

Fig.85 Painting below *jharoka* (detail of **10**).

3. Afzal Khan (*Mir Saman*)
5. Allahverdi Khan
9. (Asaf Khan) (*Wakil* and *Diwan*)
10. Asalat Khan
12. (Prince Awrangzeb)
13. A'zam Khan (*Mir Bakhshi*)
14. Bahadur Khan Rohilla
19. Raja Bithaldas
20. (Prince Dara-Shikoh)
22. Dilawar Khan Kakar
24. Gaj Singh Rathor
30. Islam Khan Mashhadi
37. Ja'far Khan
39. Jai Singh Kachhwaha
40. Jannisar Khan
43. Khan Alam
44. Khan Dawran
45. Khan Jahan Barha
47. Khan Zaman
51. Mahabat Khan
53. Mir Jumla
58. Mukhlis Khan
59. Muqarrab Khan
60. (Prince Muradbakhsh)
62. Murshidquli Khan
64. Musavi Khan (*Sadr*)
65. Mu'tamad Khan
66. Mu'taqid Khan
67. Muzaffar Khan
68. Shaykh Nazir
70. Qilich Khan Turani
72. Rao Ratan Hada
74. Riza Bahadur
76. Sadiq Khan
80. Sarafraz Khan
82. Sardar Khan
84. Sayf Khan
85. (The Emperor Shah-Jahan)
86. Shahnawaz Khan
87. (Prince Shah-Shuja')
88. Shayista Khan
90. Wazir Khan

Fig.86 Diagram of the central perspective elements of the composition in **10**.

single vanishing point but in several points on the vertical axis running through the center of the painting (Fig.86). This selective use of perspective directs the spectator's attention down the whole length of the imperial axis where illusionistic images – the solar symbol on the canopy; the globe on which two lions lie down with a lamb in front of a platform bearing a pole holding the Emperor's chain of justice and flanked by attendant shaykhs (Fig.85); and the stool supported by the traditional Indian throne animals – express the essence of Shah-Jahan's imperial ideology.[8] In Bichitr's accession painting, form is put convincingly to the service of content to create with realistic detail an abstract image, a visual manifesto of Shah-Jahan's rule as head of a cen-tralized state, 'supported politically and ideologically by a hierarchically graded corps of soldier-administrators,'[9] his amirs and *mansabdars*.[10] EK

This painting depicts one of the major ceremonies celebrating Shah-Jahan's accession to the throne, and it is the first illustration in the *Padshahnama* narrative that shows him as Emperor. In addition to his four sons, their grandfather, Asaf Khan, the *Wakil* or Prime Minister, also appears on the *jharoka* level. Asaf Khan was the son of I'timadud-dawla, the *Wakil* under Jahangir, whose tomb is one of the great monuments of Agra. His sister, Nur Jahan, married Jahangir, while his daughter was the Empress Mumtaz Mahal, for whom Shah-Jahan built the Taj Mahal. Many of the nobles in attendance were related to Asaf Khan, but no noble was more closely allied than he with the imperial family.[11]

That the scene was assigned to the artist Bichitr is appropriate, for he conveys with intense precision the sumptuousness and splendor of court life. An extraordinarily astute portraitist, Bichitr also had the technique to describe almost clinically the abundant and complicated textile patterns that enlivened Mughal ceremonies and their architectural settings. Moreover, he could place these within a composition where, while no surface was undecorated, the effect was one of clarity and perfect balance. An unusual detail – the clear shadow cast by the footstool directly in front of the Emperor, the ceremonial space for an appproach to the throne – is a testament to Bichitr's keenness of perception, as well as a visual allusion to Shah-Jahan as 'shadow of the sun, shadow of God,' as stated in the inscription above the throne. In a famous earlier work by the artist, he signed his name on the step on which Jahangir would place his foot to mount a different throne: a symbolic statement that Bichitr was worthy only to be under the foot of the emperor.[12]

Bichitr also explicates precisely the physical arrangement of the *darbar*. The top third of the painting shows the imperial family, while the middle section, within the gold rail, is restricted to those men with the highest political appointments; their placement, in fact, closely follows the verbal account given in the *Shahjahannama*.[13] Below them, the opening in the gold rail is controlled by the *bakhshis*. Shown on the left is Mir Muhammad Baqir A'zam Khan (known also as Iradat Khan), who was appointed *Mir Bakhshi* during the period leading to the formal accession, after which – at the request of his brother, Asaf Khan – he became *Wazir* (Finance Minister); and, on the right, Islam Khan Mashhadi, the second *bakhshi*. Each holds a document that presumably gives information about access and the rank of individual nobles, or other details of the day's ceremonies.

The five men standing immediately outside the opening, almost certainly awaiting a summons to enter (and new appointments of rank), are Sadiq

Khan, soon to replace A'zam Khan as *Mir Bakhshi* (he had married a sister of Asaf Khan and was therefore the Emperor's uncle); his son Ja'far Khan, another future *Mir Bakhshi*, who was married to a daughter of Asaf Khan (and therefore a sister of Shah-Jahan's wife, Mumtaz Mahal); Shayista Khan, son of Asaf Khan, and therefore Shah-Jahan's brother-in-law; Afzal Khan Allami, who was *Mir Saman* and would soon be appointed *Diwan*; and Khan Zaman, a son of A'zam Khan who also married a daughter of his uncle Asaf Khan (see the key opposite). Next to Islam Khan, but inside the railing, is Sayf Khan, the son of another of Asaf Khan's brothers and married to another of Mumtaz Mahal's sisters. Also within the enclosure, the noble standing in front of and closest to the Emperor is the Rajput Raja Gaj Singh Rathor of Jodhpur, the nephew of Shah-Jahan's mother. Such a dense series of interrelationships that united family and power cannot have occurred by happenstance, of course; it was clearly the result of careful calculation and, occasionally, intrigue. But the result was that, in a very literal way, access to the Emperor was overseen by members of his family, with Asaf Khan as the central intermediary. Family members held many of the most important appointments and were strategically placed within the space of the *darbar*. Bichitr's painting makes this immediately vivid.

The left half of this illustration, in which the *nawbatkhana* musicians are again both prominent and lively, can be attributed to Ramdas on the basis of the portion of the preceding double-page (**8**) that is signed by him. ❧ MCB

1. For Shah-Jahan's rebellion and the circumstances of his accession see *Shah Jahan Nama*, p.10ff.; Saksena 1976, pp.38–63.
2. See also **5**, **8–9**, and **38**.
3. Discussions of the *jharoka* setting can be found in **5** and **39**; and, in greater detail, on pages 133–37, above.
4. For the meaning of this flower decoration, see page 200, below.
5. Such schemes were favored in the architecture and architectural decoration of the period. See Koch 1991a, pp.93, 99, 109, 141; Koch 1988, p.17; and page 135, above.
6. At the accession of Shah-Jahan this was still the tent hall used in the time of Jahangir (see also **5**, **9**, **37**, **38**, **39** and Fig.143), but it was soon to be replaced by a wooden construction in green and red, the new *Chihil Sutun*, for which see the discussion on page 176 below, **14**, **23–24**, **32**, **43**, and **44** and Figs 144, 145 and 150; and page 135, above.
7. For the canopy and the tent ceiling, see Andrews 1987, p.155.
8. For the imagery see also pages 191, 199, and 204; for explanation of some of the symbolism, see page 160, above; Koch 1982a, in particular pp.17–21; Koch 1988, especially pp.31–35; Skelton 1988; and pages 136–37, above.
9. Eaton 1993, p.160.
10. See also pages 135–36, above.
11. The genealogy of I'timaduddawla's family is given on page 229, below.
12. For a study of this painting, see Ettinghausen 1961. Bichitr is further discussed on page 214.
13. *Shah Jahan Nama*, pp.20–21.

12–13
The weighing of Shah-Jahan on his forty-second lunar birthday

Agra, *Diwan-i Khass*, 23 October 1632
Folios 70b–71a painted by Bhola, *circa* 1635
Image area 32.3 × 22.9 cm (folio 70b);
31.8 × 21.2 cm (folio 71a)

Inscriptions: [folio 71a, below scales] *'amal-i [B]olâ* ('work of [Bh]ola')[1]

References: *Padshahnama*, 1/1, pp.440–41; *Shah Jahan Nama*, p.87 and pl.3
Published: Brend 1982–3, pp.352–55 and fig.4

The figures of two courtiers extend over the left gold marginal rulings of folio 71a. Trimmed to the original edge of the painting before it was set into its new page, earlier and distinct marginal rulings remain visible under these details, providing explicit evidence that such illustrations had previously been intended for a different volume.

THE MAN IN THE orange robe at the right of folio 71a allows us to identify this scene. He is the Safavid ambassador Muhammad-Ali Beg (Fig.87), who arrived at court on 22 March 1631.[2] The dark-skinned man in a Bijapuri turban just to his left may therefore be I'timad Khan, whose presence is noted in contemporary texts.[3] The major nobles of the Mughal court are also in attendance while Shah-Jahan is weighed against various precious materials that are then distributed to the Muslim holy men shown at the bottom right of the same folio.

The right-hand illustration, according to a recent publication by Professor Bonnie Wade,[4] gives further information on the court entertainment that was an important element of these ceremonies. At the top, the *nawbat* ensemble provides continual aural accompaniment to the Emperor's presence, but without the large kettledrums necessary at more public ceremonies. Below them, a group of male singers stands before instrumentalists who play a *rudra vina*, a typically Indian instrument with double resonating gourds; a *rabab*, a heavy lute (also called a *barbat* or rebec) descended from a type introduced into India from West Asia in the eleventh century); a *dhol*, or large Indian drum; and a slender-necked Persian *tambour*. This latter group forms a more intimate ensemble than the *nawbat* musicians, and shows a synthesis of Indian and West Asian musical traditions that parallels interrelationships between painting styles. But it also establishes the kind of ensemble basic to performances of classical Indian music today, dividing performers into the producers of the melody (singers, *rudra vina* and lute players), the keeper of time (drummer), and the keeper of pitch referent (the *tambour* player). Professor Wade also argues that, to contemporary

Fig.87 Hashim, *Muhammad-Ali Beg*, circa 1632.

2. (Khwaja Abu'l-Hasan)
3. (Afzal Khan) (*Diwan*)
9. (Asaf Khan) (*Wakil*)
12. (Prince Awrangzeb)
20. (Prince Dara-Shikoh)
28. (Hayat Khan)
32. (I'timad Khan?)
37. (Ja'far Khan)
45. (Khan Jahan Barha)
53. (Mir Jumla) (*Mir Saman*)
56. (Muhammad-Ali Beg)
58. (Mukhlis Khan)
60. (Prince Muradbakhsh)
62. (Murshidquli Khan Turkman)
68. (Shaykh Nazir)
80. (Sarafraz Khan)
85. (The Emperor Shah-Jahan)
87. (Prince Shah-Shuja')
88. (Shayista Khan)

viewers, these depictions of performers and instruments would have enriched the experience of viewing these paintings by simultaneously evoking the familiar sounds of certain ceremonies.

No other inscribed works by Bhola are known, although additional pages of this manuscript can be attributed to his distinctive hand (see the discussions of **27–28** and **45**).[5] The central pavilion and the members of the inner imperial circle, who are shown surrounded by sumptuous textiles and decorative objects, are intensely elegant in execution and design; but the attendant figures, especially the rather stiff, chorus-line arrangement of the courtiers at the bottom of the left-hand page, or the vacant expressions of the musicians, betray the inconsistent performance that is a trait of this painter. MCB

The weighing of the Emperor on the occasion of his lunar and solar birthdays, with the celebrations of his accession and of the Persian New Year,[6] were the greatest annual festivals at the court of Shah-Jahan.[7] The Emperor was weighed against gold, silver and other articles and the equivalent of the value of the entire weighing was given away as alms and charity. Although Shah-Jahan's historians Lahawri and Kanbo tell us that this Indian custom had been adopted by Akbar,[8] it was actually practiced by Humayun[9] and was continued with variations until Awrangzeb who – though concerned about weighing 'being not the practice of the country of our ancestors and of the Mohemadans [Muslims] of this country (i.e. India)' – observed it at least in the early years of his reign because it benefited 'many needy and poor persons.'[10] Apparently there was already some orthodox concern about such ceremonies in Shah-Jahan's time because Kanbo is at pains to demonstrate that the Emperor's weighing – although not strictly conforming to the charities prescribed by the *shari'a* (Islamic religious law) – was a form of alms-giving which was in accordance with the ethics of Islam.[11]

Perhaps because of its controversial status the imperial weighing was not often depicted. Bhola clearly modeled his painting on an earlier illustration made for the *Jahangirnama* whose subject is also Shah-Jahan, here as youthful 'Baba' Khurram, being weighed in his house at Kabul by Jahangir in 1607:[12] the artist adapted the composition for the formal setting of the Agra palace. The weighing takes place in a hall overlooking the Jumna River (the Yamuna or Jawn, as the Mughals called it). The sources do not tell us in which building the ceremony was held[13] but we can identify it by examining the architecture shown in the right-hand miniature, folio 70b. This shows the south-east corner of the courtyard known today as Machchhi Bhawan, with the Hall of Private Audiences, the *Dawlat Khana-i Khass* or *Diwan-i Khass* on its eastern riverfront terrace (Figs 90, 91). The hall in which the ceremony takes place originally stood opposite the *Diwan-i Khass* on the same terrace. Shah-Jahan built it as an entrance hall for the Imperial Hammam (baths) in the 1630s as part of his reconstruction of the palace, which was completed in 1637, but it was dismantled early in the nineteenth century (Fig.88).[14] While the buildings are thus positioned correctly in their context, Bhola takes more liberties with their views.[15] The hall is shown in a close, frontal view, with its longer side parallel to the river, although it actually faced the water – like the *Diwan-i Khass* – along its shorter side. At the same time Bhola

Fig.88 Colonnade from the hall in front of the Imperial Hammam. Agra fort, 1630s.

narrows the building by reducing the number and the width of its intercolumniations in order to fit it into the vertical format of the page. The *Diwan-i Khass* is subjected to a more complex manipulation. Bhola presents it correctly with its shorter side towards the court but shows not its side view but a compressed form of the more representative main façade with its five multi-cusped arcades, through which one has a faint glimpse – as in the original – of the back wall with a dado decorated with two registers of flowery plants in marble relief below blind ornamental niches (filled by the artist with flower vases; Fig.89).

The hall of the weighing is largely covered with sumptuous hangings[16] but what is visible of its faceted pillars reproduces the characteristic flickering impression created by the use of the technique of stone intarsia (*parchin kari*) for the floral scrolls of the originals. The typical elements of the new imperial marble style, which found one of its earliest expressions in these pavilions, are thus well observed in these details.[17]

Shah-Jahan's reign was an era of great architectural awareness, which is reflected in the contemporary sources. From no other Mughal period do we possess such detailed comments on architecture. Painters had to match these literary descriptions with more realistic images of architecture while, on the other hand, they were expected to observe the canon of formal hierarchy imposed on the official court style. Bhola solves this dilemma by using a frontal view for the hall where the imperial ceremony takes place while experimenting with a manipulated view and a perspectival representation in the part of the courtyard reserved for the dancers and musicians. ❧ EK

1. The reading of the artist's name presents difficulties. The placement of one dot inside the bowl of the *lâm* in *'amal* seems to indicate the reading *Bola*; the presence of what appears to be two dots above, however, would indicate the reading *Tola*. The name is almost certainly Bhola, an epithet of Shiva. (WT)
2. His ceremonial reception is depicted in its proper place later in the manuscript; see **17**.
3. *Padshahnama*, I/I, p.441. See the passages quoted on pages 39 and 42–43, above.
4. Wade 1995.
5. For further discussion of Bhola, see page 214, below.
6. See also **17**.
7. *Tawarikh-i Shahjahani*, fol.10b (re-foliated 12b). One notes that no Muslim festivals were included.
8. *Padshahnama*, I/I, pp.243–44; *'Amal-i Salih*, I, pp.267–68. Both passages are translated above, page 39. Kanbo's description of the ceremony does not take in the changes made by Shah-Jahan. *Cf. Tuzuk-i-Jahangiri*, I, p.115. The princes were weighed on their solar birthdays. Occasionally weighings were also held in thanksgiving for the recovery from an illness or to invoke good health. See also Ansari 1974, pp.118–20; Mubarak Ali 1986, pp.51–54.
9. *Qanun-i Humayuni*, p.76.
10. *Ruqa'at-i 'Alamgiri*, pp.78–79; *cf. 'Alamgirnama*, I. pp.227 ff; Awrangzeb abolished the practice in 1668, for which see *Ma'asir-i-'Alamgiri*, p.48. See also Sarkar 1972–4, III, p.56.
11. See note 8.
12. British Museum, London, inv. no.BM 1948. 10-9.069; illustrated, for example, in S.C. Welch 1978, pl.18. *Cf. Tuzuk-i-Jahangiri*, I, p.115.
13. The weighing of Shah-Jahan usually took place in the Hall of Public Audiences, the *Dawlat Khana-i Khass-u-'Amm* or *Diwan-i 'Amm*, but other locations were also possible.
14. Parts of it were brought to England and are kept today in the reserve collection of the Victoria & Albert Museum. For the story of the dismantling of the hall and an attempt to reconstruct it, see Koch 1982 c.
15. The same setting is rendered with more truth to the original on Bulaqi's double page, **25–26**.
16. These have floral patterns framed by a border of vertically arranged *chinikhana* motifs (*cf.* **5**), a demonstration that this popular wall-decoration appeared in a textile form. For a discussion of the hangings see also Andrews 1987, p.157 and fig.8.
17. See Koch 1991a, pp.93–95.

Fig.89 Façade of the *Diwan-i Khass*. Agra fort, 1630s.
Fig.90 View into the *Diwan-i Khass* and south wing of the 'Machchhi Bhawan.' Agra fort, 1630s.

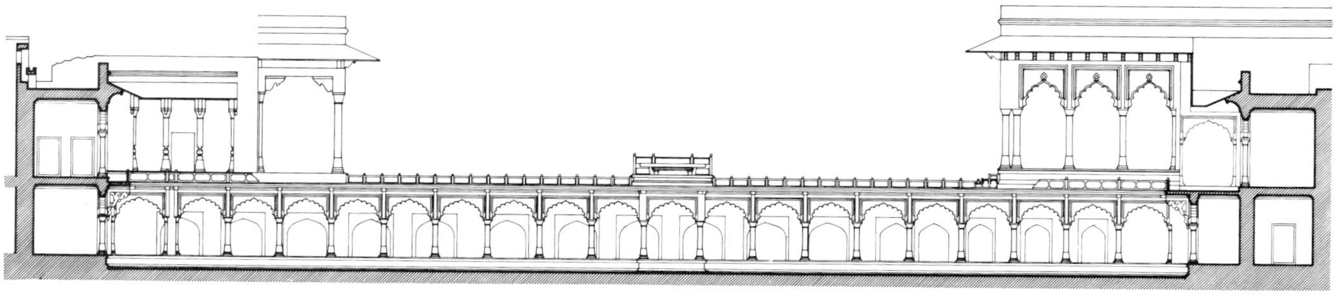

Fig.91 Elevation drawing of the riverfront terrace of the Agra fort, with the *Diwan-i Khass* and Imperial Hammam, 1630s.

14
The presentation of Prince Dara-Shikoh's wedding gifts

Agra, *Diwan-i 'Amm*, 4 February 1633
Folio 72b painted by Balchand, *circa* 1635
Image area 34 × 23.4 cm

Inscriptions: [on paper held by figure at left] *allâhu akbar banda-i dar[gâh] raqam-i Bâlchand-i pir ghulâm* ('God is great. Slave of the court, drawn by Balchand, the aged servant') [beneath painting] *shabih-i darbâr-i hazrat-i Sâhib-Qirân-i Sâni Shâhjahân Pâdishâh anâra 'llâhu burhânahu wa-sultânahu* ('a likeness of the court of His Majesty the Sahib-Qiran II Shah-Jahan Padishah, may God illuminate his proof and his rule')

References: *Padshahnama*, I/I, pp.454–56; *Shah Jahan Nama*. pp.90–91
Published: Smart 1991, p.143 and fig.10 [1]

3. (Afzal Khan) (*Diwan*)
9. Asaf Khan (*Wakil*)
12. (Prince Awrangzeb)
20. (Prince Dara-Shikoh)
28. (Hayat Khan)
30. (Islam Khan Mashhadi) (*Bakhshi*)
52. (Makramat Khan)
53. (Mir Jumla) (*Mir Saman*)
59. (Muqarrab Khan)
60. (Prince Muradbakhsh)
64. (Musavi Khan) (*Sadr*)
75. (Mirza Rustam Safavi)
80. (Sarafraz Khan)
85. (The Emperor Shah-Jahan)
87. (Prince Shah-Shuja')

THIS PRESENTATION scene seems to show the display of the *jahaz* (wedding gifts) of one of the sons of Shah-Jahan, most likely Dara-Shikoh, who is given a prominent position in the *darbar*. Shah-Jahan's wife Mumtaz Mahal had already begun to plan what she wanted to give her son Dara on the occasion of his wedding to the daughter of Sultan Parvez, when her untimely death in June 1631 interrupted these preparations. The organization of the wedding was consequently entrusted to her eldest daughter, Jahanara Begam, who succeeded Mumtaz to the position of first lady at the court.

The wedding festivities of February 1633 began with an exhibition of the imperial presents. The *Diwan-i 'Amm* was chosen for this purpose and temporarily pronounced *khalwat* (a place of 'seclusion' or 'retirement'), which meant that men had to evacuate the area to allow Jahanara and her women to arrange the articles which were afterwards displayed during a public audience given by the Emperor to the court. On the following day the bride's mother arranged her own presentation.[2]

The assignment of this *darbar* scene gave Balchand the opportunity to try his hand at the established formula of representing the Emperor holding court in the *jharoka-i khass-u-'amm*, with the inclusion of the wooden *Chihil Sutun* (the 'Forty-Pillared' audience hall).[3] His use of the pillars of the hall to divide the scene into three vertical sections is closely paralleled in the Windsor manuscript by the *darbar* scenes of Murar (**32**) and Payag (**43**). Balchand employs the division here to organize the presentation ceremony and to give it some inner logic within the stylized *darbar* image. The objects on display – gold and silver vessels, chandeliers and precious textiles, specified in detail by Lahawri – are placed in the left and middle compartments on a white, patterned cloth spread over the carpet within the railing of the hall. They are thus approximately in the direction of the Emperor's gaze (conditioned by his side view) and directly in front of the grandees in the privileged central position below the *jharoka*. The right side is also reserved as usual for the group bearing the Emperor's *qur-i khassa* (ceremonial weapons and insignia),[4] which include interesting figurative silks. EK

There is no accompanying text to explain the subject or placement of this illustration within the manuscript. The composition, however, clearly stations the nobles to the Emperor's left, from where they observe a lavish display of jewels, textiles and metalwork, among other items, presided over by three central figures: the *Wakil* Asaf Khan, the *Diwan* Afzal Khan, and Mirza Rustam Qandahari, a retired senior noble whose daughter married Shah-Shuja' in a ceremony held almost simultaneously with the marriage of the Prince's elder brother, Dara-Shikoh.[5] This, together with the prominence of Dara-Shikoh on the *jharoka*, suggests that the scene was intended to depict the presentation of the wedding gifts to Dara as organized by Jahanara Begam, his sister. To the right are the *Mir Saman* Mir Jumla and the *Sadr* (Supervisor of Religious Endowments), Musavi Khan, while Islam Khan Mashhadi, the second *bakhshi* until his appointment as *Mir Bakhshi* in 1633–4, is at the Emperor's right.[6] Outside the railing, mace-bearers and other guards provide visible protection. ❧ MCB

1. Dr Smart identifies the scene as the presentation of the wedding gifts of Shah-Shuja', but this seems unlikely as he is placed in a minor position in the *jharoka* and it is the display of Dara's gifts that is described at length in the text.
2. *Padshahnama*, I/I, pp.454–56, translated on page 46 above; *'Amal-i Salih*, I, p.437. See also Koch 1994, p.145 and n.18.
3. For the *darbar* formula, see **10–11**; and page 135, above, where the *Chihil Sutun* is discussed; for this, see also **17**, **23–24**, **32**, **43**, and **44**.
4. *Cf.* **6**, **9**, **10**, **17**, **19**, **32**, **38**, **39**, **43**, and **44**. For references regarding the insignia see page 195, note 5, below.
5. The reason for the prominence of Mirza Rustam here is uncertain. Another of his daughters had earlier married Sultan Parvez, the elder brother of Shah-Jahan and the father (by a different wife) of Nadira Banu Begam, Dara's bride.
6. This figure, who holds a paper giving the name of the artist Balchand, was wrongly identified in an earlier discussion by the present author as a portrait of the artist; see Beach 1978, p.95 and fig.6.

15
A'zam Khan captures Fort Dharur

Dharur (Deccan), January 1631
Folio 92b painted by Balchand, *circa* 1633
Image area 32.9 × 23 cm

Inscriptions: [on shield, lower left] *banda-i dargâh raqam-i Bâlchand-i pîr ghulâm* ('slave of the court, drawn by Balchand, the aged servant') [on banner finials, middle right] *yâ fattâh* ('O conqueror!'); *yâ 'Alî* ('O Ali!')

References: *Padshahnama*, I/I, pp.339–43; *Shah Jahan Nama*, pp.52–53 and pl.14
Published: S.C. Welch 1995, pp.322, 328 and fig.6;[1] Smart 1991, p.143 and fig.11

Fig.92 Raiders scaling the walls of the fort at Dharur (detail of **15**).

SINCE THE SPECIFIC text (or texts) which the paintings now placed in the Windsor *Padshahnama* manuscript were originally intended to illustrate is unknown, it is difficult to be certain of the exact relationship of text to picture. This scene nonetheless corresponds to Abdul-Hamid Lahawri's narrative.[2] The leading equestrian figure is recognizably A'zam Khan, who led the siege of Dharur. The chronicle states that 'he ordered that ... some should scale the walls by ladders and others by lassoes,' a detail which evidently siezed the artist Balchand's interest (Fig.92), as did the capture of three elephants that had wandered into the ditch around the fortress (Fig.93). Given the thoroughness of Lahawri's description of this and other events, artists were continually forced to choose particularly appealing moments of the action, or to compress sequential occurrences into a single scene. Unlike his brother, Payag (see **18**), Balchand invariably chose to concentrate on the people and their interrelationships, never allowing the action to overwhelm the actors.[3] ❧ MCB

Dharur was the first fort to fall into Mughal hands in Shah-Jahan's great Deccan campaign of the early 1630s. The campaign had been brought about by the defection of the high-ranking Mughal amir Khan Jahan Lodi, who sought support from the northernmost of the Deccani sultanates, the Nizamshahis of Ahmadnagar.[4] Shah-Jahan marched with a large army to open an offensive against the rebel and his ally, the Nizam Shah; he divided his forces, who began to reduce and conquer the most important strategic sites in the Nizamshahi kingdom. The Mughal commander A'zam Khan and his son Multafit Khan laid siege to the town of Dharur in the Balaghat Range, some hundred miles south-east – as the crow flies – of the Nizamshahi residence at Daulatabad.[5] Dharur was an important market town with a fort 'renowned in the Deccan for its impregnability and ability to withstand siege.'[6] After an embittered resistance Sidi Salim Habashi, the commander of the fort, had to surrender it on 27 January 1631.

Balchand sets the taking of the fortress by the Mughal forces in a landscape where a succession of terrain segments in varied browns creates a diagonal movement into space, from the mound in the left foreground to the fortress and to the hills in the distance. The device of superimposing landscape features in silhouette and often in different colors, like the coulisses of stage scenery, to create a sense of depth is a universally employed formula in two-dimensional representation; it had been widely and variously used in earlier Persianate painting before being adopted by the Mughals. The succession of ridges recommended itself in particular for battle scenes or sieges, because it emulated the hills or trenches in which troops took cover to advance on the enemy (Fig.98).[7] This correlation between landscape formula and subject matter is explored by Balchand in a new realistic idiom (*cf.* **36**); his interpretation, however, appears laboriously pieced together when compared with Payag's powerful *Siege of Qandahar* (**18**).

The fortress of Dharur is correctly rendered with concentric double walls giving a stepped effect.[8] The outer wall would be termed *sher hajji* in the contemporary terminology of fortification.[9] The walls and towers are equipped with *kanguras* (battlements or crenelations) in the form of small, pointed arches lined with a narrow 'piping', and *sangandaz* (embrasures or loopholes) of two types, in the form of vertical slits in the inner wall and with stone hoods in both the inner and outer walls. The details of the architecture are closely patterned on Mughal forts, so that one feels that Balchand worked some of his everyday experience of the buildings of Agra into his depiction of the Deccani stronghold.[10]

The artist took care to include actual episodes of the siege into the painting. The three elephants seem to allude to an early episode of the siege, when the Mughals captured elephants hidden by the inhabitants of the town, together with their other belongings, in the ditch (Figs 59 and 93). An important aspect of the siege of Dharur was the use of heavy guns on both sides. It is interesting to compare the effect on the depiction of the cannon of the two modes of the Shah-Jahani court style, the two-dimensional manner in which the main event was depicted and more three-dimensional illusionistic rendering. The two Mughal cannon on the left, in the area of landscape containing the

Fig.93 The capture of three enemy elephants (detail of **15**).

main protagonists, are represented in formal profile, which obscures to the modern viewer the fact they are meant to point up to the fortress. The gun in the background is rendered in perspective and positioned realistically. Moreover – like its Nizamshahi counterparts set up between the battlements of the fortress – it is shown in action, with atmospherically depicted fire and smoke emerging from the cylindrical mouth of the barrel.

The siege was eventually brought to its successful conclusion when the Mughals managed to scale the walls of the fortress with ropes and ladders. This inspired Balchand to show two figural groups in perspective: next to the 'formal' cannon, two Mughal besiegers carry a ladder and a rope past wounded soldiers; beyond them, three men scale the wall of one bastion of the fortress on another ladder, preceded by a fourth man on a rope (Fig.92). Unlike Payag, who focused on a single event in his *Siege of Qandahar* (**18**), Balchand shows several successive scenes of the siege happening simultaneously in one picture. ❧ EK

1. Welch attributes fol.92b to Payag, despite Balchand's signature.
2. See the text quoted on page 48 above.
3. For further discussion of Balchand, see page 214.
4. See also **16**, **18**, and **31**.
5. *Shah Jahan Nama*, p.52–53; Sherwani & Joshi 1973–4, I, p.271; Saksena 1976, pp.132–33.
6. *Padshahnama*, I/1, 340, as translated on page 48 above; *cf. Shah Jahan Nama*, p.52.
7. Balchand's bare segments of terrain are close to those in early Timurid landscapes of the same type; the battles and sieges in the copy of the *Zafarnama* of Sharafuddin Ali Yazdi, illustrated for Timur's grandson Ibrahim Sultan in Shiraz in 1436, express the motif in a linear, highly abstract manner. See Lentz & Lowry 1989, nos 3, 29, and pls on pp.26, 103; and Sims 1992. See also page 177 above, and Fig.98.
8. See Navarane 1995, pp.246–47, 374 (illustration) and fig.127 (plan).
9. See page 178, note 5, below.
10. See Koch 1991a, fig.37 and Nicolle & McBride 1993, p.3.

16
The death of Khan Jahan Lodi

Kalinjar, 3 February 1631
Folio 94b painted by Abid, *circa* 1633
Image area 31.8 × 20 cm

Inscriptions: [bottom, middle] *'amal-i banda-i dargâh 'Âbid* ('work of the slave of the court Abid')
[on shield, top middle] *fawj-i 'âlamgîr-i Sâhib-Qirân-i Sânî* ('the world-conquering army of the Sahib-Qiran II')
[on finial, top right] *fathun qarîb* ('victory is nigh')[1]
[on sword, middle right] *shabîh-i Sayyid Khânjahân* ('depiction of Sayyid Khanjahan')
[on figure, lower middle right] *kamîna châkar-i û bar 'adûw mansûr bâd | khudây Shâh-i Jahânrâ zafar* [*hamesha*?[2]] *dihâd* ('may his most insignificant servant be victorious over the enemy; may God always grant victory to the king of the world [Shah-Jahan]')
[on figure, lower middle, left] *Shâhjahânî* ('Shah-Jahani')
[on collar of beheaded figure] *shabîh-i besar* ('depiction of the headless')
[below left hand of figure, middle left]: *'amal-i ...mr* ('work of ...mr')

References: *Padshahnama*, I/1, pp.350–51; *Shah Jahan Nama*, p.56 and fig.15
Published: Beach 1992a, pp.132–33 and pl.G; S.C. Welch 1963, fig.4

That the illustration was trimmed along the edges is evident by the extension of the figure of a warrior over the left marginal rulings, and by the no longer legible severed head near his right shoulder. With this extension, which also includes an indication of original marginal rulings, the width of the illustration is 21 centimetres. There is also some damage just above the shield at the right, where support for the two staffs is now unexplained.

THE AFGHAN Pir 'Pira' Khan Lodi, entitled Khan Jahan, rose to power under Jahangir and became one of his most esteemed amirs. In 1625 he was made governor of the Mughal territories in the Deccan. However, he endangered his position and reputation when he let himself be bribed by the sultan of Ahmadnagar, the Nizam Shah or Nizamulmulk, to cede part of the regions entrusted to his governorship.[3] Khan Jahan fell completely out of favor with Shah-Jahan when he failed to support his succession to the throne. Despite the fact that Khan Jahan suppressed the first rebellion of Jujhar Singh Bundela in 1629, at the beginning of Shah-Jahan's reign,[4] he was recalled from his post as governor of Malwa to Agra. Seeing no way to avert his downfall, Khan Jahan fled the court on 15 October 1629 and, joined by several other defectors, made his way through Bundela territory to the court of the Nizam Shah at Daulatabad.[5] Khan Jahan's defection put Shah-Jahan's rule to its first great test because 'nearly all the tribes of Afghans from the bank of the rivers Nilab and Indus to the confines of Kabul and the surrounding countries' were prepared to acknowledge the rebellious Afghan as their leader. The Mughals were, however, able to suppress this uprising at Peshawar[6] and to deal successfully with the Nizamshahis in the Deccan campaign which Shah-Jahan supervised personally from Burhanpur.[7] When the Mughal forces closed in on Daulatabad (see **31**), Khan Jahan and his followers became distrustful of their ally, the Nizam Shah, and set off towards the north, 'hoping to achieve something in the Punjab with the assistance of the Afghans.'[8] They were hotly pursued by a detachment of Mughal troops. When the Afghans passed through Bundelkhand, Bikramajit, the eldest son of Jujhar Singh Bundela who had previously assisted Khan Jahan in his flight from Agra, wanted to atone for it and attacked the Afghans. Khan Jahan's companion Darya Khan Rohilla was killed and his head sent to court, for which Bikramajit received the title *Jagraj* ('Lord of the World').[9] Khan Jahan managed to escape but his flight finally came to an end at Sahenda (Sihanda), north of Kalinjar, where his Mughal pursuers advanced towards him on 3 February 1631. The Afghan fought valiantly but was killed in action, many of his men perishing with him. The heads of Khan Jahan and of his son Aziz Khan were sent to the court at Burhanpur, where Shah-Jahan had them mounted on the gate of the palace as a warning. The court poet Abu-Talib Kalim composed appropriate verses to commemorate the event, and Khan Jahan's captors, Abdullah Khan *Firoz Jang* and Sayyid Muzaffar Khan Barha, were suitably rewarded; Khan Jahan's title went to the latter.[10]

Abid depicts the moment when the royal macebearers sever the head of Khan Jahan from his bound body. It appears as if the renegade loses his life at this moment, though in reality he had been cut to pieces earlier in the engagement with the Mughals. The heads of his son and followers are lying on the ground; two more heads of slain Afghans are held by Mughal officers among those who watch the main scene. The leaders of the Mughal force, Abdullah Khan and Sayyid Muzaffar Khan, have dismounted their richly armored horses to preside over the action.[11]

The artist sets the figures in countryside that is recognizably the stereotypical Persianate landscape adopted by the Mughals, namely vertically disposed terrain topped by a rocky formation lined by a row of spectators (Fig.80),[12] but the scheme is now interpreted more realistically and enriched with horizontally layered segments employed to build up the landscape.[13] The ground is thus divided into three ascending zones of varied greens. Abid moves the main scene to the foreground, showing the beheading at close quarters to give it more immediacy. We look directly at the head of Khan Jahan, depicted in the profile view to which he was still entitled in this last moment of his life as a high-ranking Mughal official. However, the intersecting arms of his decapitators create a three-dimensional space around him which, as we have seen, was the domain in such miniature paintings of unofficial activity, of anonymous or lower ranking persons

Fig.94 Etienne de Laune, *Ad Locutio*, dated 1570.

Fig.95 Detail of **16**.

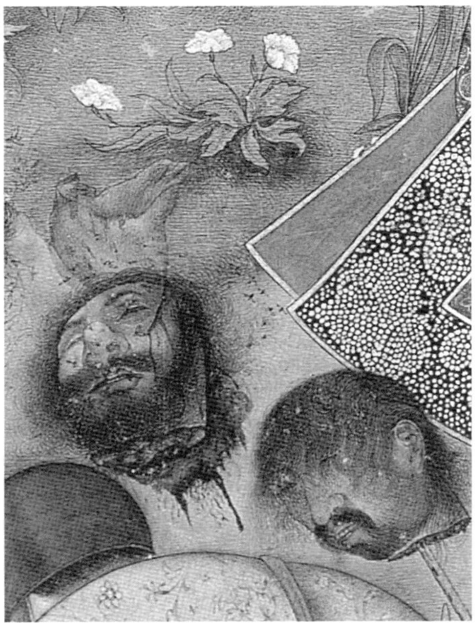

Fig.96 Severed heads (detail of **16**).

and – as in this case – of rebels and traitors.[14] This is emphasized by the other heads lying on the ground, seen obliquely and in shockingly realistic detail. The two men who mete out death and degradation are clearly designated as the agents of Shah-Jahan by small inscriptions. One has 'Shah-Jahani' written on his sleeve, the other wears a vest bordered with a beautiful inscriptional band (*tiraz*) praying for victory for the Emperor. The imperial presence is also evoked in nature, by the plane tree (*chinar*) placed above the decapitation scene. As the favorite tree of the Mughals, it is traditionally associated with imperial landscapes.

The viewer is not the only one to witness how imperial vengeance deals with the rebellious. The line of onlookers who hem the contour of the hill act as a kind of chorus. Abid adds a second line of more individualized spectators behind the flower-strewn mound that forms a tapestry-like backdrop to the key figures. One of them looks directly out at us, as if to call us to witness, and to direct our view toward the main scene.

The eye is caught by the details, generally the great strength of the Windsor manuscript,[15] but here executed with particular virtuosity. Only a few can be enumerated here, such as the meticulous studies of flowers, the sensuous treatment of drapery and armor, and the painter's ability to individualize faces. Abid poignantly captures the expression of Khan Jahan between life and death, and confronts us with the reality of extinction in the distorted features of the already severed heads, in particular those lying on the ground, over which swarm minuscule but none the less gruesome blood-filled flies (Fig.96), as if to announce decay and the transience of (rebellious) life.

Abid shared this interest in explicitly realistic beheading scenes with the followers of Caravaggio in contemporary Europe. He expressed it with the Mughal's own 'courtly miniature realism' to give the utmost immediacy to the moral lesson of his painting. ❧ EK

This painfully graphic depiction of a beheading is among the most complex and original of all Mughal pictures. We are brought inescapably close to the act of execution, isolated clearly within a complicated grouping of royalist forces, but each of the surrounding figures also demands our alert attention. The characterizations are intensely individualized, and the composition is enriched by the complicated patterns of the armor and textiles. The densely intertwined forms, and especially the stances and hand gestures of the warriors at the bottom, are self-consciously posed, however, suggesting that the artist has seen Northern European works in the style of Etienne de Laune's engraving of 1570, *Ad Locutio* (Fig.94). The balance of calculation and direct, uninhibited observation gives this work by Abid unusual power.

The Rajput whose head is emphasized below the dismounted horseman on the right is Madho Singh Hada, younger son of the Rajput Rao Ratan Hada of Bundi. (It was a spear thrust from Madho Singh that killed Khan Jahan Lodi.) Abdullah Khan *Firoz Jang*, the other horseman, became famous for this act of beheading, and a series of portraits showing him holding the head of Khan Jahan Lodi are known.[16]

Abid is in many ways the greatest of Shah-Jahan's painters, although few of his works have been identified. His compositions are always original, and tend to contain small details of riveting visual power. Even when his assignment is a *darbar* scene, the most ordinary of subjects, he includes figures who demand our personal involvement. In *Jahangir receives Prince Khurram* (**37**), for example, the line of figures at the bottom is a theatrical device, like the chorus of warriors at the top of the present scene, but the figures themselves – especially those at the extreme left and right – dissolve its artificiality through an assertion of vivid, and in this case humorous, life (Figs 39, 41).

We know through inscriptions that Abid was the son of the Persian emigrant artist Aqa Riza, who came to India from Iran in 1589 and took employment with the young Sultan Salim (the future Emperor Jahangir). Abid was also the brother of Abu'l-Hasan, who was given the title *Nadir al-Zaman* ('Zenith of the Age') by Jahangir and is described in the *Jahangirnama* as that Emperor's favorite painter.[17] Like the works of his brother, Abid's paintings show immaculate craftsmanship and retain a recognizably Persian character in such details as the solid blue sky and gold band at the top which, together with the line of distant figures, serve to flatten the space. It is uncertain what Abid was doing during the years of Abu'l-Hasan's prime, for nothing can be reliably ascribed to his authorship before Shah-Jahan's accession to the throne.[18] ❧ MCB

1. The words are found in the Koran, 61:13.
2. From the meter of the line it is obvious that three syllables are missing. *Hamesha* is suggested. (WT)
3. Prasad 1922, p.424.
4. *Cf.* **35**, **36**.
5. For the career and rebellion of Khan Jahan see *Shah Jahan Nama*, pp.34, 35–36; *Ma'athir ul-Umara*, I, pp.795–804; Sherwani & Joshi 1973–4, I, pp.270–71; Saksena 1976, pp.66–79.
6. *Shah Jahan Nama*, pp.40–42.
7. *Cf.* **15**, **18**, and **31**.
8. *Shah Jahan Nama*, p.49.
9. *Shah Jahan Nama*, pp.51–52. Yet he was soon to lose his own head in a similar situation; see **36**.
10. *Shah Jahan Nama*, pp.55–57.
11. For the armor shown in this painting, see Nicolle & McBride 1993, p.40.
12. For the landscapes of the Windsor manuscript see also pages 137–40, above.
13. *Cf.* **15**, **18**, and **36**.
14. *Cf.* **36**. For the principles of the official Shah-Jahani court style, see page 136, above.
15. See also page 141, above.
16. A nineteenth-century copy is in the Kevorkian Album; see S.C. Welch *et al.* 1987, no.90.
17. For Abu'l-Hasan, see Beach 1978, pp.86–94; and Beach 1980.
18. For further discussion of Abid, see pages 198 and 213, below.

17
Shah-Jahan receives the Persian ambassador, Muhammad-Ali Beg

Burhanpur, *Diwan-i 'Amm*, 26 March 1631
Folio 98b attributed to the 'Kashmiri Painter', *circa* 1633
Image area 30.7 × 20.2 cm

References: *Padshahnama*, I/1, pp.365 ff; *Shah Jahan Nama*, pp.62–63 and pl.16
Published: S.C. Welch *et al.* 1987, p.258[1]

THE PERSIAN AMBASSADOR, Muhammad-Ali Beg, was received by Shah-Jahan at Burhanpur, which was Shah-Jahan's base from early 1630 to the middle of 1632. Muhammad-Ali Beg is known from several individual portraits (for example, Fig.87) and appears here dressed in orange near the bottom left of the composition. The ambassador remained at court until his dismissal on 23 October 1632 (see **12–13**), and the following May Shah-Jahan sent a return embassy to the Iranian shah.

The embassy of Muhammad-Ali Beg was one of two important diplomatic missions sent by the Iranian Shah Safi to Shah-Jahan's court. The second, led by Yadgar Beg, arrived in 1638. (For a depiction of that ceremony, see Fig.150) The presentation of gifts, seen here, was an important element of these ambassadorial contacts, and manuscripts and paintings regularly formed part of the exchange. This is shown by the courtier at the left, who holds a book with painted lacquer covers similar to those in which the *Padshahnama* and many other manuscripts were bound. Other tribute-bearers are aligned along the bottom of the scene.

The painting is unsigned. Despite the sensitive treatment of the faces of the most important figures, the roughly drawn features of the attendants and the awkward proportions throughout suggest that this is the work of the artist identified as the 'Kashmiri Painter'. ❧ MCB

The Ottomans, the Safavids, and the Mughals were the three Muslim 'superpowers' of the seventeenth century. As their immediate neighbors and rivals, the Safavids were the Mughals' greater concern.[2] Their relations were peaceful over long periods, being endangered mainly by clashes of territorial interest over the possession of Qandahar (today in Afghanistan) (see **32**). From the time when a mission from Shah Tahmasp recognized Babur as *padshah* of the new Mughal empire,[3] embassies were particularly welcomed after a change of rule. Shah Safi's congratulatory embassy to Shah-Jahan on the occasion of his accession was received at Burhanpur, during the *Nawroz* (Persian New Year) celebrations in March 1631.[4] This gave the Mughal court the opportunity to show itself in its most Persian aspect. Like other Muslim rulers of India and the courts of Central Asia, the Mughals oriented themselves strongly towards Persianate culture and celebrated *Nawroz* until Awrangzeb abolished it as un-Islamic in 1659.[5]

The ambassador Muhammad-Ali Beg is received in the palace of Burhanpur[6] in a sumptuous double tent of a new, curved type called *aspak*,[7] which was set up during *Nawroz* celebrations in the courtyard of the *Diwan-i 'Amm*. The audience hall of Burhanpur is shown behind the tent; enough of it can be made out to see that – like its counterparts in Agra and Lahore – it was a wooden hall with green and red pillars. These wooden halls, which replaced the tented halls in front of the *jharoka-i khass-u-'amm* in the main palaces throughout the empire in 1628 – were proudly announced as an innovation of Shah-Jahan and called *Chihil Sutun* ('Forty-Pillared' Hall), to show their inspiration from Persia.[8] Muhammad-Ali Beg must have felt quite at home.

Otherwise, the composition is patterned on the formula of the emperor holding court in the *jharoka-i khass-u-'amm*, with the roofed throne taking the place of the *jharoka* balcony. The imperial insignia, positioned as usual on the right, include a pearl-fringed parasol (*chatr*) (see **3–4**) and a balance-shaped emblem, obviously a symbol of justice.[9] It does not seem to be a coincidence that the ambassador is shown making his obeisance outside the red railing that demarcated the inner area reserved for the highest dignitaries. Muhammad-Ali Beg was probably received inside it, but to fix him in the memory as standing outside – and with his bulging waistline realistically rendered – appears as a pictorial slight, intended to amuse the informed Mughal audience at the expense of the Safavids. The same purpose was served by the conspicuous depiction of most of the Safavid bearers of presents – arrayed in a line at the bottom and recognizable by their turbans – in the three-quarter or rear views usually assigned to persons without particular status. Unlike the historians, who had to weigh their words carefully for posterity, the painter could subtly explore the conventions of Shah-Jahani court painting to comment, in a way understandable only to the initiated, on the ambivalent feelings the Mughals harbored towards the Safavids. EK

1. S.C. Welch attributes the painting to the artist La'lchand.
2. Islam 1970 and 1993.
3. Islam 1970, pp.18–21.
4. Islam 1970, p.98.
5. *Muntakhab al-Lubab*, trans. p.84; Sarkar 1972–4, III, p.53.
6. Today the fort of Burhanpur is largely ruined. I surveyed its remains in 1984 and 1991 in the course of documenting the entire body of palace architecture of Shah-Jahan.
7. Andrews (1987, p.156, fig.7) has identified it as double *aspak*.
8. Koch 1994, in particular pp.143, 158 n.2. The wooden *Chihil Sutun* is also discussed on pages 182, 191 and 208, below, and on page 135, above; see also **43**.
9. The same emblem appears among symbolic imagery below the *jharoka* in *darbar* scenes in the Bodleian Library, Oxford, MS.Ousely Add.173, nos 13 (illustrated in Koch 1988, pls 53 and 58; *Shah Jahan Nama*, pl.5) and 18 (see also Binyon & Arnold 1921, pl.21); and see below, Figs 149 and 145.

18
The siege of Qandahar

Qandahar (Deccan), May 1631
Folio 102b painted by Payag, *circa* 1633
Image area 32.1 × 22.9 cm

Inscription: [on shield, bottom right] *raqam-i Payâg* ('drawn by Payag')

References: *Padshahnama*, I/I, pp.374–77; *Shah Jahan Nama*, pp.65–67
Published: S.C. Welch 1995, p.322 and fig.7; Dye 1991, pp.126–27 and fig.8; Leach 1981, fig.160

NOTHING SHOWS BETTER the range of individual artistic initiative within the Windsor *Padshahnama* than a comparison of this battle scene by Payag with *A'zam Khan captures Fort Dharur* (**15**) by the artist's older brother, Balchand. The scenes have very similar compositions, both taking place in a landscape defined by receding hillocks. But whereas Balchand uses a gentle light to illuminate the scene and to carefully define and emphasize the individuality of each figure, Payag has produced a dramatic spectacle of billowing clouds, clustered troops, and unearthly light. His figures, moreover, often have an identity as part of a group before any individuality is perceived. Both here and in a second, stylistically identical battle scene also intended for a historical manuscript of the Shah-Jahan period (see Fig.151), Payag emphasizes a single incident, giving the scene a dramatic impact quite different from the slow accumulation of episodes through which the viewer comes to understand Balchand's narrative. His interest in extending the viewer's ability to move into deep space is unprecedented in Mughal Indian painting. The distant troops at the upper right, and the burning building and exploding cannon beyond, are as rewarding to our attention as any foreground action. (We see this again in **36**.) The viewer's involvement with the scene, and the resulting sense of powerful forces at work, is among the most innovative contributions of Shah-Jahan's artists to the tradition of Mughal painting.

Payag's style was profoundly affected by the Northern European images and painting techniques arriving in India with missionaries and traders.[1] A sketch made in 1632 by the English traveler and amateur artist Peter Mundy showing an imperial procession (Fig.117), for example, clusters figures in groups and places them in arrangements that repeat and reinforce the rolling lines of the landscape in a way almost identical to Payag's troops in the mid-ground of this scene.[2] While Mundy is not necessarily a specific source for these innovations in Payag's work, he was utilizing a style fashionable in European illustrations of the period, the kinds of images that circulated throughout the Mughal empire and could have been seen by Payag and his contemporaries with no intermediate role being played by the court. It is also possible that in seeking out unusual images available from foreign sources, Payag and other Mughal painters came into direct contact with amateur European artists.[3] MCB

After he had taken the fort of Dharur for Shah-Jahan in the Emperor's campaign against the Nizamshahis of Ahmadnagar (see **15**), A'zam Khan – failing to take the fort of Parenda – marched on further east to support the Mughal commander Nasiri Khan in the siege of the fort of Qandahar.[4] Nasiri Khan had opened trenches all around the fort from which he had driven mines under the edifice. After A'zam Khan's arrival the mines were sprung and the explosion damaged the fort so severely that soon afterwards the garrison sued for quarter. The joint Mughal forces took Qandahar in the middle of May 1631.

Payag focuses on a single event during the siege. He shows the decisive moment when Nasiri Khan brought A'zam Khan to the Mughal trenches to witness the ignition of the mines. The explosion damaged the outer walls (*sher hajji*)[5] of the besieged fort and half of a large bastion, blowing many defenders into the air.

Payag brilliantly manipulates the two opposing modes of the Shah-Jahani court style, the linear two-dimensional type of representation, prescribed for the Emperor and nobles of rank in the main scenes, and the illusionistic three-dimensional mode which could manifest itself only in the less 'official' areas of a composition and in the detailed treatment of surface and texture.[6] The artist conforms to the linear mode in the basic structure of the landscape, for which he uses segments of terrain arranged in layers and retreating upwards to suggest depth, a compositional device that recommended itself for the setting of sieges (*cf.* **15**). The artist was certainly aware of early Timurid representations of Timur's sieges, where the scheme had been expressed in a particularly stark and expressive manner (Fig.98).[7] The two-dimensional structure of Payag's Qandahar landscape also facilitates the integration of the group of Mughal commanders on the right who, as the main protagonists, had to be shown in

Fig.97 Figures arranged behind landscape 'segments' (detail of **18**).

Fig.98 Unknown artist, *Timur conquers Baghdad*, from a *Zafarnama* manuscript dated 1436.

profile. The succession of overlapping figures reflects the 'stepping' of the terrain, indicating for both degrees of distance in depth. The observation that more distant objects move upwards, in particular when seen from an elevated viewpoint, is also incorporated into the depiction of the lines of the besiegers in the trenches. In the foreground only heads or parts of heads are visible in the main, while increasingly more of the figures becomes visible in each succeeding register of the background.

This basically two-dimensional structural system is permeated by a three-dimensional realistic mode,[8] as Payag attempts to infiltrate the linear impositions of the Shah-Jahani court style with forms of representation for which he, as a Hindu, might have felt a particular affinity, based on his own autochthonous heritage, which inclined towards the plastic arts. Payag fills the clearly outlined segments of terrain and the figures with dense modeling and microscopic stippling and hatching, to which the rich and detailed treatment of the texture of the ornamental armor of men and horses corresponds. Three-dimensional elements also invade the lines of anonymous besiegers in the form of perspectivally rendered heads and half-figures, given in oblique and frontal views and all diversified in type, character and costume.[9] The background, where illusionistic interests could always be expressed more freely without formal impositions, shows, on the right, an atmospheric landscape filled with smoke from various sources – the explosion in the fortress, the distant burning building and the cannon positioned in front of it. Similar illusionistic conditions reign in the fortress itself and expand into its immediate surroundings, which are covered with corpses and skeletons whose sketchily applied local colors are absorbed by the tonalities of the ground. The domain of the enemy was predestined for such free forms of representation, as is apparent from related subjects among the Windsor illustrations.[10] The fortress is partly obscured by a huge column of smoke from the explosion, which carries bodies into the air. The evanescent smoke rises in a succession of clearly delineated billows, reflecting the linear overlapping of the terrain and the figures (Fig.50).

The siege of Qandahar is one of the most powerful and complex depictions in the Windsor manuscript. The two contrasting idioms of Shah-Jahani painting – two-dimensional formality and three-dimensional illusionism – are here brought to a convincing synthesis while retaining fully their respective expressive powers. In addition, Payag introduces a new dichotomous dimension by exploring the emotional potential of realism. Opposite his landscape of death and destruction he places an intriguingly idyllic scene of two rabbits contemplating their reflection in a small pond surrounded by reeds and high grass (Fig.58).[11] EK

1. See also Beach 1992a, pp.133–36 and 149–50.
2. The manuscript of Mundy's account of his travels, and the drawings, are in the Bodleian Library, Oxford (Rawl. MS.A.315). The drawings are published in Mundy 1907–24, II.
3. For further discussion of Payag, see page 217, below.
4. This was Qandahar (Kandhar) in the Nander District, north-west of Hyderabad, not to be confounded with its better known namesake in present day Afghanistan. For a brief description see Navarane 1995, pp.255–56, 376 (illustration), and 482 (map). For the siege, see *Shah Jahan Nama*, pp.65–67; Sherwani & Joshi 1973–4, I, pp.272–73; Saksena 1976, p.135.
5. This term is used consistently by Mughal authors to designate the outer walls of a fortress. Muhammad Kazim explains the term for his readers as '*fasil* [fortification wall], which in the language of the common people is called *sher hajji*': *'Alamgirnama*, I, p.424.
6. For further discussion, see pages 136, 141, above.

7. See page 174, note 7, above.
8. See also pages 141–42, above.
9. Payag takes a similar approach in his *darbar* scene, **39**.
10. *Cf.* **16**, **36**, and **40**.
11. Terence McInerney has suggested that Payag inserted this motif in his landscapes much like a signature. His observation is quoted in Gamerman 1994. The two rabbits appear like miniature versions of their famous relative by Albrecht Dürer and its later imitations by the German painters Hans Hoffmann and Georg Hoefnagel, who were main protagonists of the Dürer revival before and after 1600 (Koreny 1985, pp.132–57). Perhaps one of these extremely foreshortened versions was known to Payag, who liked to perfect his naturalistic interests through the systematic study of the latest European styles. The latter point is briefly addressed by Milo Beach above, page 177 and in Beach 1992a, pp.133–36; Dye 1991, p.127; S.C. Welch 1995, pp.323.

19
Europeans bring gifts to Shah-Jahan

Agra, *Diwan-i 'Amm*, July 1633
Folio 116b painted by an unknown artist, *circa* 1650
Image area 33.8 × 23.7 cm

References: *Padshahnama*, I/I, pp.534–35; *Shah Jahan Nama*, p.117

Fig.99 Detail of **19** showing European with a lacquer box.

THE EXPLORER Vasco da Gama rounded the Cape of Good Hope in 1497, arriving on the west coast of India the next year. By 1510, a Portuguese colony had been established at Goa and for more than a century the Portuguese were an important presence on the subcontinent. This and the following illustration, however, depict the fate of a trading colony founded later at Hoogly, near present-day Calcutta, in Bengal, which had come into conflict with Mughal merchants centered at nearby Satgaon. The resulting rivalry, and the active missionary work of the Portuguese, gave Shah-Jahan justification for an attack (depicted in **20**) at which European prisoners was seized and taken to the Mughal court. They may be the group shown here.

Folios 116b and 117a, originally left blank for the addition of a double-page illustration, were instead completed by the insertion of two separate scenes. While both relate to the Portuguese presence in India, the present episode actually happened more than a year after the event in the succeeding illustration – further proof that the images were not meant to be facing compositions.

In this scene, one of the Europeans holds a black box with gold designs, evidently an object of sufficient importance and rarity to serve as an imperial gift. A very similar box is being carefully loaded into the ships at the left of **20**. Oliver Impey has suggested that these are *namban* wares, trade objects made in Japan of black lacquer and gold for both the European and Indian markets.[1]

This is the work of an unidentified and minor painter. The portraits give little sense of different appearances or personalities, and the princes, who are barely distinguished from each other, seem to be the same age as in events which happened considerably earlier (see, for example, **10–11**). This is probably evidence that the artist was careless of (or separated from) the sources which would have ensured historical accuracy.[2] ❧ MCB

Shah-Jahan receives the Europeans while holding court in the audience hall at Agra. The painter updated the architecture to take in several – but not all – of the modifications the hall underwent between what we take to be the time of the event (July 1633) and the date of the painting (the 1640s or 1650s). He retained the long-established form of the Agra *jharoka*, as it actually looked in 1633, but modernized it with floral motifs in what seems to be *pietre dure* inlay or *parchin kari*, the new and precious decoration technique used conspicuously (albeit with somewhat different motifs) in the rebuilding of the *jharoka* in the form of a loggia before 1637. The elements of the hall – white, faceted pillars framing the composition, the *chhajja* (eaves) and the ornamental parapet on top – refer clearly to the permanent construction of the same date, which is still standing (Fig.104). These sparse details represent the only verifiable pictorial reference to Shah-Jahan's new permanent audience hall in the entire Windsor manuscript.[3]

At first glance the painting looks like any other *darbar* scene, the special event being a delegation of Europeans – *firangi*s, or 'Franks', as they were generically termed by the Mughals – bringing gifts to Shah-Jahan, an activity that, incidentally, takes up much of the reports of European visitors, who felt, not unjustifiably, that it provided the key to their access to the Mughal court. The Europeans are rendered in unprestigious three-quarter views,[4] and they are positioned outside the opening of the railing of the audience hall which set off the area reserved for the highest ranks and specially favored persons. The *firangi*s are not only not allowed 'inside,' they are moreover supervised closely by a threatening group of mace-bearers, who do not usually appear in such numbers in *darbar* scenes. This could indicate that the Europeans are not here of their own free will, but that they are indeed the Portuguese prisoners brought by force from Hoogly to the court of the 'Great Moghul' at Agra.[5] The fact that they are represented as unharmed and healthy testifies to the 'selective realism' of Shah-Jahani painting. The perfect world of the Emperor's *darbar* could not be flawed by misery.[6] The Portuguese seem to owe their voluntary appearance here to the same embellishing line of thought which claimed that the game offered itself willingly to the imperial hunter, Shah-Jahan.[7] ❧ EK

1. A shield made of Indian leather, but lacquered and decorated in Japan, is in the Ashmolean Museum, Oxford. I am grateful to Oliver Impey for this reference.
2. A later version of this painting, formerly in the Sitaram Sahu Collection, Varanasi, gives greater stress to the presence of women among the Europeans; Mehta 1926, pl.39. I am grateful to Robert Skelton for this reference.
3. Freer references to them are made in **23–24** and **27–28**. On the problem of the audience halls see also Koch 1994; and page 135, above.
4. For the hierarchical use of views in the Shah-Jahani court style see page 136, above.
5. For the fate of the prisoners at Agra see *Shah Jahan Nama*, p.117; Maclagan 1990, pp.103–106.
6. See also Robert Skelton's argument on page 167 above, note 18.
7. See page 193, below.

20

The capture of Port Hoogly

Hoogly (Bengal), June–October 1632
Folio 117a painted by an unknown artist, *circa* 1634
Image area 34.1 × 23.5 cm

References: *Padshahnama*, I/I, pp.433–39; *Shah Jahan Nama*, pp.86–87
Published: Ashton 1950, no.773 and pl.138

Fig.100 A distant encampment (detail of **20**).

Fig.101 Unknown artist, battle scene from a *Jahangirnama* manuscript, 1620s.

HOOGLY, ESTABLISHED AS a Portuguese trading center in the later sixteenth century, directly rivaled the nearby Mughal port of Satgaon. This produced friction, which was further aggravated by Portuguese activity in slave-trading, their support for local rulers against Mughal authority, and the forced conversion of local peoples to Christianity. In frustration, the Mughals attacked the trading settlement in the middle of 1632, a matter complicated for the Europeans by the presence among the imperialists of Martin Alfonso de Mello, a disaffected Portuguese freebooter who sided with the Mughals and designed weapons and tactics that were especially effective against his countrymen.

In an extraordinarily descriptive scene, the Portuguese are shown loading their ships to escape down a river that is blocked by a mass of boats that have been roped together. This was a tactic of Martin Alfonso, who was ordered (in the words of Father John Cabral) 'to contrive engines and machines wherewith to obstruct the channel. Afonso did it better than the Moor had asked him. His first invention was a pontoon of large boats, which the Moors called *patalas*, connected by means of enormous beams, nails, cramps, and a number of chains.'[1] The correspondence between details of this scene and contemporary historical accounts is shown by the carefully depicted explosion, which may refer to the mine described in the *Padshahnama* text accompanying the miniature.

Among the most fascinating and innovative details of the scene is the inclusion of a barren landscape and the encampment of the surrounding Mughal troops. Before Shah-Jahan's reign, few Mughal paintings activated distant space so dramatically, or compelled the viewer to understand the broad spatial context within which the narrative takes place. In the contrast between the far-off tented enclosures and the European colony (complete with a Gothic church) closer to the foreground, we can understand the statement in the *Padshahnama* that Hoogly contained 'dwellings of the utmost splendor and strength.'[2] One can also imagine that the Mughals were startled by European costumes, as well as by the visible participation of European women in the events of the siege. MCB

The representation of the sea battle off Hoogly is based on Western sources: such undisguised Europeanism was legitimized by the subject. The cityscape of Hoogly and the ships of war could have been inspired by geographical works of the period or by engravings or paintings in the Flemish landscape tradition which the Mughals had collected, studied, copied, and used as vehicles for their own cityscapes and landscapes since Akbar initiated a systematic interest in European art.[3] The Lilliputian landscape with charming genre scenes in the distant background is a telling example of such Mughal adaption, which underwent its own evolution during which it became increasingly microscopic. It appears earlier in the background of Govardhan's famous *Rustic Concert* of *circa* 1620 (Fig.34);[4] and again on a smaller scale in a battle scene of Jahangir's period, a previously unpublished illustration of the *Jahangirnama* (Fig.101).[5] This battle shows other elements that reappear in *The capture of Port Hoogly*, such as perspectivally rendered boats and a similar 'cubistic' cityscape.

The streets of Hoogly are filled with fleeing Portuguese and with Mughal besiegers, invading the city from behind. Other illusionistic effects are brought about by the explosion which darkens the surrounding buildings. Prominence is given to the large church with two towers,[6] meant to represent one of those seen at Hoogly by the Mughal writer Khafi Khan, who noted perceptively that 'Their places of worship – contrary to the idol houses of the Hindus – are perfectly clean. Wax candles can be seen there burning during the day. According to their depraved belief, they have formed statues of Hazrat Isa [Jesus] and of Maryam [Mary] – who are our prophets[7] and peace be upon them – of wood and in different shapes, and they have painted them all over. But in the churches of the Angrez [the English] who are also Nazarenes [Christians] there are no idol figures.'[8] In July 1633 these statues came with the Portuguese prisoners from Hoogly to Agra, where Shah-Jahan had many of them thrown into the Jumna, to the silent regret of his historian Kanbo, who found much in them to admire.[9]

Another object which seems to have made its way to Agra – at least pictorially – is the lacquer box brought out of the exploding building onto a nearby ship; a similar box appears among the presents in **19** and might have been intended as a link between the two pictures (Fig.99). EK

1. Quoted in Manrique 1927, II, p.410.
2. Manrique 1927, II, p.410.
3. See page 138 above.
4. Now in the Chester Beatty Library, Dublin, Minto Album, MS.7A.11; illustrated in color in Leach 1995, cat. no.3.21, pl.57.
5. It is owned by Terence McInerney who has kindly brought it to my attention and provided me with a photograph. McInerney attributes the battle scene to Bishandas and identifies it as the defeat of Prince Khusraw's army near Govindwal on the Beas in 1606, for which see *Tuzuk-i-Jahangiri*, I, pp.63–65; Prasad 1922, pp.143–44.
6. Perhaps this is the church of Our Lady of Mercies which was stormed by the Mughal forces on 2 July 1632. See Maclagan 1990, p.102.
7. 'Isa ibn Maryam ('Jesus son of Mary') is acknowledged in the Koran as a great prophet; his mother Maryam (Mary) is generally held in Muslim tradition to be one of the four best women that ever existed. See Anawati 1978; Wensinck & Johnstone 1991.
8. *Muntakhab al-Lubab*, I, p.469; translated by Ebba Koch.
9. *'Amal-i Salih*, I, 509; *cf. Shah Jahan Nama*, p.117. Since the Mughals usually adopted a tolerant stance toward religion (see page 73) this seems to have been a political gesture directed at three-dimensional religious representations, particularly offensive to orthodox Islamic thinking. Otherwise the Mughals, including Shah-Jahan, were happy to ignore such concerns and collected pictures of Jesus and Mary which they had copied by their painters. Jahangir went so far as to associate himself openly with Christian imagery by having his *jharoka* decorated with it. See page 200, below, where further literature is cited.

21–22
The delivery of presents for Prince Dara-Shikoh's wedding

Agra, the riverfront, November–December 1632
Folios 120b–121a attributed to Bishandas, *circa* 1635
Image area 33 × 23 cm (folio 120b); 33.1 × 23 cm (folio 121a)

References: *Padshahnama*, I/1, pp.452–53; *Shah Jahan Nama*, p.90

Fig.102 Attributed to Bishandas, *A procession*, circa 1625.

Fig.103 The riverfront at Agra (detail of **21**).

ON 17 JUNE 1631, Mumtaz Mahal, the beloved wife of Shah-Jahan, died at Burhanpur, and plans then underway for the wedding of Dara-Shikoh, the eldest of the imperial princes and the designated heir, were delayed until late the next year.

Marriage alliances within the imperial family were arranged within a very small circle, as can be shown best by the genealogical chart on page 229. This ceremony united Prince Dara-Shikoh with Nadira Banu Begam. Her father was Sultan Parvez, Shah-Jahan's elder brother, and her mother was Jahan Banu Begam, daughter of Prince Murad (the brother of Jahangir and therefore Parvez's uncle). Bride and groom were thus both grandchildren of Jahangir. The ceremony depicted here is the delivery of presents from the imperial household to the mansion of Jahan Banu Begam, who at that time was a widow. The gifts are accompanied by musicians and flower bearers, all under the escort of Asaf Khan, Sadiq Khan, and Afzal Khan Allami. The procession moves along one bank of the Jumna River, with the gardens of the nobility seen clearly on the farther bank. The density of the plantings within these walled enclosures, and the shade and cool air they must have provided, are in expressive contrast to the barren, dry riverbed. No amount of verbal detail can equal the direct visual awareness of pomp and ceremony, and of luxurious life, with which the painter invests the scene.

While unsigned, both halves of this painting can be reliably attributed to Bishandas, an artist who worked first for the Emperor Akbar.[1] Lauded by Jahangir for his skill in portraiture, he was sent to Iran with the embassy of Khan Alam, remaining away from India between 1613 and 1620. Upon his return, he was honored with the gift of an elephant.[2] A second processional scene (Fig.102), the remaining left half of a similar double-page composition, and usually thought to be from an illustrated *Jahangirnama* manuscript, can be attributed to the same artist at virtually the same time; and since it is the same size as well, it may actually have been made to illustrate an event from Shah-Jahan's histories.[3] The figures in these pages share certain traits (whether as individual types, in their arrangement in rows receding into depth, or in distinctive raucous action in the background) with those in a double-page from a *Masnavi* of Zafar Khan manuscript that can be dated to the early 1640s.[4] This both suggests an extension of Bishandas' working life beyond the period hitherto accepted, and also finally makes credible the inscriptions on a series of late portraits derived from his stay in Iran. Central to this group are two depictions of *The meeting of Shah Abbas and Khan Alam*.[5] Seen in relation to Bishandas' earlier achievements, the late paintings are convincingly the work of an old man, an artist whose recognizably personal figure types and compositions remain intact, but whose technique has become weak.[6] ❧ MCB

The wedding of Dara-Shikoh, whom Shah-Jahan favored as his successor, to the daughter of his uncle Sultan Parvez, was performed with exceptional splendor. It was the first major celebration after the court had spent over a year in mourning for Mumtaz Mahal, the Emperor's favorite wife and mother of the Prince. The ceremonies began in late November 1632 with the formal engagement expressed in the delivery of the *sachiq*,[7] presents sent on the part of the bridegroom's family to the house of the bride.[8]

The men carrying the *sachiq* on large round trays covered with precious cloth were accompanied by high dignitaries, female relatives of the Prince (not shown), by the imperial insignia and standards, and by members of the *naqarakhana*, the imperial orchestra, carried on elephants. They were preceded by musicians and singers in horse-drawn chariots and riding elephants, and followed by men on foot holding sticks strung with marigolds (*gul-i ja'fari*), still the most popular flower in India for festive garlands.

The pageant is shown proceeding along the bank of the Jumna to the dwelling of Sultan Parvez, which was on the other side of the river, immediately north of the tomb of I'timaduddawla (Fig.132, no.11).[9] The procession must have crossed the river in boats of the type visible in the painting. The *sachiq* procession had a highly ceremonial character, and the artist responded to the task of depicting a formal outdoor scene by setting it on an unstructured flat surface. He took care, however, to designate it as the barren bank of the Jumna by giving it a greyish sandy color and by topping it with a typical riverfront view of Agra, with its garden buildings and residential towers placed near the water (Figs 63, 103).[10] The arrangement of the procession, which is composed of clearly defined columns of human and animal figures, corresponds to this steeply rising picture plane. The layers are distinguished by the use of alternating light and dark colors, used studiedly in the succession of horsemen on the right-hand page. The painter takes care to make the groups more lively by introducing small irregularities, such as figures turning in all directions, rendered in perspective. In addition, the artist reduces the size of the figures at the top of each 'column' to express distance not only by means of vertical layering, but also through perspective. Here, and generally in Shah-Jahani painting, decreasing size not only conveys an optical perception but can also indicate decreasing rank. The dignitaries are shown at exaggerated size in relation to the other participants of the procession, to enhance their status and to express, literally, that they are 'the great ones' of the empire. ❧ EK

1. This attribution was proposed in Beach 1978, pp.110–11, and is based on the styles of a signed painting, *The house of Shaykh Phul*, from the Bharat Kala Bhavan, Varanasi, and an attributed work, *The birth of Jahangir*, in the Museum of Fine Arts, Boston (Beach 1978, no.15).
2. *Tuzuk-i-Jahangiri*, II, p.117.
3. See Brown 1975, pl.XXXI, where the colors in the reproduction are far richer than the cool colors of the original.
4. Royal Asiatic Society, London, fols 11v–12r; see Pinder-Wilson 1957.
5. One such depiction is in the Museum of Fine Arts, Boston (acc. no.14.665; see Beach 1978, no.36, where the work is dated *circa* 1650 and proposed as a copy after Bishandas). The other was recently auctioned (Sotheby's 1996, lot 6). For an important

discussion of these works, see Robinson 1972, pp.58–63. Related works are in the St Petersburg Album (Akimushkin *et al.* 1996, pls 79 and 102).
6. For discussion of Bishandas, see page 214, below.
7. For the weddings of Shah-Jahan's sons see also **14**, **23–24**, **25–26**, **27–28**, and **45**. For Muslim marriage customs in India see Herklots & Sharif 1972, pp.57–88; Asad 1988, pp.59–74. For Mughal weddings in particular see Chopra 1976, pp.182–84 and Mubarak Ali 1986, pp.106–108; both authors, however, hardly consider the changes which the ceremonies underwent over the years.
8. *Padshahnama*, 1/1, 452ff, translated on page 60 above. See also *Shah Jahan Nama*, p.90ff.
9. This site is indicated on the early 18th-century map of Agra in the Maharaja Sawai Man Singh II Museum, Jaipur, cat. no.126. For an overall view of the map see Gole 1989, pp.200–201. I am indebted to Dr. B.M. Jawalia, keeper of manuscripts, for reading the inscriptions on the map with me.
10. For a discussion of this urban scheme, see Koch 1991a, pp.33–34, 86–87, 99; Koch, forthcoming (d). See **29** for a more realistic rendering of the same landscape. The formula is discussed on page 138, above.

23–24
The wedding procession of Prince Dara-Shikoh

Agra, palace courtyard, 12 February 1633
Folio 122b painted by Murar, *circa* 1635, and folio 123a by an unknown artist, *circa* 1640
Image area 33.9 × 23.5 cm (folio 122b); 34 × 24.1 cm (folio 123a)

Inscriptions: [folio 122b, middle] *'amal-i Murâr* ('work of Murar')

References: *Padshahnama*, 1/1, pp.457ff; *Shah Jahan Nama*, p.91 and fig.18

At the right edge of folio 122b, the original lines defining the space for a gold ruling around the illustration remain, but the band was never colored. This may indicate that the illustration was never used in a final copy of the manuscript for which it was first intended.

UNLIKE EITHER THE preceding (**21–22**) or the following illustrations (**25–26**), the two halves of this composition are not continuous, and they are clearly by different artists. It is therefore uncertain that they were originally intended to form a single composition. Moreover, the tiny signature 'Murar,' placed just to the right of the equestrian Dara-Shikoh's waist, may refer simply to the figure of the Prince. The tight, crisp execution of the Prince's portrait, similar to that seen in other signed works by Murar (for example, **38**), is not found in the more loosely rendered faces of the various attendant courtiers on the left-hand page – suggesting that only that portrait is by Murar; nor do Murar's other works show the interest in shadows so evident in the right-hand page here.[1]

Despite contemporary textual descriptions of the event,[2] Awrangzeb is not shown accompanying his brother, for reasons that are unclear. On the other hand, all the major ministers at court are in the equestrian procession. Dara-Shikoh is accompanied by Shah-Shuja' and Prince Murad-bakhsh, with (moving clockwise) Asaf Khan, Afzal Khan, Khan Jahan Barha, Mir Jumla, Musavi Khan, Shayista Khan, and – in a particularly prominent position – Ja'far Khan.[3] MCB

The wedding festivities of Dara-Shikoh in February 1633 consisted of a series of events. The display of the wedding presents (*jahaz*) in the *Diwan-i 'Amm* on two consecutive days in early February (**14**) was followed, on the 11th of the same month, by the *hinabandan* ceremony.[4] On the evening of the following day the actual marriage ceremony took place. The bridegroom was escorted by his brothers in a solemn procession from his house to the palace (Fig.132, nos 5 and 4), where he was received by the Emperor in the *Diwan-i 'Amm* and given ceremonial presents. Later, in the evening, the Emperor veiled the face of the bridegroom in the Indian manner with a *sehra* (**25–26**), after which the sermon and signing of the marriage contract (*nikah*) were performed in the Shah Burj pavilion. A week later the Emperor paid an official visit to the house of the newlyweds.[5]

The painting of Dara's wedding procession on folio 122b is paired with a page showing standard bearers and musicians on elephants with splendid, partly figurative saddle cloths, as well as men on foot carrying lamps and trays similar to those in the *sachiq* procession. The figures are arranged in vertical and horizontal layers and move toward one of the gates of a complex inside the Agra fort.[6] The gate is faced with sandstone and topped by ornamental battlements (*kanguras*), like the present-day gates of the courtyard of the *Diwan-i 'Amm*; however, it shows a different style of ornamentation with geometrical patterns of the earlier period. The wing pierced by the gate presents not the characteristic arcades of the inner courtyard of the *Diwan-i 'Amm* (*cf.* Fig.104) but a blank wall that corresponds to the outer southern wall of the enclosure. This places the procession outside the courtyard, though Dara – who would have been preceded by such a group (*cf.* **27–28**) – is already inside in front of the audience hall. It is thus doubtful that the supporting page was originally conceived in this context.

Since the scene takes place in the evening, the hall is illuminated with *chiraghs* (oil lamps) placed on top of the eaves along the roof parapet, with *fanus* (lanterns of colored glass) hung on a *chub bandi* (scaffold) in front of the arcades, and with *mash'alas* (torch-lights) stuck into the ground below; these, together with the *sham'dans* (candlesticks) held by some men in the procession, give us a detailed picture of the various forms of illumination of the period.[7]

The depiction of the hall poses some problems. This is not the green wooden *Chihil Sutun* which Shah-Jahan built in front of the *jharoka-i khass-u-'amm* after his accession and which represents the audience hall in most of the *darbar* scenes of the Windsor manuscript, including that recording the display of Dara-Shikoh's wedding presents.[8] The white masonry building seems to refer to the later, permanent hall, which replaced the wooden con-

Fig.104 Elevation drawing of the façade of the *Diwan-i 'Amm* and adjoining wings, Agra fort, 1630s.

3. (Afzal Khan) (*Diwan*)
9. (Asaf Khan) (*Wakil*)
20. (Prince Dara-Shikoh)
37. (Ja'far Khan)
45. (Khan Jahan Barha)
53. (Mir Jumla) (*Mir Saman*)
60. (Prince Muradbakhsh)
64. (Musavi Khan)
87. (Prince Shah-Shuja')
88. (Shayista Khan)

struction at some time before January 1637.[9] Thus the painting must have been executed after the construction of the hall. However, in his attempt to actualize the building, the painter took little care to model it closely on the original, showing solid masonry arcades instead of its elegant faceted pillars and multi-cusped arches. EK

1. For a related equestrian procession in the Victoria & Albert Museum, London (inv. I.M.12-1925), see Archer 1957, pl.v. While convincingly ascribed to Manohar in the marginal notation, a very similar inscription, 'work of Murar,' is placed next to the figure of Shah-Jahan, suggesting that Murar worked on the central portrait.
2. See the text quoted on page 61 above.
3. Related scenes, later in date but also identified as showing Dara-Shikoh's wedding procession, are reproduced in Sotheby's 1986, lot 23; Colnaghi 1978, no.36; and Carroll 1972, p.80.
4. This event is not represented in painting, probably because it was performed by high-ranking women of the palace. Henna from the bride's house was applied to the hands of those admitted to this exclusive ceremony; *Padshahnama*, I, p.456; *'Amal-i Salih*, I, p.439; *Shah Jahan Nama*, p.91.
5. *Padshahnama*, I/1, pp.452–60; *'Amal-i Salih*, I, pp.435–45; *Shah Jahan Nama*, pp.90–92; Saksena 1976, pp.311–12.
6. For a plan of the fort based on new measurements see Koch 1991a, fig.36, and pp.106–109. The most comprehensive description is still that in Husain 1937; see also Asher 1992, pp.182–89.
7. Most of these terms are used by Kanbo of the illumination of the *Diwan-i 'Amm* on this occasion (*'Amal-i Salih*, I, p.441). The term *chub bandi* as a lantern support is in Lahawri's description of the illumination of Shalimar gardens in Kashmir in 1640 (*Padshahnama*, II, p.199). See also **27–28**.
8. See **14**, **32**, **43**, and **44**.
9. See Koch 1994, p.143; and page 135, above.

25–26

Shah-Jahan honoring Prince Dara-Shikoh at his wedding

Agra, *Diwan-i Khass*, 12 February 1633
Folios 124b–125a painted by Bulaqi, son of Hoshang, *circa* 1635
Image area 38.8 × 24.1 cm (folio 124b); 38.8 × 24.1 cm (folio 125a)

Inscriptions: [folio 124b, on wall below Shah-Jahan; and repeated on folio 125a, on book held by second figure from left] *'amal-i kamtarîn-i khânazâdân [Bu]lâ[q]i walad-i Hoshang* ('work of the most insignificant of the slaves, [Bu]la[q]i son of Hoshang')[1]

References: *Padshahnama*, I/1, pp.458ff; *Shah Jahan Nama*, p.91 and fig.19.
Publication: Beach 1992a, pp.130–32 and figs 97–98

The preceding double-page of text (folios 123b–124a) bears only three rather than the usual twenty-one lines of text on each of its pages. This is explained by the close sequence of illustrations within the narrative and the wish nonetheless to maintain the placement of illustrations at the points where they are textually appropriate.

KNOWN CONTEMPORARY TEXTS tell us little about the early career of Ja'far Khan, the son of the *Mir Bakhshi* Sadiq Khan, but his rise to power is evident from his increasingly prominent placement in paintings. In the coronation *darbar* of Shah-Jahan by Bichitr (**10**), for example, he is standing alongside his father and among the men awaiting entrance to the inner enclosure, while in *The weighing of Shah-Jahan on his forty-second lunar birthday* (**13**) he is placed next to both Prince Murad and Asaf Khan. In *The wedding procession of Prince Dara-Shikoh* (**23**), he leads the nobles accompanying the Prince to greet Shah-Jahan in the *Diwan-i 'Amm* or Hall of Public Audience on the eve of Dara's wedding; and here, he is given the extraordinary distinction of standing next to Prince Awrangzeb, within the inner imperial circle.[2] His robes and jewelry are virtually identical to those of the princes, a highly unusual honor, given that neither his title nor any special status at this time are recorded in either the *Padshahnama* or the *Shahjahannama*.[3] Two further works in the St Petersburg Album should also be noted: a portrait of Ja'far Khan with female attendants, and a depiction unprecedented for any nobleman, in which Ja'far Khan is aureoled and placed facing the Emperor (Fig.105).[4] While Mughal historical texts are not helpful about his rank or role, paintings leave no uncertainty about Ja'far Khan's prominence at court: this underlines their importance as independent historical documents.

A possible explanation, however, is provided by Niccolao Manucci, a Venetian traveler in India between 1656 and 1680, who writes: 'All the world knows that the Mahomedans, following the example of their master, Muhammad, are very licentious; wherefore there are men among them, some more and some less, chiefly the nobles and the kings, who do not content themselves with a few wives, but seek for every method of gratifying themselves in this particular. It may be asserted that Shahjahan was not superior to others in this respect, for, not contenting himself with the women that he had in his palaces, he forfeited the respect of the nobles at his court by intrigues with their wives .... The chief of the women, one that he thought a great deal of, was the wife of Jafarcan [Ja'far Khan], and from the love he bore her he wished to take her husband's life, but she saved him by praying that he might be sent as governor

Fig.105 Unknown artists, *Ja'far Khan* (left) and *Shah-Jahan* (right), circa 1635.

to Patana [Patna], as was done .... The intimacy of Shahjahan with the [wife] of Ja'far Khan was so notorious that when they went to court the mendicants called out in loud voices to Ja'far Khan's wife: "O Breakfast of Shahjahan! Remember us!" '[5] Ja'far Khan was married to Farzana Begam, a daughter of Asaf Khan and therefore sister to the Empress Mumtaz Mahal. While Manucci's account is little better than rumor, the relationship of Ja'far Khan's increasing prominence at court, so amply recorded in paintings, to his wife's possible indiscretion remains to be explored.

Little is known about the artist Bulaqi, whose earliest known paintings are in a *Razmnama* manuscript of 1598.[6] Other than a single page in Jahangir's album, his only other identified work is this scene and a series of related Shah-Jahan period history paintings (see Figs 157, 158).[7] Based on the Windsor illustration and a hitherto unnoticed inscription, however, a page presumably intended for the imperial *Jahangirnama* manuscript can also be assigned to Bulaqi (Fig.15).[8] It shows Jahangir greeting Muqarrab Khan, who returned to court from an ambassadorial mission to Goa in 1612 with interesting exotica, including a monkey, a turkey, and a pheasant, to present to the Emperor. This scene and his other late works reveal Bulaqi's particular skill in creating a majestic architectural setting within a visually shallow space, and in using broad areas of white pigment both to clarify artful arrangements of people and to intensify the effect of color. Bulaqi is a distinctive and original artist. ❧ MCB

The most conspicuous Indian element in the wedding ceremonies of Shah-Jahan's sons was the ceremonial covering of the bridegroom's face with a *sehra* or *sihara* (see also **45**), a custom intended to avert the ill effects of the evil eye. While a common bridegroom's *sehra* would usually be made of a fillet with miniature flower garlands, more precious materials were used by the well-to-do.[9] Dara's *sehra* consisted of strings of pearls interspersed with gems; it was a family heirloom which had already served for Shah-Jahan's own wedding.[10]

The ceremony took place before the actual marriage (*nikah*) on the evening of 12 February 1633, on the riverfront terrace between the *Diwan-i Khass* (shown on folio 124b) and the portico of the Imperial Hammam (shown on folio 125a) of the Agra fort (Fig.91).[11] In attendance among the men aligned in front of the Hammam portico is a white-bearded and turbaned *mulla* who must be Qazi Muhammad Aslam, ready to perform the *nikah*. The Emperor, who holds the *sehra* against his son's face, sits on a white marble platform hidden by a balustrade. In reality this seat is situated in the middle of the riverfront terrace and with the halls on both sides forms a bilaterally symmetrical composition, conforming to the period ideal of *qarina*;[12] Bulaqi's double-page composition therefore reflects the symmetry of the actual setting. The imperial platform, however, could not be shown in its true position because this would have meant cutting it in half. Bulaqi solves the dilemma by moving it closer to the *Diwan-i Khass*, to place Shah-Jahan and Dara in the center of the right-hand composition.[13] The Emperor's seat is covered by a white 'night canopy' – characterized by its ornamentation of silver stars – which is supported by ropes and by four poles; the flanking halls are furnished with splendid hangings of gold-brocaded velvet and extended by tented porticoes.[14] In the courtyard below, dancers and musicians perform by the light of large, two-armed candlesticks. In the background, on the bank of the Jumna, elaborate fireworks give the painter occasion to show atmospheric night effects. With the help of a magnifying glass we can make out the 'Artificiall trees' and 'ranck [row] of great Eliphants whose bellies were full of squibbs, Crackers, etts.' seen by the English observer Mundy on this occasion.[15] ❧ EK

1. For the reading of this artist's name, see the inscription on a page of the 1598 *Razmnama* (S.C. Welch 1959, no.4 and fig.6) where the name is written *plene* and with dots, *Bûlâqî walad-i Hushang*. (WT)
2. Standing on the upper terrace on the right-hand page are (from the left) Gaj Singh Rathor, Khan Jahan Barha, Afzal Khan (*Diwan*), Asaf Khan (*Wakil*), Muradbakhsh, Dara-Shikoh, Shah-Jahan, Shah-Shuja', Awrangzeb, Ja'far Khan, Hayat Khan, and three unidentified men.
3. An inscription on a detached frontispiece *shamsa* in the Art and History Trust Collection, Houston, notes that it was once in an imperial album presented by Shah-Jahan to Ja'far Khan. The Emperor wrote 'This album ... on [5 May 1630], in Burhanpur, was given from the royal library to [my] true disciple of perfect understanding, Ja'far Khan' (see Soudavar 1992, no.130).
4. This combination of images was composed in Iran when the St Petersburg Album was assembled. For the most recent discussion of this album, see Akimushkin *et al.* 1996.
5. Manucci 1889, I, pp.186–87.
6. S.C. Welch 1959, no.4 and fig.6.
7. For further discussion of Bulaqi, see page 214, below. Nothing is known of his father, Hoshang.
8. The page is in the Raza Library, Rampur (MS.I, fol.7a). A badly rubbed signature appears to credit the work to Bulaqi.
9. Aziz 1947, pp.116–17; Herklots & Sharif 1972, p.73.
10. *Padshshahnama*, I/I, p.458, translated on page 66 above; *cf.* *'Amal-i Salih*, I, p.441.
11. The architectural setting is explained on page 170, above.
12. Koch 1991a, pp.93, 141. See also page 135, above.
13. *Cf.* the discussion of the composition of **45**.
14. For a discussion of this tentage see Andrews 1987, in particular p.160, fig.10.
15. Mundy 1907–1924, II, p.202, quoted more fully on page 66 above.

27–28
The wedding procession of Prince Shah-Shuja'

Agra, palace courtyard, 4 March 1633
Folios 126b – 127a attributed to Bhola, *circa* 1635
Image area 33.2 × 23.5 cm (folio 126b);
33.4 × 23.3 cm (folio 127a)

Inscription: [folio 127a, on center finial]
Allâh 'Alî Muhammad

References: *Padshahnama*, I/1, pp.460ff;
Shah Jahan Nama, p.93

The original painting on the right-hand page (folio 126b) has been unevenly trimmed and two figures at the right edge extend over the marginal rulings. With this extension, the width of the illustration is 24 centimetres.

THIS PROCESSION DIRECTLY corresponds to the ceremony held for Dara-Shikoh less than a month earlier (**23–24**), although fewer of the major imperial ministers accompany Shah-Shuja'. This, and the fact that the marriage ceremonies are depicted in one double-page illustration rather than three, is an indication of his junior status.

Both halves of this composition can be attributed to the artist Bhola, who signed the left-hand page of *The weighing of Shah-Jahan on his forty-second lunar birthday* (**13**). Bhola was an inconsistent artist whose skills seem to have increased with the hierarchy of the person portrayed. The princes and major nobles here, for example, are skillfully and perceptively depicted, whereas the minor attendants often seem weightless (note the figures at the right edge of the left page) and their eyes seldom focus. While the artist was interested in the phenomenon of shadows (as distinct from the long-held Mughal interest in shading), the treatment is inconsistent and the effect is more like smudges of dirt. Judging from the various scenes by Bhola in the Windsor manuscript, the inclusion of birds along the roofline of court buildings and evocative blue skies are personal traits of Bhola's style. MCB

The wedding of Shah-Shuja' was celebrated in March 1633, immediately after that of his elder brother Dara-Shikoh. The Emperor pressed his astrologers to find a date for the festivities before Ramadan, the Muslim month of fasting. The arrangements were again made by Jahanara Begam in the same manner as a month before, but on a more modest scale.[1] The same is true of the depiction of the wedding which shows merely the procession of the bridegroom. These two pages – unlike those used in the manuscript to illustrate his elder brother Dara's procession (**23–24**) – were undoubtedly conceived as a single composition.

The prince and his escort have entered the courtyard of the *Diwan-i 'Amm* surrounded by musicians and preceded by bearers of the imperial insignia, who prominently display the *panja* (outspread hand) and a dragon emblem.[2] The figures in the procession form compact groups, differentiated, as is usual in formal scenes, by contrasting local colors.

Several of the mahouts, the drivers of the tamed elephants, and footmen on the left-hand page look back to the group of princes on the right-hand page and thus establish a visual connection between the two miniatures. The architecture in the background also creates a unified space. We recognize on the right-hand page the northern part of the hall of the *Diwan-i 'Amm*, connected by arcades to the gate leading to the Zanana Mina Bazar (Fig.104),[3] and on the left-hand page the courtyard arcades as they continue north of the gate. For as yet unknown reasons the painter shows the architecture with pointed arches instead of the present engrailed ones (*cf.* **23–24**). He nevertheless displays his skill for close observation, also characteristic of **13** in the rendering of ornamental detail, such as the minutely painted parapets above the slanting eaves of the arcade, or the flower-filled vases on the panels of the gate. The white color of the courtyard wings and the painted floral motifs, which also appear in other images of the *Diwan-i 'Amm* (see **11**), seem to be based on the contemporary appearance of the architecture. Some evidence to this effect is preserved inside the courtyard wings, whose front arcades present themselves today with no facing over their red sandstone construction.[4] EK

1. *Padshahnama*, I/1, pp.460–65; *'Amal-i Salih*, I, pp.447–52; *Shah Jahan Nama*, pp.92–93.
2. For the insignia see the literature cited on page 195, note 5, below.
3. For the palaces of the Agra fort, see the literature cited on page 183, note 6.
4. In 1981 I discovered traces of flowery motifs painted on stucco on the back wall of the arcades north of the gate to the Zanana Mina Bazar. They appear to be part of the original decoration.

29
Prince Awrangzeb facing a maddened elephant named Sudhakar

Agra, the riverbank, 7 June 1633
Folio 134a painted by an unknown artist, *circa* 1635[1]
Image area 24.6 × 40.3 cm

References: *Padshahnama*, I/1, pp.489–92;
Shah Jahan Nama, pp.95–96 and fig.20
Published: S.C. Welch *et al.* 1987, p.288 and fig.15[2]

FROM ANCIENT TIMES elephants were the ultimate status symbol of southern Asia. Akbar's historian Abu'l-Fazl knew that because elephants added 'materially to the pomp of a king,'[3] rulers 'have always shown a great predilection for this animal, and done everything in their power to collect a large number.'[4] Indian rulers used elephants in war and for transport; they also required that their elephants be depicted in certain ways, and the animals played an important part in representing their power. Elephant fights already had a long history in India when they were adopted – like many Indian customs – by the Mughals,[5] and became a jealously guarded imperial prerogative.[6] At the court of Shah-Jahan elephant fights were usually held on several days of the week except Monday – the day of the *julus* (accession) – and Wednesday – the death day of Mumtaz Mahal on which all entertainments were forbidden.[7] Because elephant fights needed a great deal of space, and because it was preferable to set the animals to fight when they were in rut (a state called *mast* in which they were extremely difficult to control), the spectacle was held outside the palace on the riverfront below the *jharoka-i darshan*, the window in the outer wall of the palace in which the Emperor appeared to the public (see Figs 157 and 158).[8]

The historic elephant fight commemorated in the Windsor manuscript, however,[9] took place at a slightly different location on the river, to the north of the fortress palace and below the *jharoka* of the building where the Emperor had lived when still a prince (*jharoka-i mubarak-i manazil-i ayyam-i padshahzadagi*).[10] In the course of their fight the two elephants moved away from this site down the riverbank to engage each other again below the *jharoka-i darshan* of the fort (Fig.132, no.4). To continue to watch the spectacle at close hand the Emperor and the princes mounted their horses and followed the beasts. In this situation one of the elephants, named Sudhakar, turned from his opponent to attack Awrangzeb, who faced him bravely. It is this moment on which the depiction of the fight focuses.

This image stands out among the illustrations in the Windsor manuscript in that it is composed with only the barest reference to a landscape formula,[11] in an attempt by the painter to render the topography of the riverfront at Agra as it presented itself to his eyes. The oblong 'landscape' format, used only once in the Windsor manuscript, corresponded to the particular urban planning of Mughal Agra, which consisted of a series of gardens and garden palaces lining both sides of the river (Fig.132).[12] The river bank is shown as it appears in June; it is wider then than at other times of the year, because just before the outbreak of the monsoon the water is at its lowest. Consequently, the river in the painting is reduced to a narrow watercourse flowing in front of the façades of the garden residences aligned along the bank opposite the fort, until the point where it makes a great bend, changing its course from south to east, which brings into view the gardens on the southern side of the river, just before the site of the Taj Mahal, which was then under construction (Fig.61).[13]

This riverscape provides an ideal setting for a large figure composition following the strict principles of Shah-Jahani court painting. The flat stretch of terrain made it easy for the painter to observe the necessary hierarchical considerations, which meant that the Emperor and the princes had to be shown in profile. As a result, the confrontation between Awrangzeb and the elephant Sudhakar is set parallel to the picture plane in the foreground. The group is framed in both the lower corners by two opposing riders on dark horses, functioning as repoussoir elements to give depth. The horseman on the left is the Emperor's eldest son, Dara-Shikoh, surrounded by a group of footmen carrying *charkhi*s and other fireworks to ward off the elephant.[14] The mahout perched on the back of Sudhakar – who is wearing the barest minimum of his harness[15] – has already used his main means of slowing the progress of his mount. He has dropped the *loh langar*, a long chain tied to the elephant's right forefoot with a thick log at its other end meant to make the beast trip.[16] The mahout has failed to throw over the elephant's eyes the *andhiyari*,[17] a red cloth weighted with bells – it is still lying in front of him. Sudhakar's tusks are cut and fortified with *bangri*s (brass rings)[18] while his opponent is tusk-less. According to the English observer Mundy, this did not make the match uneven because elephants with small or no tusks were of a more strongly built breed that 'com[m]only overcome the others.'[19]

As usual in the *Padshahnama* landscapes, the Emperor appears in the middle ground. He is sitting on the piebald horse that appears in some of his other equestrian portraits, and is protected from the front by a line of footmen with shields and a *charkhi*. The men behind him carry his insignia of kingship, the parasol (*chatr*)[20] the *sayaban* or *aftabgir* (sunshade)[21] and his *'alam*s (standards).[22] Shah-Shuja', on a prancing dun horse, advances to the left of the Emperor. His mount is rendered almost in three-quarter view, announcing the unrestrained illusionism of the background, where a stampede of white-clad, almost transparent figures rendered in free perspectival views and with an intensity similar to those storming through the jungles of Orchha (**35**) indicates the confusion which could arise on such occasions.

Freely painted realism is not confined to the illustration of the chaos in the background. It also permeates other areas of the painting, largely through the smoky fireworks which are exploding around Sudhakar's opponent on the right and in which his mahout seems to have put his entire confidence, having dropped neither his *loh langar* nor the *andhiyari* on which he still firmly kneels. The most revolutionary of the illusionistic effects is, however, the consistent use of cast shadows. This quintessential device of naturalistic painting, never popular in Asia,[23] is seen only sporadically elsewhere in the Windsor illustrations.

The illusionistic tendencies are intensified by the use of Europeanizing color and aerial perspective. There are three main color areas, beginning with greyish beige for the fore- and middle grounds, followed by green and blue in the background, to indicate recession into space.[24] The meticulously painted stretch of the far river bank at the left, which despite its microscopic format conveys so much of the experience of Indian gardens that may still be felt today, is masterfully differentiated in treatment and color from the stretch of bank on the right. To indicate that this area is further away, namely on the other side of the turning river, the white façades and the trees are rendered in such thin layers of paint that they are almost absorbed by the atmospheric tonalities of the blue sky. ❧ EK

Mughal paintings continually tell us of the important role which the Rajput princes played at court. Several illustrations in the *Padshahnama* relate to the rulers of Mewar, their defeat and eventual service to the Emperor (for example, **5**, **6–7**, and **38**). Among the most prominent figures in *The death of Khan Jahan Lodi* (**16**) is Madho Singh Hada, son of the Raja of Bundi, who was granted a new state, Kota, in return for his support of the imperialists. And junior members of the Rathor family of Jodhpur are shown in attendance on the Emperor alongside their father, Raja Gaj Singh Rathor, in *The departure of Shah-Shuja' for Kabul* (**32**). That illustration includes Raja Jai Singh Kachhwaha of Amber, who is also shown in this scene at the lower right, riding swiftly to the aid of the young Awrangzeb. This interrelationship and occasional camaraderie had considerable impact on the way of life – and the artistic aspirations – fashionable at the courts of the princes' homelands.[25]

In *A'zam Khan captures Fort Dharur* (**15**), the artist Balchand meticulously described each individual, even those at a distance or in a group; his brother Payag, however, presented groups as masses, their communal identity more important than their individuality, as seen in *The siege of Qandahar* (**18**). Just as the earliest Mughal techniques for depicting physical mass, spatial depth, and human individuality were profoundly influenced by Northern European images imported into India and collected by the emperors as early as the mid-sixteenth century, so the different interests later developed by Payag and the unknown artist of this work were paralleled by seventeenth-century European images of the school of Jacques Callot and his contemporaries.[26] Among these new interests were the depiction of shadows created by visible illumination, and the effect of individual insubstantiality when humans were active in groups. This new style, of which Payag was the most successful exponent, may well have been transmitted in part by amateur artists such as the adventurer Peter Mundy, whose travel account is accompanied by informative sketches (Fig.117).[27] Thus while the Rajput princes aspired towards the taste of their Mughal overlords, the Mughals themselves looked towards more exotic foreign styles in their own cultural explorations. MCB

1. A drawing based on this composition, dateable to the mid-seventeenth century, is reproduced in Sotheby's 1988, lot 16.
2. Welch attributes the painting to Govardhan.
3. Abu'l-Fazl gives a detailed treatment of the elephants of the imperial establishment in Akbar's time: *A'in-i Akbari*, I, p.123.
4. *A'in-i Akbari*, I, p.130.
5. Elephant fights (*hasti-yuddham*) were singled out as part of the work of war elephants by the ancient Indian manual on statecraft, the *Kautiliya Arthasastra*, which might date to the Mauryan period of the 4th century BC, although the version we know seems to be later. See *Kautiliya Arthasastra*, II, p.177. A brief treatment of Mughal elephant fights can be found in Ansari 1974, pp.168–69; Mubarak Ali 1986, pp.79–81.
6. Lahawri makes this explicit in his description: 'The elephant fight which is a prerogative of the sovereigns of India and worthy as a spectacle for rulers of elevated rank' (*jang-i fil ki khassa-i padshahan-i Hindustan ast wa darkhwar-i tamasha-yi farman-rawayan-i wala-shan*), *Padshahnama*, I/I, p.145. See also the translation of the whole passage by Nur Bakhsh 1903–1904, p.188; *cf. Shah Jahan Nama*, pp.567–68. Jahangir's concern in this respect is evident from the stern reminder he sent at the end of the sixth year of his reign (1612) to his amirs on the borders of the empire, who – far away from imperial control – had engaged in unauthorized actions, which included the staging of fights; see *Tuzuk-i-Jahangiri*, I, p.205, as quoted on page 73 above. Not even royal princes were allowed to hold elephant fights; see Khosla 1976, pp.77–78.
7. *'Amal-i Salih*, I, p.461.
8. In this location elephant fights were also more easily accessible to European observers. Mundy (1907–24, II, pp.127–28) gives a description of Shah-Jahan's elephant fights from a European point of view. See also Bernier 1891, pp.276–78.
9. An earlier elephant fight in which royal princes were involved is recorded at the end of Akbar's reign; *A'in-i Akbari*, I, pp.520–21.
10. The garden palace was situated 'outside the fort on the bank of the river Jawn' (*'Amal-i Salih*, I, pp.460ff.). Lahawri describes Shah-Jahan's princehood palace as the dwelling of Shah-Shuja' during his wedding in March 1633 (*Padshahnama*, I/I, p.463; see page 67 above) but eventually the building came to Awrangzeb (*Padshahnama*, I/2, p.268; see page 108 above). According to the 18th-century plan in the Maharaja Sawai Man Singh II Museum, Jaipur, it seems to have been been the fifth and sixth garden complex north of the fort (Fig.132, nos 6, 7).
11. For a discussion of this formula, see pages 138–39.
12. See page 182, note 10.
13. While the buildings and gardens on some stretches of the Agra riverfront are reasonably well preserved, little of them survives in the areas shown in the painting. The Jaipur plan (Fig.132 and note 10 above) designates the three gardens opposite the fort (corresponding to those in the background at the left of the painting) as imperial gardens (*vag patisahi* or *bagh-i padishahi*), of which the middle one has a more specific name, Moti Bagh. The more distant buildings which come into view on the southern bank of the river (at the right in the painting), between the fort and the Taj Mahal, most probably belonged to the garden complex of A'zam Khan.
14. A *charkhi* was shaped like a polo stick: its loosely fixed head had hollow ends filled with gunpowder. When both ends of the head were lit it rotated rapidly and made a frightening noise. For a description of this contraption, which was invented by Akbar, see *A'in-i Akbari*, I, pp.134–35. When elephants are shown in the Windsor manuscript, men with *charkhis* are often near them; see **5**, **6**, **8**, and **11**.
15. This is exactly as described in *A'in-i Akbari*, I, pp.134–37.
16. Also invented by Akbar (*A'in-i Akbari*, I, p.134.)
17. *A'in-i Akbari*, I, p.135. The bells were also Akbar's idea.
18. *A'in-i Akbari*, I, p.137.
19. Mundy 1907–24, II, p.128.
20. See **3–4**.
21. See **41–42**.
22. See **34**.
23. Schäfer 1986, pp.73, 272.
24. See also pages 139 and 143, note 42.
25. For this development, see Beach 1975.
26. See Beach 1992a, pp.149–50 and figs 115–16.
27. Twenty-nine of the drawings are reproduced in Mundy 1907–24, II.

30
Prince Khurram attacking a lion

Bari, late 1610
Folio 135b painted by Balchand, *circa* 1640
Image area 34.3 × 23.3 cm

Inscription: [on shield, lower right] *Allâhu akbar / raqam-i Bâlchand / pir ghulâm* ('God is great. Drawn by Balchand, the aged servant')

References: *Padshahnama*, I/I, pp.493–95; *Shah Jahan Nama*, p.5 and pl.12; *Tuzuk-i-Jahangiri*, I, p.186; *Jahangirnama*, folios 74b–76a
Published: Smart 1991, p.144 and fig.12

WHILE JAHANGIR WATCHES, Khurram strikes a lion which has attacked the Hindu Anup Singh Badjugar during a lion hunt. He is helped by I'timad Rai and a Hindu, the elderly Ramdas Kachhwaha. While the text describes the active participation of Hayat Khan in this episode,[1] he is placed here at the lower right as an observer. Hayat Khan (Fig.107) was the head of the Emperor's domestic servants and a personal attendant on Shah-Jahan; in **44**, for example, he is placed in his more usual role as a whisk-bearer standing at the Emperor's side.

Like many of the painters whose work is included in the Windsor manuscript, Balchand began his career under Akbar's patronage, although his early maturity coincided with Jahangir's reign. In *Jahangir receives Prince Khurram on his return from the Mewar campaign* (**5**), Balchand's warm and compassionate view of humanity is strongly in evidence, as it is in his often reproduced depiction of Shah-Shuja' and his wife, dateable to the time of their wedding.[2] The date proposed for the execution of this Windsor page, three decades after the episode shown, may explain the uncharacteristically cool, rather detached quality of the human depictions; however, this effect is countered by the intensely naturalistic, loving depiction of the grasses.

Several alternative versions of this hunting scene, which became emblematic of Shah-Jahan's bravery as a prince, are known – none, however, contemporary with the event.[3] MCB

Fig.106 Hayat Khan (detail of **30**).

Fig.107 Murar, *Hayat Khan*, circa 1630.

Fig.108 Unknown artist, *Khurram slays the lion attacking Anup Rai*, circa 1610.

The episode of the famous lion hunt of 1610 was inserted by Lahawri as a flashback in the narrative, to counter the description of the bravery of Prince Awrangzeb in the elephant fight of June 1633 (**29**) with an even more heroic deed of Shah-Jahan's own princehood. Jahangir, then the reigning emperor, had not planned to meet a lion at close quarters when he took his son and a small party on a chase hunt with cheetahs near Bari in Rajasthan, south-west of Agra. Lahawri even embeds in his account some criticism of Jahangir for going so foolhardily after a lion, with night closing in and with his hunting party mounted only on horses and not on elephants.[4] But even if it had been planned, the hunt could not have been staged more successfully to display, in an almost programmatic way, kingly courage and the bond which hunting established between the Mughals and their Rajput subjects. A Hindu noble (Anup Rai) saving the life of a Muslim emperor (Jahangir) and having his life saved in turn by the designated successor to the throne (Khurram, later Shah-Jahan) at the risk of his own life during a personal combat between lion and men – the hunt indeed had epic dimensions. Moreover, Prince Khurram's achievement against the lion put him among the ranks of the great oriental hunters of the past, in particular the Sasanians and Achaemenids of Persia, popularized and mythologized in Firdawsi's *Shahnama* and deliberately emulated by the Mughals as model kings and model hunters.[5] As in the distant past, the lion hunt in Mughal India was an imperial prerogative, symbolizing the emperor's power to protect his people and to overcome the forces of evil, often meant in a political sense: Mughal lion hunts were frequently staged to ensure a good omen in connection with a military campaign.[6] The representation of Khurram slaying a lion with his own hand testifies to his ability to fulfill his kingly functions and retrospectively legitimizes the Prince as the rightful successor to the throne.

The image of Khurram slaying the lion that mauled Anup Rai was repeatedly represented in Mughal art, no doubt because of the appealing associations it evoked.[7] It is particularly interesting to compare Balchand's representation in the Windsor manuscript with an earlier drawing in a private collection (Fig.108) which shows a more dramatic rendering of the scene.[8] Balchand minimizes the action as far as possible and gives the scene a detached formal character, clearly to underline the commemorative purpose of the representation and to draw attention to its symbolic connotations. Khurram (who is shown haloed in reference to his present status as Shah-Jahan), his fellow hunters, the lion and Anup Rai, who has been knocked down, seem frozen in their movements. The main focus is on the Prince, who raises his sword to strike the lion. Jahangir is removed completely from the action, his role reduced to watching the scene in the pose of an official portrait that has been adapted to the circumstances of the hunt with what seems a touch of humor at his expense. Jahangir's head is surrounded by a green 'hunting halo,' and a lit matchlock fuse has been placed in his right hand instead of the jeweled object customary in such representations. To emphasize that the combat with the lion took place in the past and that it is linked symbolically to an even more distant Persian past, the artist places it in an old-fashioned Persian landscape. The terrain has been tilted up to create a vertical plane with little suggestion of depth; it is dotted with tiny tufts of grass and rises steeply from a watercourse at the bottom to a rocky formation at the top which is lined with a row of spectators, deliberately evoking the favorite Safavid land-scape formula of the sixteenth century (Fig.80). Balchand nevertheless links the landscape to the Mughal present by bringing significant innovations to this time-honored scheme. He reinterprets the countryside in a painstakingly naturalistic manner, inserting into it minute nature studies of plants and birds, one of the hallmarks of his style.[9] His minuscule depictions of sarus cranes at the bottom of the illustration on the left[10] and of the two ducks in the swampy crevice of the rocky plateau at the top are particularly delightful (Figs 55, 56). EK

1. See page 76 above.
2. Beach 1978, no.31.
3. See Khandalavala 1953–5, pp.1–5. There are also two illustrations in the St Petersburg Album; see Akimushkin *et al.* 1996, pls 114 and 118.
4. Lahawri's account (*Padshahnama*, 1/1, pp.493–95) is a shorter version of Jahangir's own detailed narrative (*Tuzuk-i-Jahangiri*, 1, pp.185–88, where *sher* is wrongly translated as 'tiger'). Both passages are translated on pages 76–78 above. Lahawri's critique is shaped on *adab* literature, for which see pages 191 and 210, below.
5. Shah-Jahan is eulogized as 'lion-capturing Khusraw' (*Khusraw shirgir*, *'Amal-i Salih*, 1, p.200); or compared to Bahram Gur (Kalim,

Padshahnama, fol.276a, margin).
6. Kanbo, for example, explains that Shah-Jahan hunting lions was a good omen for his Udaipur campaign of 1613–15; *'Amal-i Salih*, I, p.58. See also **46**.
7. In addition to the examples cited by Milo Beach above, note 3, see also the fragmentary 18th-century version in the Millionenzimmer of Schönbrunn Palace in Vienna; Strzygowski & Glück 1923, no.28. The scene also exists as an intaglio; see Victoria & Albert Museum 1982, no.377.
8. S.C. Welch 1985, no.117.
9. Smart 1991, p.146.
10. For the Mughal viewer of the miniature, these birds (*Grus antigone*) would have a specific Jahangiri connotation, for that emperor was the first to observe and describe their behaviour. The relevant passages in his *Jahangirnama* are acknowledged as an original contribution to ornithology; see, for example, Alvi & Rahman 1968, pp.67–77.

31
The siege of Daulatabad

Daulatabad (Deccan), April–June 1633
Folio 144a painted by Murar, *circa* 1635
Image area 34.1 × 23.9 cm

Inscription: [on banner, bottom] *'amal-i banda-i dargâh Murâr-i umedvâr* ('work of the slave of the court, Murar the hopeful')

References: *Padshshnama*, 1/1, pp.484ff; *Shah Jahan Nama*, pp.105–106 and pl.4

Fig.109 Detail of **31** showing the fortifications of Daulatabad.

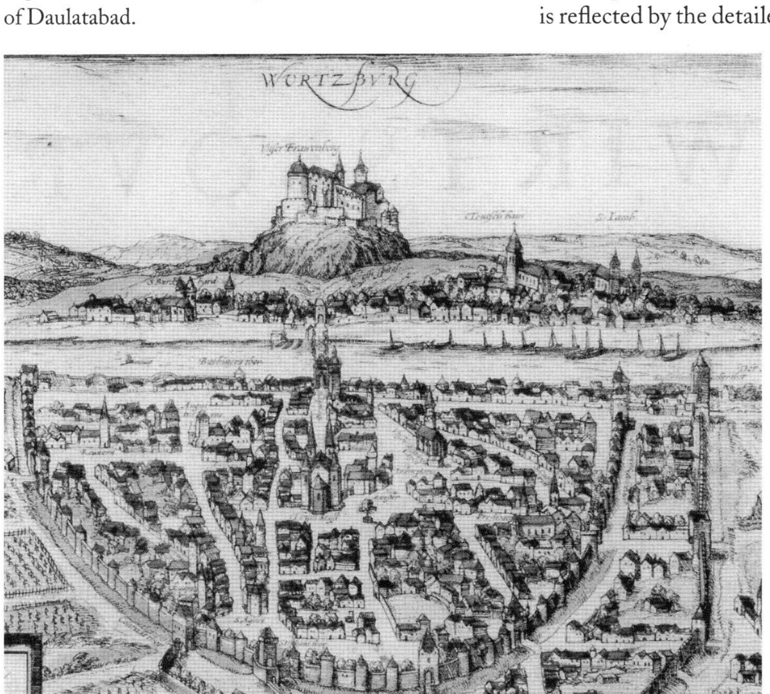

Fig.110 Georg Braun and Franz Hogenburg, Map of Würzburg, from the *Civitates orbis terrarum*, 1572–1618.

ON 27 JUNE 1633 the Mughal forces under the command of Mahabat Khan finally succeeded in conquering the fortress of Daulatabad, the great Deccani stronghold northwest of present-day Awrangabad. This was the key victory in a series of successful sieges that consolidated Mughal power in the Deccan.[1] The fall of its capital Daulatabad brought the kingdom of the Nizamshahis of Ahmadnagar to the verge of extinction and induced the formal submission of the two remaining Deccani sultanates of Bijapur and Golconda. Shah-Jahan was immensely proud of this military and political success.

The old Hindu fortress of the Yadava dynasty called Devagiri (Deogiri), or 'Hill of the Gods,' had become the object of the ambitions of the rulers of Delhi at an early stage: Sultan Ala'uddin Khalji took and plundered it in 1294, and Muhammad ibn Tughluq made the fort, now renamed Daulatabad, his second capital in 1328. When the weakening of the Tughluqs' hold on the Deccan led to the formation of independent Deccani sultanates, Daulatabad became the most powerful stronghold among their possessions in the northwest, and was deemed impregnable until – as one of his historians puts it – Shah-Jahan's forces 'broke the spell of this ancient fortress.'[2] The importance of the conquest for the Mughals is reflected by the detailed descriptions of Daulatabad and its siege in the *Padshahnama* and other contemporary histories, and also by the fact that Shah-Jahan made a point of visiting it in spring 1636. He then confessed that he had always taken the descriptions of the fortress as exaggerations but now – seeing it with his own eyes – he found that the reality surpassed even what he had heard about it and that people were right in calling it a work of the demons (*kar-i diw*, an allusion to its old name, Devagiri), for it certainly could not be the work of men![3] It seems that on the occasion of his visit Shah-Jahan also ordered the depiction of the fortress because one year later, in May 1637, the Emperor included a topographical image of the fort (*naqsha-i qil'a-i Dawlatabad*) among the diplomatic gifts he sent with an envoy to Shah Safi of Iran, to impress the Shah not only with the conquest but also with its artistic representation.[4] We can assume that Murar's view was prepared in close connection with this *naqsha*.

Murar's birds-eye panorama represents the earliest depiction of Daulatabad known so far; it is also the most impressive of the topographical illustrations in the *Padshahnama*. The viewer seems to be suspended above the landscape looking toward the hill of the fortress and the mountain range beyond; the triple apron of fortifications spread out on the sloping ground beneath the eastern side of the hill is depicted in a map-like fashion. The manner of representation changes again in the foreground to show the conquering Mughal forces as conventional groups of figures, mainly in profile, on the hilly terrain. This ingenious merging of views, which was a form of representation characteristic of topographical European maps of the 16th and 17th centuries (Fig.110),[5] enabled Murar to provide the spectator with a ready survey of the most salient features of the fortress and of its surrender. The three rings of fortification (Anbarkot, Mahakot, and Kalakot)[6] are compressed into concentric half-circles; the areas in between them contain houses and minuscule figures to indicate the settlement of Daulatabad. The fortified hill in the background is surrounded by the moat cut from the rock that features prominently in all contemporary accounts. Murar also carefully depicts the characteristic scarping of the black rock of the mountain base, which is smoothed in such a way 'that even the sight cannot get hold on it'[7] (Figs 26 and 109). The artist took more liberties with the rendering of individual buildings. The group of buildings in the center of the second ring is perhaps meant to represent the mosque or Jami'

Masjid of the Khaljis (built in 1318), the large reservoir next to it, and the conspicuous Chand Minar (a minaret built in 1445), but all are actually situated in a different order in the outer ring.[8] There Murar places a second white minaret which has a rounded top and bears a greater resemblance to the Chand Minar than the inner minaret, which is shaped on Mughal models of the period. It is also not clear if any of the palace buildings represent the pavilions which Shah-Jahan added to the fortress in 1636.[9]

Murar's painting represents a splendid attempt to reconcile the imperial interest in Daulatabad with a true report of its topography, while at the same time both conforming to the conventions of Shah-Jahani court painting and exploring its margin of artistic freedom. EK

Instead of depicting a specific moment of the siege of Daulatabad, Murar has instead chosen a more general view showing Nasiri Khan *Khan Dawran* (at the right) gesturing as if to deliver the fort to Mahabat Khan *Khankhanan*, the Commander-in-Chief of the Mughal armies (at the left). Amanullah *Khan Zaman*, the son of Mahabat Khan, who was given major responsibilities during this campaign, is the equestrian figure in the center. In the background, the portion of the outer defense wall that had been mined on 19 April 1633 is shown clearly and as if the mine had only recently exploded. The orange flags flying from the parapets of all the defensive walls, however, reveal that the fortress was completely in Mughal hands; its capitulation took place on 27 June following. Despite the explicit and varied historical references, the detailed depiction of Daulatabad, and the precisely recognizable portraits of the commanding figures, Murar is clearly interested in providing a more inclusive depiction of the siege.

Inscriptions reveal that Murar painted six illustrations in the *Padshahnama*, more than any other artist (the others are **9**, **23**, **32**, **38**, and **44**). This is the only scene that includes a broad topographical view. Murar is best known for his reliably exact, carefully crafted images of courtiers in *darbar* and the impact of these works, which are based on established formulas, is achieved through his variation and refinement of a known theme, rather than by novelty and fresh invention. Some artists, like Abid, were able to view each assignment as if it were an unprecedented subject (for example, **37**); Murar was not. This scene, therefore, reveals an otherwise unknown facet of Murar's talent. We continually see in Mughal painting that new subjects and unusual artistic challenges could bring forth correspondingly sensitive and original responses from even the most routine artist. *The siege of Daulatabad* must have been such an assignment for Murar, and the result is one of the greatest views of architecture in all of Mughal painting and a worthy depiction of a thrillingly dramatic site. MCB

1. See also **15**, **18**.
2. *'Amal-i Salih*, I, p.469. For the history of the fort, see Sherwani 1983.
3. *'Amal-i Salih*, II, p.131.
4. *Muntakhab al-Lubab*, I, p.543.
5. For the fusion of maps and landscape, or maps and architectural views, see Gibson 1989, pp.52–53. The Mughals' interest in European geographical works is discussed in more detail on page 133.
6. *'Amal-i Salih*, I, p.469; *Shah Jahan Nama*, p.99.
7. *'Amal-i Salih*, I, p.468.
8. For the topography see Burton-Page 1983; Burton-Page 1986, pp.17–25.
9. *'Amal-i Salih*, II, pp.130; *Shah Jahan Nama*, p.169.

32
The departure of Prince Shah-Shuja' for Kabul

Agra, *Diwan-i 'Amm*, 16 March 1638
Folio 147b painted by Murar, *circa* 1640
Image area 34.3 × 23.9 cm

Inscription: [on scroll, center] *'amal-i banda-i dargah kamtarin Murar* ('work of the slave of the court, the most insignificant Murar')

References: *Padshahnama*, II, pp.35–37; *Shah Jahan Nama*, p.224
Published: Begley 1987, p.143 and fig.4

THE FIGURES OUTSIDE the railing include Khan Dawran, Wazir Khan, and Bahadur Khan, who were all appointed to accompany Shah-Shuja' – the prince shown being embraced by Shah-Jahan – on a mission to Kabul during the takeover of Qandahar in 1638. A number of Rajput princes were also prominent in this campaign: Jai Singh Kachhwaha of Amber, the man in yellow in the foreground, who would be given the title *Mirza Raja* the following year; and Amar Singh Rathor, placed just inside the dividing rail, on the right. His younger brother, Jaswant Singh, had been brought to court for the first time by his elderly father, Raja Gaj Singh, whom he would succeed as Maharaja of Jodhpur. Gaj Singh is a major figure in *Padshahnama* illustrations, and when his appearance here, at the center-left of the scene, is compared to his depiction a decade earlier, when he attended the coronation *darbar* for Shah-Jahan (**10–11**), or to his portrait at the wedding ceremonies in 1633 for Dara-Shikoh (**25**), one can see how attentive the painters could be to the changes in the physical character of the men at court. Gaj Singh, a vital presence in the earlier scenes, is shown to have aged considerably here, and he died two months after this *darbar*.

Nonetheless, as in *The siege of Daulatabad* (**31**), time is a flexible element for Mughal artists. The *Shahjahannama* ends the description of the departure ceremonies on 16 March by describing how 'the illustrious Prince loaded with favors took leave along with the nobles and *mansabdars*, appointed to accompany him ...' and only later adds 'After some days, Khan Dauran *Bahadur Nusrat Jang* arrived at court ... he was appointed to join the retinue of Prince Muhammad Shah Shuja' ....'[1] It was at this time, too, that Bahadur Khan and Wazir Khan were ordered to take part in the Kabul campaign. The painting therefore accurately reflects historical fact, but not the actual physical event. This group of people did not all attend the same ceremony on one specific day. MCB

Fearing reprisals by Shah Safi I of Persia, on 16 March 1638 Shah-Jahan dispatched his second eldest son, Muhammad Shah-Shuja', with a large contingent of troops from Agra to Kabul, to back up the Mughal forces who had just brought Qandahar under their control. The takeover had been made possible by the defection to the Mughal side of Alimardan Khan, the Persian commander of the fortress. With the recovery of Qandahar – which had always been, and was to remain, a bone of contention between the Safavids and the Mughals (see **17**) – Shah-Jahan had realized the second of his three objectives in foreign

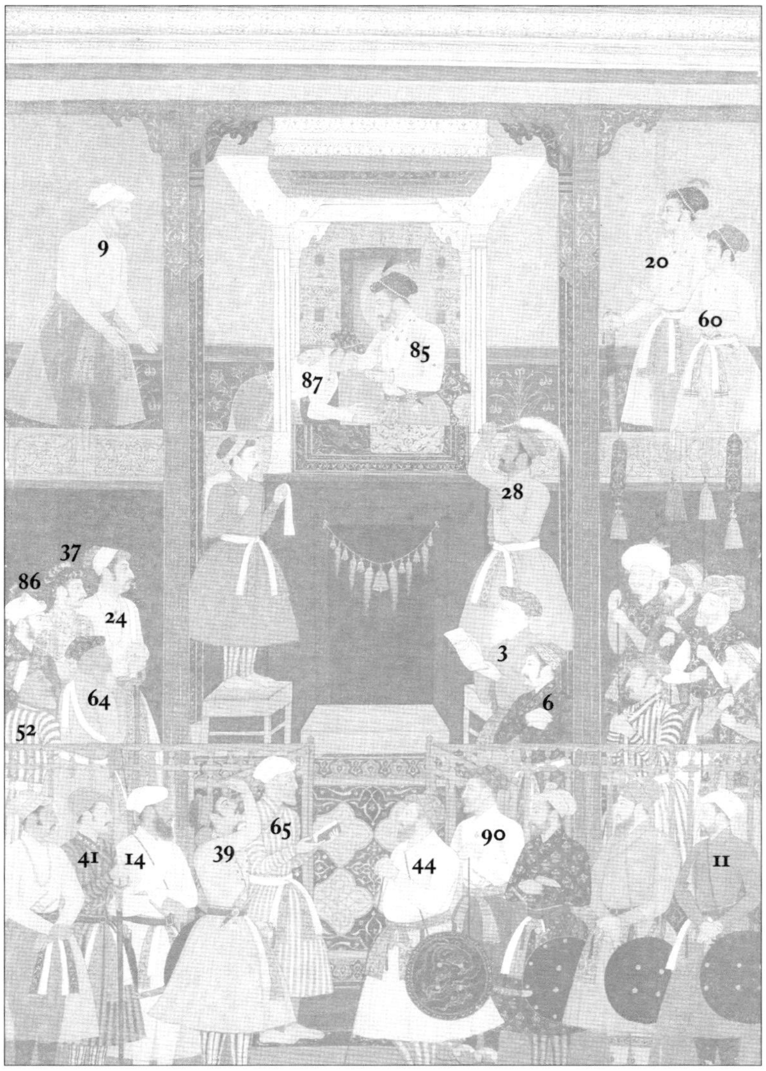

3. (Afzal Khan) (*Diwan*)
6. (Amar Singh Rathor)
9. (Asaf Khan) (*Wakil*)
11. (Asalat Khan)
14. (Bahadur Khan Rohilla)
20. (Prince Dara-Shikoh)
24. (Gaj Singh Rathor)
28. (Hayat Khan)
37. (Ja'far Khan)
39. (Jai Singh Kachhwaha)
41. (Jaswant Singh Rathor)
44. (Khan Dawran)
52. (Makramat Khan) (*Mir Saman*)
60. (Prince Muradbakhsh)
64. (Musavi Khan) (*Sadr*)
65. (Mu'tamad Khan) (*Mir Bakhshi*)
85. (The Emperor Shah-Jahan)
86. (Shahnawaz Khan)
87. (Prince Shah-Shuja')
90. (Wazir Khan)

Fig.111 Throne-*jharoka* of Shah-Jahan in the *Diwan-i 'Amm*. Red Fort, Delhi, completed 1648.

policy, the first being the assertion of his suzerainty over the Deccan. (He was to fail in the third, the re-establishment of the power of his dynasty in his ancestral lands in Central Asia).[2]

This *darbar* is one of four painted by Murar for the Windsor manuscript.[3] The composition adheres to the established scheme,[4] even though the event depicted took place after the Agra *jharoka* had been reconstructed in an entirely different form as a loggia; at this time, the green-and-red wooden audience hall was also substituted by the present hall of monumental masonry (Fig.104).[5]

As usual in the architecture of *darbar* scenes in the *Padshahnama* we find creative variations mainly in the ornament, where the artist could express himself most freely. The motifs are usually taken from real architecture, but the highly eccentric, Europeanizing grotesque patterns of the golden railing flanking the *jharoka* (Figs 127, 128), which include lyre-shaped cartouches formed by winged dragons with pendulous breasts who spit out acanthus scrolls, were an extravagance certainly only admissible in the context of painted architecture. Here the motif obviously met with approval, because Murar repeated it in his depiction of the *jharoka*-railings at Ajmer and Lahore (**38, 44**).

Another intriguing feature of the decoration is the wall behind the *jharoka*. The dark panels embedded in flowery arabesque patterns topped by a register of flower vases refer to elements typical of the extraordinary decoration – with figural Italian and Mughal *pietre dure* – of the *jharoka* at Delhi, built between 1639 and 1648 (Fig.111).[6]

Below the *jharoka*, a symbol of Shah-Jahan's justice features prominently. This is the 'chain of justice' formed of bells alternating with yak's or horse's tails.[7] The chain of justice is one of the most ancient Asian symbols of rulership and can be traced back to the Achaemenids and Sasanians.[8] Its memory was kept alive in Islamic *adab* literature (literally 'politeness' or 'good upbringing' recommended in books of advice such as the *Mirrors for Princes*),[9] in which the ancient Persian kings were included as model rulers.[10] The rulers of India – both Hindu and Muslim – revived, or continued, the tradition and put up real bells in their palaces.[11] Their function is summed up poetically in the *Rajatarangini*, the Sanskrit chronicle of the kings of Kashmir: 'At the palace gate (*simhadvara*) he [king Harsa, 1089–1101] hung up great bells in all four directions, to be informed by their sound of those who had come with the desire of making representations. And when he had once heard their plaintive speech, he fulfilled their desire [as quickly] as the cloud in the rainy season [fulfils] that of the Cataka birds.'[12]

Shah-Jahan's chain of justice seems to have been primarily a literary and pictorial symbol, but here it is shown as a real chain, meant perhaps as a 'topographical allusion' to evoke the memory of the chain of justice which Jahangir had put up at the Shah Burj tower of the Agra palace.[13] EK

1. *Shah Jahan Nama*, p.224.
2. Islam 1970, especially pp.99 ff.; Islam 1993, p.318.
3. *Cf.* **38** and **44**.
4. Analyzed on pages 162, 167–68 and 203, and on pages 133–37.
5. *Cf.* page 135. See also **14, 17, 23–24, 43**, and **44**.
6. Koch 1988, in particular pp.10–22.
7. The Turko-Mongolian yak's tail (which could be substituted by a horse's tail) also appears in a Mughal context as a fly whisk, as an ornament of horses and elephants, and in the insignia of the highest orders, the *tuman tugh* and the *chatr tugh*, which consisted of yak's or horse's tails attached to a pole. See *A'in-i Akbari*, I, p.52; Huart 1993; and the literature quoted below, page 195, note 5.
8. For references see Koch 1988, pp.34, 46, n.134; see also page 137.
9. On the concept of *adab* see Gabrieli 1979.
10. Busse 1968, pp.12–19.
11. Ibn Batuta describes the chain of justice which Sultan Shamsuddin Iltutmish (r. 1210–1236) put up between two marble statues of lions on towers in his palace at Delhi; Ibn Batuta 1993, III, p.630.
12. *Rajatarangini*, book 7, verse 879; I should like to thank Joachim Deppert for bringing this passage to my attention.
13. *Tuzuk-i-Jahangiri*, I, p.7; Skelton 1988, p.182.

33
Shah-Jahan hunting

Perhaps near Rupbas, after 1640
Folio 165a painted by an unknown artist, *circa* 1645
Image area 32.9 × 21.6 cm

Published: *Shah Jahan Nama*, fig.7;[1] S.C. Welch *et al.* 1987, p.291 and fig.21;[2] Barrett & Gray 1963, pp.112–114[3]

Fig.112 Unknown artist, *Asalat Khan* (detail), circa 1645.

Fig.113 Asalat Khan (detail of **33**).

Fig.114 Chitarman, *Ja'far Khan* (detail), circa 1645.

Fig.115 Ja'far Khan (detail of **33**).

IF THE DEPICTION of a precise historical episode is intended here, the event is now unknown. Given the ages of the Emperor and his son Dara-Shikoh, and the whiteness of Shah-Jahan's beard, it must have taken place after 1640. In addition to the Prince, the two most prominent attendants are Salabat Khan (d. 1644) and Asalat Khan (d. 1647; Figs 112, 113), who served together as *Mir Bakhshi* and second *bakhshi* between 1639 and 4 August 1644, when Asalat Khan succeeded as *Mir Bakhshi*. It is unlikely that they would be shown together in such an intimate imperial group during any other period. Standing directly to the Emperor's left is Dara-Shikoh, while the man behind, with a fresh rifle, is Ja'far Khan (Figs 114, 115), who would succeed Asalat Khan as *Mir Bakhshi* in 1647. Shah-Jahan is therefore shown hunting with a group of particularly powerful nobles.

Perhaps the most surprising new element of Shah-Jahan period painting is the frequent incorporation within the picture of vast spaces, and the realization that landscape can independently contribute important expressive elements. In *Prince Khurram attacking a lion* (**30**), for example, the landscape acts as a backdrop to emphasize the foreground action, which the distant figures serve to frame and contain. Here, by contrast, we move back into the landscape and observe the extent and the peacefulness of the world over which the Emperor reigns. No longer restricted to the imperial party, we also observe farmers at work and distant herds of deer. If the hunting party itself proclaims Shah-Jahan's personal power, the golden glow of both sky and landscape leaves no doubt of the prosperity this power will continue to bring. The time, after all, is dawn. MCB

This miniature was inserted in the manuscript to illustrate the account of Shah-Jahan's hunt of black buck antelope (*ahu siah*) at the imperial hunting-ground (*saydgah-i muqarrar*) of Palam near Delhi, where the Emperor halted in March 1635 on his return to Agra from Kashmir and Lahore.[4] At that time Shah-Jahan would have been forty-three years old and his son, Dara-Shikoh, only twenty; since both look much older in the painting, it would seem to depict a later hunt. This could well have taken place at Palam too, because that particular hunting-ground was Shah-Jahan's favorite place to shoot *ahu siah*. During each of his visits to Delhi in the years before he moved his residence there in 1648, the Emperor tried to outdo himself in shooting at Palam ever greater numbers of *ahu siah* with his royal gun, named *Khassban*.[5]

The hunt takes place on a plain; the rising sun, low and bright on the distant horizon, indicates that the time of day is early morning. On the plain, which is covered with trees and bushes, a large herd of antelope has gathered on both sides of a winding rivulet (Fig.54).[6] The Emperor, whose halo echoes the sun to symbolize his power as a 'sun king' (*cf.* **3**–**4**), has taken cover behind a bush near the water. He is sitting on a small carpet spread out on the ground, and with the finger of his right hand he pulls the trigger of his matchlock to bring the tiny lit fuse down on the powder-filled pan that will fire the charge. With his left hand supported by the elbow on his raised knee he steadies the gun by means of a string fastened to a ring on the barrel. However, the barrel is mainly supported on the shoulder of the huntsman who sits in front of the Emperor. Behind Shah-Jahan stand three of his nobles and Prince Dara-Shikoh. A huntsman in the foreground keeps the game within the range of the imperial gun. The whole party wears green as camouflage, necessary for this particular type of hunt.

The Emperor had obviously taken position before sunrise to await the antelope coming to the water; to make sure of his success, huntsmen would have driven the game slowly and carefully in his direction. This hunting technique – in which the drivers would also employ tame antelope or cows as decoys – had long been used by the Mughal emperors, certainly since Akbar.[7] Although its connotations were less noble, the antelope drive had the advantage of being simpler and less exerting than the dangerous lion hunt (see **30**, **46**) or the war-like *qamargah* hunt, the great battue in the Mongol tradition in which thousands of drivers would enclose the game within a vast circle and the Emperor would bring the animals down from on horseback. Having the game driven towards one's gun was a practice well-suited to the new, courtly habits of the Mughals, who had become increasingly sedentary and formal. It was also a subtler form of the hunt in which firearms – then still unreliable and slow to handle – could be used to greater advantage.[8] Finally, this style of hunting brought the main hunter into close and prolonged contact with nature. While he lay quietly in waiting, he had leisure to observe his surroundings; consequently the painters, who had to meet the interests of their patrons, had to give more attention to the landscapes in which such hunts took place. Thus we owe to the antelope drive the most naturalistic landscapes of Mughal painting, where illusionistic background features are extended to the fore- and middle grounds. The earliest preserved depictions of the drive hunt date from the time of Shah-Jahan and show royal princes as the main hunters.[9] The antelope hunt in the Windsor manuscript is the first painting so far known in which the Emperor himself is engaged in this type of activity. The role of the men driving the antelopes to the ambush is minimized, however, being merely symbolized by the waving hunter in the foreground. Ruses, violence and killing are absent; the scene almost has the character of a *darbar* or imperial reception. Its rendering reflects the embellishment of the royal hunt by the court poets, who depict the game as so eager to be hunted by the Emperor that it offers itself voluntarily to his gun, like the moth that seeks annihilation in the flame of the candle in the favorite metaphor of love in Persian poetry.[10]

The harmonious poetic mood of Shah-Jahan's hunt is reinforced by the landscape. The plain, seen from above, is structured with a multitude of little mounds (on which stand tiny trees and animals), diminishing in size and set closer to each other towards the distance to create the impression of depth.[11] This compositional device successfully fuses the very far and the very near in a gliding recession into space without abrupt breaks. This uninterrupted impression of depth is subtly intensified by the atmospheric tonalities of the landscape; the dark green of the trees in the fore- and middle grounds changes into a light, sharp green in the forested ridges of the background, and the soft pinkish light of the rising sun invests the most distant regions on the horizon with an insubstantial quality. The hierarchical impositions of Shah-Jahani court painting,[12] which require that in landscapes the main characters should, ideally, be placed in profile in the middle ground and on a larger scale than those in the foreground, are integrated so skillfully into the optical logic of this composition that we hardly notice that the size of the objects and figures varies not according to their distance from our viewpoint but according to their distance from the Emperor. Even the trees are included in these hierarchical considerations: the largest, acting as a framing repoussoir motif at the left, is placed in the middle ground on the 'imperial plane.' Behind it the rivulet winds diagonally towards the background to link the landscape of the royal hunt with that of the peasants watering their fields (Fig.62). More minuscule scenes of rustic life can be seen beyond and in the background to the left. In the West we know such juxtapositions of aristocratic life and rustic occupations from calendar illustrations of the 'Labors of the Months,' a main vehicle of landscape representation in Renaissance Europe.[13] In seventeenth-century Mughal India, the juxtaposition of princely with peasant life had the specific function of legitimizing the imperial hunt. It was intended to express the idea that the hunt brought the royal hunter in close contact with his subjects so that he could see their conditions with his own eyes and dispense justice – the foremost duty of the sultan – on the spot, without intermediaries.[14] This age-old and widely employed argument to morally justify the royal hunt as a serious social duty was much emphasized in India, where royal hunters since the time of the Buddhist emperor Asoka (r. 270–232 BC) had run into conflict with the principle of *ahimsa*, non-violence towards living beings.[15] ❧ EK

1. Begley suggests that the scene is a hunt held near Rupbas in November 1643 or February 1644.
2. Welch has, without explanation, identified the scene as the work of Hashim. While no signed paintings by Hashim show major landscape elements, Welch has also assigned **46** and an additional scene showing the *Emperor Awrangzeb hunting nilgai* (Chester Beatty Library, Dublin, MS.11, no.27) to the same artist (S.C. Welch 1985, no.176). The latter work seems instead to relate very closely in its landscape depiction to *Shah-Shuja' hunting nilgai* (Museum of Art, Rhode Island School of Design, 58.068), which has been attributed to Payag by Dr Ellen Smart (Beach 1995a, Fig.10).
3. Barrett and Gray proposed the no longer acceptable theory that the illustration is dateable to the reign of Bahadur Shah (1707–1712).
4. *Padshahnama*, I/2, p.73.
5. For Shah-Jahan's hunts at Palam during his visits to Delhi see Koch 1991b, pp.101–102; Koch 1993, pp.9–11, 14–15. See also page 84 above.
6. They are black bucks (*Antilope cervicapra Linn.*; 'black buck' is a direct translation of the Persian *ahu siah*), once the most numerous of the larger wild mammals of India. That the painter took great care to render the antelope true to nature is evident when we compare his depiction with a zoological description of the species. The females are a 'light brownish tan on the upper part of the body, tail, outer flanks of hind legs, face and neck, with a similar coloured strip running down the forelegs. The underparts of the body and tail and remainder of the legs are white, with a similar

coloured patch around each eye. The males, which are similar to the females in colouration when immature, assume a darker pelage with age, and in certain individuals become almost black in sharp contrast to the white on the nose and lower muzzle, eyes, chest, belly and legs .... The male blackbuck carries straight, shallow-spiralled horns with circular rings' (Ranjitsinh 1989, pp.29–30; the Mughal emperors as hunters of black buck are discussed on pp.1–3.) The herd represented in the painting consists mainly of females; a dark male drinks from the rivulet below the tree in the left middle ground and a few males are minutely depicted in the background.
7. *A'in-i Akbari*, I, p.302–303. For this point and the following discussion of the antelope drive, see also Koch, forthcoming (b).
8. This is the suggestion of Lisa Golombek with whom I discussed Mughal hunts in New York on 21 June 1995.
9. *Dara-Shikoh shooting nilgai* (Indian blue antelope), Arthur M. Sackler Gallery, Washington, DC, first published by Beach 1995a, fig.1; and *Shah-Shuja' hunting nilgai* (see above, note 2; illustrated in Smart 1979, figs 16 and 17.)
10. Kalim, *Padshahnama*, fol.282b. In the painting this parallel is further emphasized by the fact that most of the antelopes depicted are female. For a discussion of the hunt as a metaphor of love in Persian poetry see Hanaway 1971, pp.29ff. See also Schimmel 1992, p.286.
11. For a discussion of this landscape type see pages 139–40, above.
12. See pages 135–38, above.
13. Pächt 1950, in particular pp.37ff. See also Koch, forthcoming (c).
14. Kalim, *Padshahnama*, fol.282a. *Cf.* Hanaway 1971, pp.27–29, with further examples of Persian poetry. See also Koch 1991b, p.102, and Koch, forthcoming (c), with further references.
15. Koch 1991b, pp.102, 106, n.55.

34
A royal procession

Unidentified city, *circa* 1655
Folio 166b attributed to the 'Kashmiri Painter', *circa* 1655
Image area 33.1 × 22.9 cm

WHILE THIS ILLUSTRATION is placed to correspond with a passage describing the imperial caravan moving between Lahore and Agra in March 1634, no member of the ruling family is included and, judging by the composition, one can be certain that the scene was originally the left (and less important) half of a double-page scene in which the Emperor or some surrogate would have been shown in the now missing right half. There are few significant details in this remaining section, other than the city in the background, which has not yet been identified, although the presence of the cheetah suggests that it was intended originally to show a royal hunting expedition.

This is rougher, less precisely detailed painting than is usual in this manuscript – the figures are types, rather than individuals, for example – although it is almost certainly by the artist who executed the left half of **6–7**, as well as **17** and **41–42**, dubbed the 'Kashmiri Painter.'[1] Since the latter double-page shows an elderly Shah-Jahan, and the episode can be dated to 1654, that scene could not have been painted earlier than about 1655. The similarity in style of that work and this painting suggests the same general date for both.

While the identity of the artist is unknown, several important facts emerge from the various illustrations by him included in the Windsor *Padshahnama*. The first, of course, is that his work is quite distinct in style and quality from that of other illustrators of the manuscript. Because of this difference, it has been proposed that these scenes were added much later, when the manuscript was in Lucknow.[2] However, *A royal procession* is only half of the original composition, and the two halves of *Shah-Jahan visits the shrine of Khwaja Mu'inuddin Chishti at Ajmer* (**41–42**) seem not to have been intended for a single composition either; they are slightly different in overall size and in the scale of the figures. This would indicate that they and other scenes were paintings in stock at the time the miniatures were added to the manuscript; being both incomplete and, in this case, inappropriate for the text opening at which they were placed, they clearly were not painted specifically for inclusion in the Windsor volume. On the other hand, the left half of *The submission of Rana Amar Singh of Mewar to Prince Khurram* (**6–7**) was certainly made to complete La'lchand's composition on the right; unlike other double-pages, where the lesser half is a stock composition with no necessary relationship to the main scene (for example, **8–9**), these pages depicting the Rana with Khurram were intended to match both compositionally and in terms of the pigments chosen. There is every evidence therefore that the painter was active in the middle decades of the seventeenth century. ❧ MCB

The Mughals honored their nomadic heritage by maintaining a largely peripatetic life style, alternating between their capitals Agra, Lahore, and Delhi, and their summer residence in Kashmir. The emperors often led their military campaigns in person and went on extended hunting expeditions to consolidate their hold over their large empire. Akbar had responded to this situation by transforming the Mughal camp into an elaborate bureaucratic institution which effectively moved the imperial household and the administration of the empire to wherever the emperor was traveling.[3] This painting shows that on such journeys the imperial appurtenances were displayed prominently. We see the Emperor's hunting animals, represented by the falcon on the hand of its falconer and the cheetah (*yuz*) on the bullock cart;[4] the horses, whose elaborate saddle cloths seem to mark them as the Emperor's own mounts; and the imperial insignia which followed the Emperor wherever he went.[5] They consisted of the imperial standards (*'alam)* with the emblem of the lion and the rising sun (*sher-u-khurshed*),[6] and the *qur-i khassa*,[7] a collection of ceremonial arms, flags, and other insignia, carried here by the elephants, and observed by the English eyewitness Peter Mundy in 1632, who described them as 'sundrie figures of gold and silver upon long staves.'[8] Such items formed an element of the ruler's magnificence (*shukoh-i dawlat*) and contributed 'full distinction and spectacular display.' Shah-Jahan's historian Kanbo goes on to tell us that pomp and outward show should not be dismissed as 'beautiful and external.' They were necessary to enhance the status of the sultan in the eyes of rival rulers and of his subjects, and would, in turn, induce the people to follow his commands and obey the law.[9]

The imperial train is shown traveling through open country, passing a small town and a fortress in the background. The composition is clearly related to that of the antelope hunt (**33**). Both paintings show the same type of landscape seen from an elevated viewpoint and across a diagonal arrangement of figures in the fore- and middle ground. The 'Kashmiri Painter' interprets this formula in a more tectonic manner, however, with hardened contours and sharper colors. The figure group is so closely knit together that it acts like a compact diagonal segment within the terrain, beyond which we look onto a plain structured by little mounds similar to the landscape of the hunt. In the previous painting the decreasing size of and intervals between the mounds towards the distant horizon indicate that the plain continues more or less on the same level, while in *A royal procession* the intervals between them become wider in the background to create the impression of ascending mountainous terrain. The mounds, painted with mechanical regularity, are in turn structured by other, miniature mounds, and all are topped by trees or bushes, the size of which is not proportionate to their distance from the viewer. This manner of presenting landscape is clearly inspired by Western pictorial maps (Figs 35–37).[10] In addition, the terrain is dotted with a multitude of minute figures who appear behind the ridges, so that often only their heads are visible.[11] With the help of a magnifying glass one can make out that most of them seem to belong to the imperial train, dispersing across the area for all sorts of activities, while the local farmers go about their usual occupations, for instance, threshing with bullocks in three round, makeshift enclosures. The town in the background at the left consists largely of stereotypical pink houses; several architectural features are highlighted in white. This color dualism, borrowed from imperial Mughal architecture, is varied in the little fortress on the right horizon; its pink fortifications enclose white residential buildings. The more conspicuous features of the city, such as the large reservoir and the caravanserai beyond, populated by the microscopic figures of travelers and merchants, seem to indicate that the painter was attempting a reference to actual topographical features, though it has as yet not been possible to identify the place.

The painter's greatest strength was clearly his talent for detail; this is also evident in the group of the imperial train. While the figures are stereotypes, the imperial textiles are rendered with such attention to texture that – looking at them in close-up – one is almost tempted to touch the flowered brocades and velvets of the horses' saddle cloths, and their jeweled harnesses. The detailed realism culminates in the lions on the standards, which are shown in much livelier postures than their 'real' counterparts in the scenes of lion hunts (**30** and **46**). EK

1. Robert Skelton called the artist of these pages the 'Kashmiri Painter,' and has suggested a general stylistic relationship between his work and that of Muhammad Nadir Samarqandi, an artist working in Kashmir early in the second half of the 17th century. For this argument and a list of related works, see Leach 1987.
2. S.C. Welch 1963, p.103.
3. Richards 1981, pp.258 ff; Andrews, forthcoming.
4. An almost identical bullock cart, though in reverse, appears on a late sixteenth-century illustration of the *Akbarnama*, inscribed to L'al and Sanwala; Victoria & Albert Museum, London, inv. no.I.S. 2/1896, acc. no.92/117, illustrated in Sen 1984, pl.10.
5. The histories of Shah-Jahan tell us little about the imperial insignia, and there is as yet no comprehensive study of the insignia of the Mughals. Brief treatments are provided by Irvine 1962, pp.31–35; Aziz 1947, pp.77–100. The insignia of Akbar's period (several of which continued to be used by Shah-Jahan) are assembled in Verma 1976, pp.41–45. See also Burton-Page 1991, p.536–37. Some information about insignia of the late Mughal period can be found in Delhi 1911, pp.27–28. Mughal insignia also relate to the Persian tradition, for which see Ackerman 1964–5.
6. See 'Shir u-Khurshid,' forthcoming. The Mughals shared this ancient solar emblem with the Safavids; it had been used earlier by Timur. See Ackerman 1964–5, pp.2777–78. For the sun-rulership of the Mughals, see **3–4**.
7. For the *qur* see in particular *A'in-i Akbari*, I, pp.52, 115–16.
8. Mundy 1907–24, II, p.199.
9. *'Amal-i Salih*, III, p.18. For a more extensive quotation see page 131, above.
10. For a discussion of this landscape type and its European sources see pages 139–40 above.
11. This manner of showing figures in a landscape is ingeniously explored by Payag in *The siege of Qandahar* (**18**).

35
The capture of Orchha by imperial forces

Orchha, October 1635
Folio 174a painted by Tezdast, *circa* 1637
Image area 34.2 × 23.2 cm

Inscription: [bottom, middle] *'amal-i Tirdast* ('work of Tirdast [Tezdast]')[1]

Published: *Padshahnama*, I/2, pp.94–100; *Shah Jahan Nama*, pp.155–58

THE THREE GREAT COMMANDERS of Shah-Jahan's campaigns in Central India and the Deccan are shown together in this scene, indicating that it relates to their joint assignment to pursue the rebellious Jujhar Singh and his son Bikramajit. The exact moment shown here does not clearly relate to known historical texts, however.

This is the only work known by an artist who signed himself Tezdast. That he was an important painter is shown by his skill in portraying the victorious procession about to enter the captured fortress, heralded by trumpeters over the gateway; we can sense both the individuality of each man, and his place within an informal and spirited crowd. The clarity of the distant fortress creates an unusually crystalline space, in which the distant armies, although only lightly sketched, show the artist's meticulously careful draftsmanship. These

Fig.116 The imperial army storming through the jungle of Bundelkhand (detail of **35**).

figures are very different to those in the background of the next illustration by Payag (**36**). The distant figures here are so miniaturistically depicted, however, that they fail to open the space as successfully as Payag's more assertive spatial manipulations. It was recently suggested that this illustration was the work of an eighteenth-century painter at Lucknow,[2] but there is no reason to doubt a seventeenth-century date. MCB

The painting depicts the taking of one of the fortresses of the rebellious Raja of Orchha, Jujhar Singh Bundela, during Shah-Jahan's Bundelkhand campaign of 1635. About one hundred years earlier, at the beginning of the sixteenth century, the Bundela Rajputs had asserted themselves with the foundation of a capital at Orchha (also spelled Urchha or Ondchha), which was situated a few miles south of Jhansi in central India; their territory became known as Bundelkhand. The Bundela ruling house had accepted Mughal sovereignty under Akbar but a crisis developed when Bir Singh Deo, a member of the Bundela royal lineage, murdered Akbar's favorite, Abu'l-Fazl, at the request of Sultan Salim (later Jahangir), who saw Abu'l-Fazl as the main reason for the estrangement between himself and his father. When Jahangir came to the throne in 1605 he rewarded Bir Singh Deo with grants of titles and territories.[3] To emphasize his position as ruler of the Bundelas, Bir Singh Deo embarked on ambitious building activities; particularly outstanding are his grandiose palaces at Orchha and Datia, which merge ancient Indian shastric ideas with Mughal concepts of regular planning.[4] The good relationship between the Bundelas and the Mughals was not carried on by the following generation: Bir Singh Deo's son, Jujhar Singh, rebelled against Shah-Jahan in the first year of that Emperor's reign; he was subdued and pardoned.[5] In 1635 he again embarked on unauthorized actions which prompted Shah-Jahan to send the high-powered punitive expedition shown here. The Mughal forces were led by Abdullah Khan, Sayyid Khan Jahan Barha and Khan Dawran under the nominal command of the Emperor's son, Awrangzeb, who was then only seventeen years old.[6] Despite their difficulty in advancing through the dense jungle of Bundelkhand the Mughal forces succeeded in taking the fortress of Orchha on 14 October 1635, and soon afterwards the fortresses of Dhamoni and Chawragarh. Pursued by the Mughals Jujhar fled with his sons, wives and treasures towards the Deccan, where he met his ignoble end.[7]

On 1 October 1635, when the Bundela campaign was drawing to its close, Shah-Jahan set off from Agra to see Daulatabad and to further secure his hold on the territory of the Nizamshahis in the Deccan (see **31**). At the request of Awrangzeb, who described the beauty of its scenery in vivid terms, the Emperor made a detour to visit Bundela territory, in particular newly conquered Orchha.[8]

The illustration of the Bundela campaign shows the three army leaders surrounded by their troops in front of a city and its fortress, and about to enter it in a victory procession. The route they are going to take is indicated by the foot soldiers who form the right arm of the half-circle of troops. To approach the fortress, these standard-bearers, matchlockmen, and lance-bearers turn their back to the viewer; the artist was free to use such a view only in the rendering of persons of no rank.[9] The compact column of foot soldiers is preceded by more loosely grouped footmen and riders making their way through the streets of the city, over two bridges of the moat and into the gates of the fortress, one of which has been severely damaged in the siege. The other, the main gate, is topped by a *naqarakhana* with a band playing to provide the event with the proper accompanying music.

The composition fully subscribes to the conventional principle of Mughal landscape representations, which is that, when seen from a higher point, figures and objects which are actually on the same level appear higher the more distant they are. The Mughal troops, the buildings and trees are projected onto the vertical surface of the painting with little exploration of depth. The middle ground is, as usual in landscapes by Shah-Jahani court painters, reserved for the main protagonists; the three army commanders appear as a diagonally layered group in the strict side view obligatory for their rank. Their importance is also underlined by their larger proportions. The compositional device of a closed circle used for the arrangement of the troops and the houses of the city is unusual; it seems to take its cue from the fortification lines seen in *The siege of Daulatabad* (**31**), transforming, as it were, a pattern of defense into one of attack. Contrary to the Daulatabad composition, however, which stands out because of its topographical realism, the Orchha fortress is rendered in such general terms that we have to take recourse to the textual sources and the overall topography to establish its identity. The texts mention the victorious entry of all three commanders only with regard to Orchha, and it seems plausible that the taking of the most important fortress would be chosen to illustrate the Bundelkhand campaign. In a general way, the juxtaposition of a city and a fortified castle surrounded by an irregular watercourse applies to Orchha and its fortress, which is built on an island in the Betwa River, but there is nothing in the buildings which would allow us to make a connection to the actual Orchha, nor can we make out any of the features which Shah-Jahan's authors pointed out as peculiar to its architecture, such as the surrounding walls, without battlements, that were built of blocks of quartz (*sang-i chini*) without the use of mortar.[10]

The vague treatment of the architecture, the formal composition, and the stiff figures might cause us to dismiss the painting as one of the least-inspired illustrations of the *Padshahnama* were it not for the stunningly 'modern' background scenes. The atmospheric landscape into which the terrain beyond the fortress develops and a patch of jungle closing up on the city behind the elephants to the left show the artist Tezdast (whom we have no reason to doubt was responsible for these parts of the painting) as one of the great masters of the illusionistic background mode of Shah-Jahani painting. Particularly striking are the Lilliputian troops led by an elephant storming through dense dry jungle shrubs in a rising cloud of dust; we see nothing like it again till the impressionists of the nineteenth century. Tezdast evokes here – with microscopic brushwork in light airy tonalities that dissolve local color, and with a sense of drama entirely contradicting his formal 'main event' mode – an actual aspect of the campaign, which features prominently in all accounts, namely the thickly forested terrain in Bundela country and the achievement of the imperial troops in overcoming this formidable obstacle.[11]

The capture of Orchha by imperial forces emerges as a highly complex painting. More than any other illustration in the Windsor manuscript it demonstrates the antagonistic potential of the Shah-Jahani court style: its highly formal official idiom and the wild, almost shocking freedom accorded to its painterly backgrounds. EK

1. As written, *Tirdast* would mean 'arrow-handed,' not a satisfactory name or epithet. Amending the reading with a single missing dot, we obtain Tezdast, which means 'swift of hand,' a more suitable appellation for an artist. (WT)
2. Verma 1994, p.365. The character of the author's remarks suggest that they might have been based on a misidentification. For further brief discussion of Tezdast, see page 217, below.
3. *Shah Jahan Nama*, pp.27, 156. For this and the following historical outline, see also Saksena 1976, pp.79–91.
4. Tillotson 1987, pp.71–79; Asher 1992, pp.162–64.
5. *Shah Jahan Nama*, pp.27–30. See also **16**.
6. *Shah Jahan Nama*, pp.149–51, 154–59.
7. *Shah Jahan Nama*, pp.162–67. See also **36**.
8. *Shah Jahan Nama*, pp.155, 161–62.
9. The principles of Shah-Jahani court painting are explained in more detail on page 136, above.
10. *Padshahnama*, I/2, p.122.
11. The painted vegetation also corresponds to the actual flora of the Orchha state, at least as it survived until the early twentieth century, when it was described in the British Gazetteers as consisting of 'small trees and rough undergrowth;' Luard 1989, pp.348, 354.

36
Khan Dawran receiving the heads of Jujhar Singh and his son Bikramajit

Deccan, January 1636
Folio 176b attributed to Payag, *circa* 1638
Image area 35.9 × 23.8 cm

References: *Padshahnama*, I/2, p.116; *Shah Jahan Nama*, p.165 and pl.22
Published: S.C. Welch 1995, pp.322, 328, and fig.8

AMONG THE MOST ORIGINAL contributions by painters to these *Padshahnama* illustrations is the occasional incorporation of scenes within a distant space: these include the gardens seen across the Jumna River at Agra (**29**), the barren lands surrounding the Portuguese colony at Hoogly (**20**), and far-off battles (**35**). Here, the distant armies and elephants, with covered howdahs in which ladies were transported in privacy, are integral to the narrative.[1] So vital and so innovative is the execution of these vignettes that our interest is often actually drawn away from the foreground. In contrast to the expectedly meticulous detail, perceptive portraiture, and painstaking craftsmanship found in the foreground scene of Khan Dawran accepting the heads of the captured rebels, Jujhar Singh and his son Bikramajit, the background figures are quickly and deftly sketched, with individual gestures and group movements skillfully caught. We sense a directness, the artist's immediate reactions to the subject and the movement of his brush, which is often lacking in the mainstream of earlier Mughal painting. And once again, there is a direct visual correspondence between the ladies' elephants with their attendants and the grouped elephants and warriors at the right of the sketch made in India in 1632 by Peter Mundy (Fig.117). Payag, to whom this scene has been attributed by Robert Skelton and S.C. Welch,[2] was able to extend this immediacy of execution throughout his best paintings; in *The siege of Qandahar* (**18**) and *Jahangir presents Prince Khurram with a turban ornament* (**39**) especially, we can sense the action with which he applied pigment to paper, and this gives the images the vividness of a direct encounter with the artist. The foreground figures of the present scene, and those in the stylistically conservative *The arrival of Prince Awrangzeb at the court at Lahore* (**44**), on the other hand, show a meticulous, rather dry execution that is considerably less spirited and engaging. ❧ MCB

Fig.117 Peter Mundy, *A royal progression*, dated 1632.

Fig.118 The capture of the rebels' elephants (detail of **36**).

This painting shows the end of Jujhar Singh Bundela and his son Bikramajit, or *Jagraj*,[3] as well as several of their followers (see **35**). In their flight towards the Deccan the Bundelas had been pursued by the Mughal forces under Abdullah Khan and Khan Dawran into the region of the Gonds, a forest people, in Chanda territory, in early January 1636. When the Mughals caught up with the fugitive Rajputs, Jujhar Singh and his son managed to escape once more into the woods, but were put cruelly to death by a troop of Gonds.

Upon finding their corpses, Khan Dawran had their heads cut off and sent to the court, then encamped at Sehore in Malwa, en route to Burhanpur. The Mughal troops also captured elephants carrying the women and treasure of the Bundelas (Fig.118).[4]

The gruesome scene takes place in a landscape which parallels the one used by Balchand in *A'zam Khan captures Fort Dharur* (**15**). The artist, probably Payag, uses this landscape formula, in which recession is established by means of segments of terrain, to create a well-ordered open-air scene according to Shah-Jahani principles of representation. He divides the landscape into a very shallow foreground, a large middle ground containing the main scene of the beheading, and a background which shows the taking of the Rajput elephants, subdivided in turn by forested ranges. The contours of the individual segments of terrain are lined with bushes and trees rendered in minutely observed, naturalistic detail. The vegetation does not, however, obscure the line of the ridges, but emerges neatly from behind them; the same is true of the figures. This creates clearly delineated areas of landscape in different shades of brown. This method of ordering the scenery, in which distance is expressed by the addition of elements to each other, echoes the way in which formal groups of figures are arranged in Shah-Jahani compositions, with different layers distinguished by the use of alternating light and dark colors. The group containing the figures of the Mughal commanders is represented according to this principle. The Mughal forces occupy the larger, right half of the painting, both in the middle ground where the decapitation takes place and in the background, which shows the parallel scene of the capturing of the Rajput elephants.

The two-dimensional approach changes into a three-dimensional one in the narrower, left half of the composition. Here, more realistic forms of representation deemed appropriate for the rebels (who had literally fallen out of the ordered ranks of the Mughals), gave the painter the opportunity to show distorted bodies, heads seen at oblique angles and studies of facial expression. This change of mode actually extends over the whole of the left half – the 'rebel' half – of the painting: it even affects the Mughals who behead the Rajputs. Their group includes a horse shown in three-quarter view (Fig.51). Accordingly, the terrain around the rebels is also more deeply modeled and the shallow vegetation on the ridge behind them develops into a small grove. In the background this difference in manner is intensified by illusionistic effects. The elephants surrounded by soldiers are set into a hazy clearing of the forest. The gradually diminishing size of the figures conveys distance, as does the treatment of their colors, which merge progressively with their surroundings.

The painting is not signed but the clear silhouettes of the landscape and the carefully outlined and detailed treatment of the vegetation in the fore- and midground point to Balchand (*cf.* **15**). The three-dimensional views and experiments with facial expression in an open brush technique, on the other hand, reflect the interests of his brother, Payag, as does the intricate juxtaposition of the texture of faces and of ornamental armor in the closely knit group of the Mughal commanders (compare **36** with **18**). ❧ EK

1. See page 90.
2. Welch 1995, pp.322 and 328.
3. Bikramjit was awarded the title *Jagraj* ('Lord of the World') by Shah-Jahan after he had sided with the Mughals against the rebellious Khan Jahan Lodi; *Shah Jahan Nama*, p.52. See also page 174.
4. Saksena 1976, pp.88–89.

37
Jahangir receives Prince Khurram

Ajmer, *Diwan-i 'Amm*, April 1616
Folio 192b attributed to Abid, *circa* 1635
Image area 35.8 × 24.2 cm

References: *Jahangirnama*, folios 132b – 133a; *Tuzuk-i-Jahangiri*, 1, p.322

THERE IS LITTLE TO identify the event being depicted, since only a small number of the usual courtiers are shown, but unlike the majority of *darbar* scenes, including the following illustration (**38**), it seems to be a more intimate, less publicly ceremonial occasion. However, since this is the only illustration in the *Padshahnama* which shows Jahangir wearing a jeweled armlet, it may refer (with some artistic license) to a prominent passage in the *Jahangirnama* account of the eleventh regnal year, which began on 19 March 1616.[1] The narrative describes a particularly important ruby-and-pearl armlet created for the Emperor, and provides extraordinary evidence for Prince Khurram's sensitivity to and knowledge of jewels. Abid, a painter known for his artistic independence, actually shows Jahangir wearing a pair of armlets, in which single rubies are flanked by paired pearls, rather than the single armlet described in the text. On the other hand, among the attendants is Karan Singh of Mewar, who had attended court between 15 February and 5 June 1615 in homage to the Emperor after Khurram had defeated Karan Singh's father, the Rana Amar Singh (see **5**). Karan Singh returned to the court only once, from 19 March to the end of April 1616, and could therefore have been present when the armlet was first worn.

Abid frequently portrayed noblemen and sages with lively and quite unorthodox poses and expressions; in this case, note especially the grinning courtier at the bottom left (Fig.39).[2] A second man, depicted in the circle of nobles to the Emperor's right in the inner enclosure is almost certainly Asaf Khan, in a very unusual, but not unique, three-quarters profile portrait (Figs 119, 120). The unidentified figure at the lower right shows that the artist could rival his brother, Abu'l-Hasan, in his extraordinary mastery of the techniques of modeling and the evocation of three-dimensional form (Fig.41).[3] ❧ MCB

This painting is the first in a group of three *darbar* scenes inserted at this point in the narrative of Shah-Jahan's move to Daulatabad in 1635, illustrating 'flashbacks' that evoke his princely achievements and early triumphs in the Deccan.[4] The *darbar* scene, which has not yet been identified with certainty, was used here to illustrate the account of Prince Khurram's departure from Ajmer for his first Deccan campaign in November 1616, when Jahangir awarded him the title *Shah*.[5]

The Mughals followed a careful policy with regard to the award of honorific titles appropriate

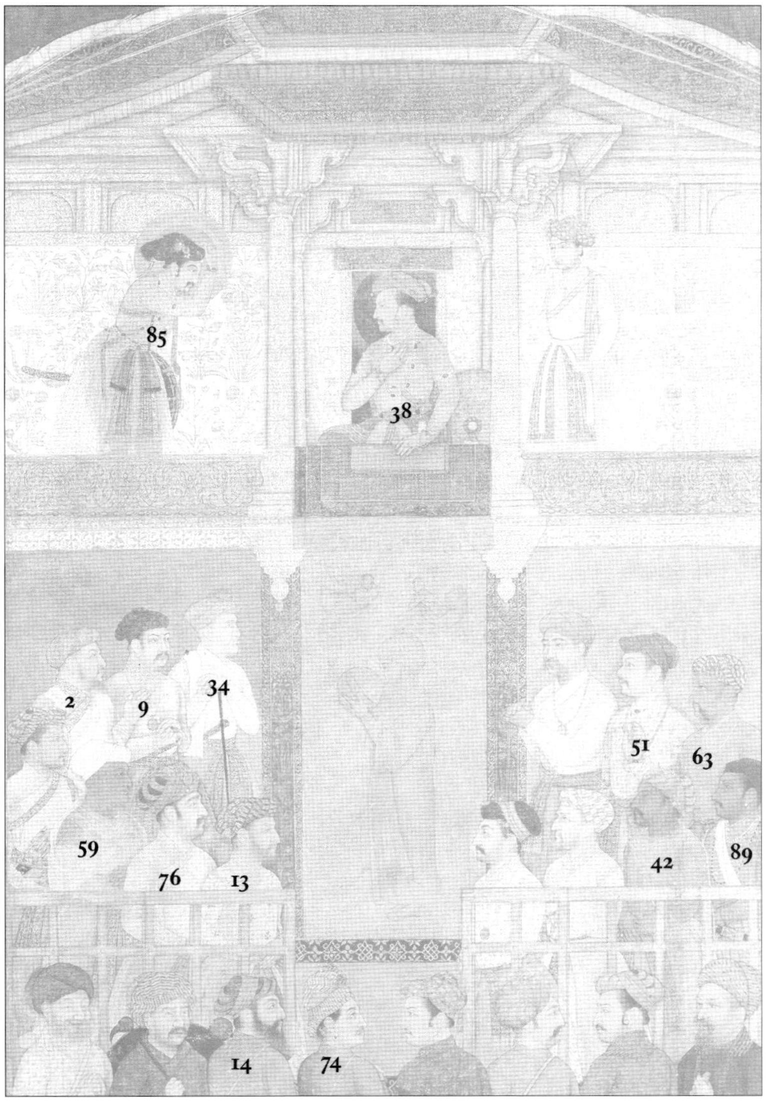

2. (Khwaja Abu'l-Hasan) (*Mir Bakhshi*)
9. (Asaf Khan)
13. (A'zam Khan) (*Mir Saman*)
14. (Bahadur Khan Rohilla)
34. (I'timaduddawla) (*Wakil* and *Diwan*)
38. (The Emperor Jahangir)
42. (Karan Singh Sisodia)
51. (Mahabat Khan)
59. (Muqarrab Khan)
63. (Murtaza Khan)
74. (Riza Bahadur)
76. (Sadiq Khan)
85. (Shah-Jahan [Prince Khurram])
89. (Suraj Singh Rathor)

Fig.119 Asaf Khan (detail of **37**).

Fig.120 Unknown artist, *Asaf Khan*, circa 1616.

to services rendered or expected.[6] The extraordinary distinction of the title of *Shah*, which 'no king during his own lifetime had ever bestowed on a son,'[7] shows the high favor in which the Prince stood with his father at this time and the expectation that he would subdue Malik Ambar, the Abyssinian general of the Nizamshahis, who challenged Mughal power in the Deccan.[8]

This scene has been convincingly attributed by Milo Beach to Abid, an artist with unorthodox tendencies who resisted total conformity to the conventions of the official court style. Here, Abid also takes his own approach to the established formula of the emperor holding court in the *jharoka*. He renders the court session in a minimalistic manner with fewer architectural elements and participants. The wall with the *jharoka* is almost identical with the picture plane; this brings us closer than usual to the imperial level of the *jharoka* and leaves only a narrow strip at the bottom of the page for the courtiers, who are therefore arranged in a very steep formation. To create some depth, Abid reduces the size of the figures proportionally towards the background, a rare instance among the Windsor *darbar* scenes where this illusionistic approach is placed above hierarchical considerations in the scale of figures. On the other hand, the painter conforms fully to the established *darbar* formula in the way he observes mirror symmetry, in the architecture as well as in the figure arrangement. No other Windsor *darbar* – not even Bichitr's accession *darbar* (**10–11**), which is a paradigm of Shah-Jahani symmetry – shows such a perfectly balanced figure arrangement. He breaks the rhythm of hierarchically ordered profiles, however, with a few three-quarter-view portraits, including that of Asaf Khan. It is hard to find a satisfactory reason why the most high-ranking noble of the empire should be shown in this less prestigious form, usually reserved for characters with low or no rank.[9]

The *jharoka* and its gallery are covered with fine gold ornaments in the same manner as the monochrome margins (*hashiyas*) of contemporary books or albums. It is as if the painter wanted deliberately to draw attention to the parallels between architectural decoration and such margins – the flowery plants are indeed a motif common to both areas of the decorative arts.[10]

The image below the *jharoka* – a drawing of a shaykh on a turqoise ground – makes the usual symbolic comment on Mughal rule. It refers to a specific allegorical interest of Jahangir, the then ruling emperor. Jahangir and his artists created Mughal allegory, for which techniques of European Mannerist allegory were adopted to express, in painting, concepts of rulership which had hitherto been expressed only in writing.[11] Jahangir explored in particular the theme of spiritual and worldly authority by associating himself with the images of Sufi shaykhs.[12] Abid evokes this connection with his large image of a shaykh holding up to the Emperor and his son a globe as a symbol of world rulership.[13] Divine blessings are guaranteed by the angels hovering above him. EK

1. See page 92 above.
2. He may be Muhammad Riza Kashmiri, who is portrayed with a very similar expression by Bichitr (Arnold & Wilkinson 1936, pl.60). See also the Shaykh in *Shah-Jahan enthroned with Mahabat Khan and a shaykh* (Fig.142).
3. For discussions of Abu'l-Hasan, see Verma 1994, pp.47–55; Beach 1978, pp.86–92; and Beach 1980, pp.7–33.
4. *Cf.* **8–9**, **38**, and **39**.
5. *Tuzuk-i-Jahangiri*, pp.337–38; Prasad 1922, pp.270–71.
6. Bosworth 1986, p.629.
7. *Shah Jahan Nama*, p.7.
8. Saksena 1976, pp.20–21. See also **15**, **16**, **18**, **31**.
9. See page 136, above.
10. For this motif, see page 200, below
11. Koch 1982a, p.21; see also pages 136–37, above.
12. For the role of shaykhs in Jahangiri allegories see Skelton 1988, pp.183–84; he identifies the shaykh shown here as Mu'inuddin Chishti of Ajmer. The same imagery was adopted by Shah-Jahan in **10–11**, **41–42**, and **44**.
13. For this symbol, see pages 200 and 203, below.

38
Jahangir receives Prince Khurram on his return from the Mewar campaign

Ajmer, *Diwan-i 'Amm*, 20 February 1615
Folio 194b painted by Murar, *circa* 1640
Image area 34.2 × 23.4 cm

Inscriptions: [on wall beneath Emperor] *Chu ârâsta kardam în bazm-i shâh / zi dargâh-i wâlâyam ummedvâr / chu altâf-i shah sarfarâzam kunad / taraqqî numâyad yadam dar nigâr* ('When I adorn this assembly of the king, I am hopeful of the exalted court. When the king's kindnesses ennoble me, my hand advances in depiction')
[on lower wall] *'amal-i zarra-i beqadr-i bekas / shâgird-i Nâdiru'z-zamân Murâr* ('work of the destitute, worthless speck of dust, pupil of Nadiruzzaman, Murar')

References: *Padshahnama*, 1/1, pp.174ff; *Shah Jahan Nama*, p.6; *Jahangirnama*, folios 107a–108a; *Tuzuk-i-Jahangiri*, 1, pp.276–78

Fig.121 Attributed to Murar, *Jahangir receives Prince Khurram on his return from the Mewar campaign*, circa 1640.

WHILE IT IS PLACED within the *Padshahnama* volume to correspond to text describing the granting to Prince Khurram of the title *Shah-Jahan*, the presence of Karan Singh of Mewar at a full court ceremonial suggests that the scene actually shows (once again) the return of Khurram from the Mewar campaign. This illustration depicts a slightly different moment during this extended audience to the other painting of the event used elsewhere in the manuscript (5). There we see the moment where Karan Singh arrives and is greeted; here, however, he has moved to join the nobles placed on the Emperor's right, as described in Jahangir's own account: 'After prostrating and salutations were completed, at the request of my son Khurram, I ordered them to place him in front on the right hand of the circle.'[1]

A drawing closely related to this illustration is in the Museum of Fine Arts, Boston (Fig.121). While a few minor adjustments of posture or placement have been made, the drawing (which has been trimmed) is so similar to the painting in individual expressions and execution that it too must be by Murar, and may be a preliminary study. And, as in Murar's *Jahangir receives Prince Khurram on his return from the Deccan* (9), the depiction of Karan Singh in both this scene and the related drawing follows the artist's adoption of the portrait formula developed by Abu'l-Hasan (see page 165). MCB

Since the *darbar* shows an event in Shah-Jahan's princehood, the generic *jharoka* setting is characterized with typical Jahangiri elements. Murar conforms to the other Jahangiri *darbar* scenes in the manuscript,[2] providing us with a view of awnings at the ceiling to indicate that, in Jahangir's time, the hall in front of the *jharoka* was merely a tent. The architectural decor is also historicizing. The *jharoka* is flanked by copies after European paintings, put up between the brackets supporting the roof of its gallery. To the right of the imperial group is the image of the Virgin Mary, and on the left that of Christ holding a globe. The latter image much fascinated Jahangir – it was, after all, a pictograph of the name he adopted at his accession, *Jahan-gir*, ('the Worldseizer') – and he borrowed it and other European globe imagery to symbolize his own world rulership.[3] Murar's retrospective setting also corresponds almost exactly to the decoration of the *jharoka* at Agra described by the Jesuit father Francis Xavier in a letter of 24 September 1608: 'Above ... on the right was a painting of Christ Our Lord with the globe of the world in His left hand, and on the other side Our Lady very well painted ... on either side of Our Lord and of Our Lady were various Saints in a posture of prayer.'[4] Such blatant 'Europeries' were not confined to Christian imagery. They were part of extensive programs of figural decoration in all of Jahangir's palaces that included life-size portraits of Jahangir and his family,[5] his nobles and foreign personalities.[6] European-style images served also as vehicles to express Islamic concepts. The outermost figure of the series to the right of the *jharoka* holds a peacock and thus alludes – as a part representing the whole – to a distinct pictorial program which had a considerable impact and was first realized for Jahangir in the palace of Lahore. This is the theme of Solomon on his flying throne, accompanied by his retinue of birds and spirits; it was meant to glorify Jahangir as a second Solomon.[7] The imagery with which Murar decorates the *jharoka* wall thus refers to identifiable historical decorations. We do not know if his particular scheme was fictional or if such a permanent mural decoration indeed existed in the fort at Ajmer, where the scene is supposed to have taken place.[8] Shah-Jahan largely refashioned Jahangir's palaces because he disapproved of his father's taste as 'old fashioned and of bad design [*kuhnagi wa bad-tarhi*].'[9] As Emperor he took a more orthodox stand and reverted to the floral decoration sanctioned by Islamic law.[10] The use of figural images, in the paintings as in real architecture, was confined to the symbolic imagery of the *jharoka-i khass-u-'amm*.[11] A major feature of Shah-Jahani floral decoration, which was not confined to architecture but extended to all aspects of the court, including dress, was naturalistically observed flowers.[12] These carried a strong symbolic message: they were to illustrate the flourishing of the country under Shah-Jahan's just and benign rule when 'Hindustan became the flower garden of the earth.'[13]

This is one of the paintings in the Windsor manuscript on which the artist puts his signature in the most prominent position imaginable, right below the Emperor in his *jharoka*.[14] In addition, Murar brings to the attention of the Emperor – in the place usually reserved for the symbolic imagery proclaiming the justice and peace of the Great Moghul's rule – his hopes for remuneration (for which he offers in return a satisfactory artistic performance), couched in polite court language. This sheds a significant light on the special relationship between the imperial patron and his painters. The painters were part of the royal household with direct access to the ruler, even in the formalized court of Shah-Jahan. The Emperor met his artists daily to supervise their work, acting as his own artistic director.[15] The iconography and the style of the *Padshahnama* illustrations must be understood as the product of this close collaboration. EK

1. *Tuzuk-i-Jahangiri*, 1, pp.276–77. According to Wheeler Thackston, the Persian text (*Jahangirnama*, fol.108a) actually reads, 'I ordered that they should have him stand forward to the left of the *jharoka*,' or possibly '... forward of the left-hand *jharoka*.' This is quite different to the translation quoted, and contradicts both what is shown in the pictures and the 1864 edition of the *Jahangirnama*'s Persian text (I am grateful to Ebba Koch for this reference). In any case, the two paintings discussed show two different moments of the ceremony, and contemporary texts imply that Karan Singh was moved from one side of the assembly (where he was greeted) to the other (where he then stood).
2. *Cf.* 5, 8–9, 37 and 39.
3. Of the images so far known which show Jahangir holding a globe, the most striking is the life-size portrait sold recently; see Sotheby's 1995, lot 85, pp.74–83, where the globe as Mughal symbol is discussed, with further literature; see also

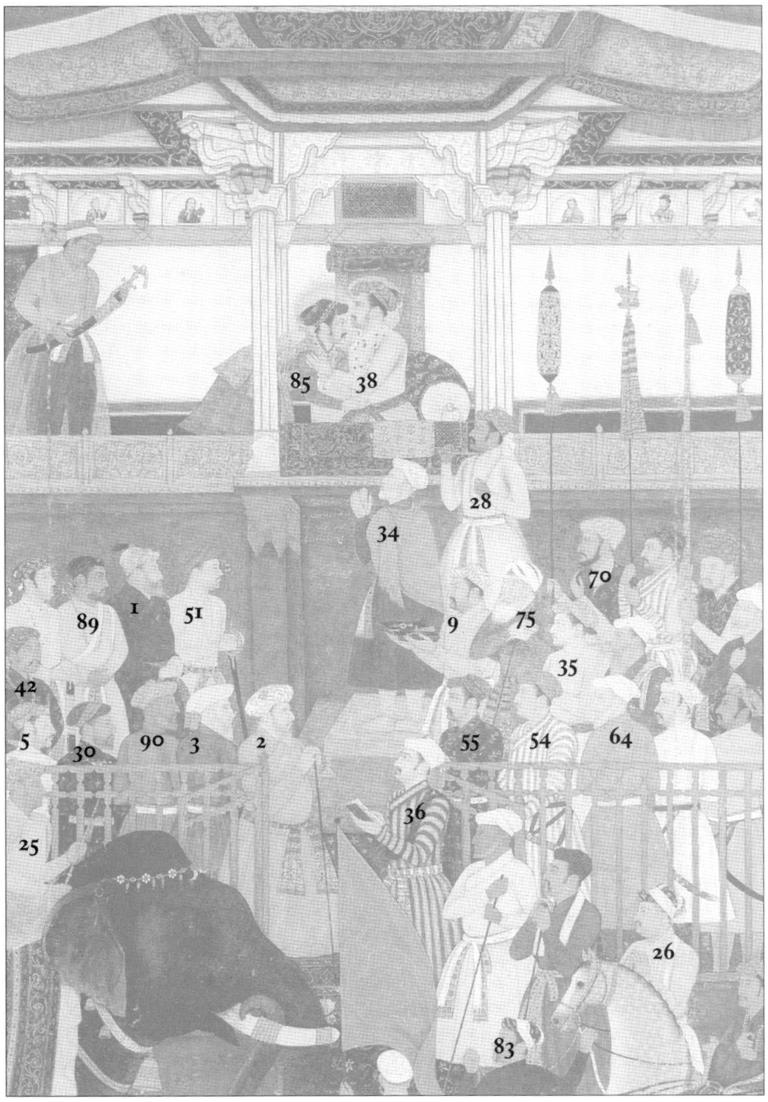

1. Abdullah Khan
2. Khwaja Abu'l-Hasan (*Mir Bakhshi*)
3. Afzal Khan
5. Allahverdi Khan
9. Asaf Khan
25. Gajpat Khan
26. Gridhra La'l
28. Hayat Khan
30. Islam Khan Mashhadi
34. I'timaduddawla (*Wakil* and *Diwan*)
35. I'tiqad Khan
36. Jadu Khan
38. (The Emperor Jahangir)
42. Karan Singh Sisodia
51. Mahabat Khan
54. Mirmiran
55. Mu'azzam Khan
64. Musavi Khan
70. Qilich Khan
75. Mirza Rustam Safavi
83. Sarla (?)
85. (Shah-Jahan [Prince Khurram])
89. Suraj Singh Rathor
90. Wazir Khan

page 137, above. For Jahangir's identification with Christ, see Koch 1982a, p.26, pl.5. For Muslim saints holding globes, and associated symbolically with Jahangir and Shah-Jahan, see **10–11**, **37**, **39**, **41–42**, and **44**.

4. Published by Fernao Guerreiro as *Relaçam*, v (1607–8, Lisbon 1611) and quoted in Maclagan 1990, p.239. William Finch, who was in India from 1608 to 1611, mentions images of Christ and Mary in the palace of Lahore (1985, p.163). For the role of Christ and Mary in the Islamic tradition, see page 180, note 7.

5. See note 3 above.

6. See the literature quoted on pages 189–190 below. See also Koch 1982a, in particular pp.25–26. For the surviving evidence see Koch 1983; Koch 1986; Cooper 1993. See also pages 136–37, above.

7. Koch 1983, especially p.186 and Fig.21. The flying retinue of spirits carry those members of Jahangir's court who cannot fly, including peacocks, through the air. See also **44**.

8. For the palaces of Ajmer, see page 162 above, note 5. The temporary decoration with English portraits of Jahangir's *darbar* at Ajmer was described by Thomas Roe in March 1616; the paintings were apparently taken to Mandu because they feature again in Roe's description of *Nawroz* celebrated there one year later. Roe 1990, pp.125–26, 357. See also Milo Beach's comments on page 202, below.

9. *Padshahnama*, 1/2, p.51.

10. Koch 1983, p.186. This is reflected in the *Padshahnama* illustrations; see **10–11**, **19**, and **32**.

11. The only figural program ever realized in a palace of Shah-Jahan's seems to have been the decoration of his *jharoka* in Delhi. Here the concept of Solomon as Koranic prophet-king and 'patron saint' of the figural arts served as an excuse for the depiction of animals, birds and Orpheus. See Koch 1988, in particular p.28.

12. Skelton 1972.

13. *'Amal-i Salih*, III, p.24; Koch, forthcoming (a).

14. For further examples of similarly placed signatures, see **5** (also below the *jharoka*); **8–9** (on the stepped stool leading to the platform below the *jharoka*); **10–11** (above the lamb within the symbolic painting); **12–13** (below the Emperor on the platform); **25–26** (below the throne platform); and **43** (on the platform below the *jharoka*).

15. The historian Qazwini gives the fullest list of the artists seen daily by Shah-Jahan, in which he puts the painters (*musawwir wa naqqash*) first; *Padshahnama*, fol.139a (refoliated 140a). See also pages 131–32, above.

39
Jahangir presents Prince Khurram with a turban ornament

Mandu, *Diwan-i 'Amm*, late 1617
Folio 195a painted by Payag, *circa* 1640
Image area 30.6 × 20.3 cm

Inscription: [on book held by figure, lower left] *Allâhu akbar. Raqam-i Payâg, barâdar-i Bâlchand* ('God is great. Drawn by Payag, brother of Balchand')

References: *Padshahnama*, 1/2, pp.187–88; *Jahangirnama*, folio 162a; *Tuzuk-i-Jahangiri*, I, pp.393–401
Published: S.C. Welch 1995, fig.1; Dye 1991, p.124 and fig.3; Beach 1980, pp.14–15 and figs 8–9

THREE FEATURES INDICATE that this is a particularly important presentation: the ceremonial gift of a jeweled turban ornament (*sarpech*); the fact that I'timaduddawla (who as *Wakil* and *Diwan* was the most important minister at court) is presiding over the event, rather than one of the *bakhshis*; and the presence of Bijapuri courtiers at the bottom right (suggesting that the event shown followed the fall of that Deccani kingdom). It could, however, be either of two ceremonies held within successive months following Khurram's return in triumph from the Deccani campaign (see also **8–9**): the award to the Prince of the title *Shah-Jahan* ('King of the World') in October 1617, or possibly the presentation to him, at the end of November, of an important family jewel, 'a good luck charm for this eternal dynasty.'[1] No detail has been isolated that might identify the scene more specifically, but it definitely does not correspond to the accompanying text of the Windsor manuscript, which describes events in 1622. On the other hand, a well-known portrait of the prince by Abu'l-Hasan shows him holding a turban ornament, and is inscribed both with the title *Shah-Jahan* and the information (in Shah-Jahan's own hand) that it was 'a good likeness of me in my twenty-fifth year.'[2] This dates that portrait also to 1617,[3] the year the title was given, and it almost certainly commemorates the same event. The association of the title *Shah-Jahan* with the gift of a dynastically important jewel seems clear; and no matter the exact incident depicted, confirmation of the title must be the actual subject.

According to Sir Thomas Roe's account of his

2. (Khwaja Abu'l-Hasan) (*Mir Bakhshi*)
3. (Afzal Khan)
9. (Asaf Khan)
31. (I'tibar Khan)
34. (I'timaduddawla) (*Wakil* and *Diwan*)
38. (The Emperor Jahangir)
59. (Muqarrab Khan)
69. Payag
76. (Sadiq Khan)
80. (Sarafraz Khan)
85. (Shah-Jahan [Prince Khurram])

Fig.122 Abu'l-Hasan, *Shah-Jahan in his twenty-fifth year*, circa 1618.

part in the *Nawroz* (New Year) celebrations on 21 March 1617, the throne at Mandu (the location for this event) was decorated with European paintings similar to those seen here: 'The Throne was the same vsed the last yeare, and all the other furniture. At the vpper end was set the King my Soueraignes Picture, the Queenes, my Lady Elizabeths, Sir Thomas Smiths and some others, two pieces of good and fine Tapestrie below them that came from Persia; a Throne of Gold set all ouer with Rubies, Emeralds and Turqueis; and the old Musicke of singing Whores.'[4]

While Northern European prints had been brought to the Mughal court by missionaries and traders from the mid-sixteenth century, English paintings came to be available around the time of Roe's embassy.[5] An official portrait of James I, almost certainly a gift from Roe to the Emperor, was copied by the painter Bichitr,[6] while Jahangir is shown here wearing among his necklaces an oval portrait medallion of the type associated with Isaac Oliver and Nicholas Hilliard (Figs 123, 124).[7] Roe had brought with him to India a picture of a woman by Oliver, and it was shown to the Emperor in an often-quoted incident: 'Asaph chan [Asaf Khan] asked mee for my little Picture and presented it to the King. He tooke extreame Content, showeing it to euerie man neare him; at last sent for his Cheefe Paynter, demanding his opinion. The foole answered he could make as good. Whereat the king turned to mee, saying: my man sayth he can do the like and as well as this: what say yow? I replyed: I knew the Contrarie.'[8]

This exchange led to a contest, with Roe being asked to recognize the original among five copies: 'At night hee sent for mee, beeing hastie to triumph in his woorkman, and shewed me 6 Pictures, 5 made by his man, all pasted on one table, so like I was by candle-light troubled to discerne which was which; I confesse beyond all expectation; yet I shewed myne owne and the differrences, which were in arte apparent, but not to be iudged by a Common eye.'[9]

The very sketchy and direct quality of execution visible in the background figures of such scenes as *Khan Dawran receiving the heads of Jujhar Singh and his son Bikramajit* (**36**), which has direct counterparts with European techniques then being introduced into India by traders and missionaries, is here extended to the depictions of courtiers. Unlike the *darbar* scene by Murar (**38**), in which the figures have carefully controlled silhouettes within which textile patterns are meticulously defined, Payag's work allows us to sense the independent texture of the pigment as well as the movement of his brush. The visual liveliness which results is matched by the variety of poses in which figures are placed, and the artist's intense interest in unusual physiognomies. MCB

Shah-Jahan's rise as a prince culminated in his victorious return from his first Deccan campaign in October 1617, after he had made a favorable peace agreement with Jahangir's arch enemy, Malik Ambar, and with the Adil Shah (see **37**). The Deccanis agreed to pay a rich tribute and to return territories they had taken from the Mughals, in particular the coveted Ahmadnagar. Jahangir, who had awaited the outcome of the campaign at Mandu, was overjoyed and showered his son with presents and honors. The Prince, already styled Shah Khurram at the onset of the campaign, was elevated to an even grander title, *Shah-Jahan* ('King of the World'),[10] which he was to keep after his accession as a claim to universal rule.[11] Shah-Jahan was confirmed as never before as the next heir to the Mughal throne. The ceremonial presentation of a turban ornament to the Prince, a special mark of distinction, apparently served the same purpose.[12]

This *darbar* scene shows Payag at the peak of his art as a realistic miniaturist, exploring the artistic freedom which an idiosyncratic painter of his power could carve out for himself within the confinement of the prescribed formulas of official court representations. His innovations are innovations of degree, yet they suffice to give his *darbar* a totally different quality from any other audience scene among the *Padshahnama* illustrations, including his own (**43**). Payag's *darbar* is the antithesis of Bichitr's canonical version; it focuses on the human element. In Bichitr's painting (**10–11**) the courtiers appear as a collection of self-contained, idealized individual portraits, unified not psychologically – they look straight ahead and their views meet only because they are placed in two groups facing each other – but solely through their subordination to the dominant figure of the Emperor.[13] This subordination is brought about by mechanical means in the form of the architectural framework, the function of which is not to create naturalistic space but to provide vertical hierarchy and centralizing symmetry.

Fig.123 Jahangir wearing a miniature portrait medallion (detail of **39**).

Fig.124 Isaac Oliver, *An unknown lady*, circa 1615.

Payag eliminates architecture and perspective as far as possible. He also minimizes symmetry: not even the flowers flanking the *jharoka* on the back wall are arranged symmetrically.[14] The participants at the *darbar* are divided into two unequal groups, allowing us an oblique glance to the center and the symbolic imagery below the *jharoka* where, perhaps in allusion to Shah-Jahan's new title, the globe receives detailed attention.[15] Instead of relying solely on mechanical means to relate the figures to each other, Payag tries to introduce an inner logic into the abstract *darbar* image and to link the nobles psychologically to the imperial protagonists, to each other, and to us, the viewer. Most of them look upwards. Payag could not, of course, go so far as to have them look directly at the Emperor and his son because this would have distorted their faces and postures,[16] and they therefore look only approximately upwards, casting their glances across the picture.

While Payag strives for psychological unification, his characters become more self-conscious and autonomous. Several of them, not only the nameless standard- and mace-bearers, even turn their attention out of the painting towards the beholder, as if openly to acknowledge his presence. Payag's recognition of the individuality of those taking part in the *darbar* gives him occasion to bring to these superb psychological studies his interest in variations in physiognomy, and his experiments with the effect of emotion on facial muscles.[17] At the same he gives his portrait figures new physical weight by modeling their faces and bodies within the confinement of their contours. In this treatment of figures Payag had no peer.[18]

The artist focuses on the individuality of his subjects not only to give them a stronger, more self-willed presence in the group; he also explores the potential of portraiture to support the prescribed hierarchical content of the painting through expressive means. While the likenesses of Jahangir and Shah-Jahan 'above' are recognizable but strongly idealized 'to show the ruler as a figure who existed at once on a real and an ideal plane,'[19] the more realistic portrayal of the nobles 'below' distances them visually from the imperial protagonists and assigns them to a more human, and less elevated sphere. EK

1. See page 96.
2. Guy & Swallow 1990, no.56.
3. Khurram was born in January 1592. While the phrase 'in my twenty-fifth year' could indicate the months following his twenty-fourth birthday, both the composition and the inscriptions suggest that he is referring instead to the period following his twenty-fifth birthday in January 1617.
4. Roe 1967, p.394.
5. See Beach 1980.
6. See Beach 1981, no.17a; and Beach 1980, figs 5–6. English pictures, including images of James I, arrived in India before Roe's embassy. In 1613, a fleet of five ships was outfitted by the East India Company, and included as presentation items were oil paintings of the King, Sir Thomas Smythe and others (Downton 1939, pp.xii–xiii.) The presence of Smythe's portrait in the decorations at Mandu, as noted by Roe, suggests that many of these works actually reached the court.
7. A variety of early seventeenth-century English portraits were certainly known in India. One, in the style of Isaac Oliver but mounted on an eighteenth-century Indian album page, is in the Victoria & Albert Museum, London (inv. no. I.S. 60-1978).
8. Roe 1967, p.214.
9. Roe 1967, p.225.
10. *Tuzuk-i-Jahangiri*, I, pp.380–81, 393–401; Prasad 1922, pp.281–86; Saksena 1976, pp.21–23.
11. Some aspects of the Mughal concept of world rulership are discussed in Koch 1982b, pp.261–62.
12. Soudavar 1992 (p.311, no.129a and b) argues that the transfer of the turban ornament corresponds to the transfer of the handkerchief, a Mongol insignium of kingship. *Cf.* Soudavar, forthcoming, Fig.25.
13. This interpretation is based on Alois Riegl's characterization of Dutch group portraiture (Riegl 1902; for a critical resumé in English, see Iversen 1993, chapter 6; see also Binstock 1995). Although Riegl's masterly study was developed for an entirely different cultural context, I feel justified in turning to it for hermeneutic inspiration, because the systematic use of the group portrait in Shah-Jahani painting is as unique in Islamic or Indian art as is its Dutch counterpart against the background of seventeenth-century European art.
14. For the floral and figural decoration, see page 200, above.
15. For the symbolic imagery see pages 168 and 199, and pages 136–37, above.
16. This point is discussed in more detail on page 136, above.
17. This even includes the symbolic animals in the 'painting within the painting.' Payag explores their physiognomies to comment ironically on imperial propaganda. The expression of the ox leaves us in no doubt how it feels about having to lay down with the ferocious lion to illustrate the justice of Shah-Jahan!
18. *Cf.* **18**.
19. Warnke 1993, p.215; his insights into European court portraiture (pp.212–24) are also valid in this context.

40

The surrender of the fort at Udgir to Khan Dawran

Deccan, October 1636
Folio 204b painted by Dhola, *circa* 1650
Image area 34.8 × 24.8 cm

Inscription: [on quiver, bottom left]
'amal-i Dola ('work of D[h]ola')[1]

References: *Padshahnama*, I/2, pp.217–19; *Shah Jahan Nama*, pp.192–94

An inner gold ruling around all edges of the illustration remains within the usual marginal rulings found surrounding this and other *Padshahnama* scenes. This would have been the original margin for the illustration in the manuscript for which it was first intended.

WHILE THE JUXTAPOSED text relates to the capture by Khan Dawran of the forts at Udgir and Ausa in the Deccan, there is nothing in this scene that corresponds specifically to the *Padshahnama* description of either event. Nonetheless, the illustration *is* specific: the captured courtiers are recognizably Deccani because of their costumes, and they are being brought out of the fortress with wrists tied together for presentation to Khan Dawran. (The Rajput courtier in orange at the bottom may be Mirza Raja Jai Singh, who is known to have taken part in the Deccani campaigns.) This scene must therefore refer to the known details of a particular event, although this is presently unidentifiable.

Only two other works by Dhola are known, and both indicate that he was painting late in the Emperor's reign.[2] His style introduces a movement away from the concerns and achievements of imperial painting under Shah-Jahan. Comparison of the skies here and in *Shah-Jahan hunting* (**33**), for example, shows the light, atmospheric effects so deftly refined in the somewhat earlier work being replaced by a solid, decorative sky that closes in, rather than extends, space. Similar skies of solid gold streaked with brilliant orange became the norm in painting under Awrangzeb and later emperors such as Muhammad Shah (r.1719–1748), and were also adopted by Rajput artists. ❧ MCB

In January 1636 Shah-Jahan again led a campaign against the Deccan to bring about the final subjugation of the kingdom of the Nizamshahis of Ahmadnagar, which, in the hands of the Maratha leader Shahuji Bhonsla, had recovered from its great defeat of 1633 (see **31**). After a series of Mughal victories the rulers of the adjoining sultanates of Bijapur and Golconda made a peace treaty with Shah-Jahan. In July of the same year, when the Emperor had taken up quarters at Daulatabad, Awrangzeb was appointed governor of the Mughal possessions in the Deccan, being entrusted in particular with the task of reducing Shahuji. He succeeded in doing so when Khan Dawran conquered the forts of Udgir and Ausa in October 1636. This completed the extinction of the Nizamshahi kingdom, which became a *suba* (province) of the Mughal empire.[3]

Dhola's painting is traditionally identified as the surrender of Udgir. His treatment of the subject is similar to that of the capture of Orchha (**35**), though with a less dramatic naturalistic landscape in the background. The pinkish Deccani fort is rendered freely with much attention to detail,[4] minutely depicted in thin layers of paint. One notices in particular the monumental open stairway, on which four almost insubstantial Deccani defenders descend to join their companions in misfortune assembling behind the main gate before leaving the fort in surrender. While the figures of the Deccanis gain progressively more substance as they move forward to meet the Mughal commanders, they are shown largely in frontal and oblique views, in stark contrast to their Mughal conquerors who are rendered in opaque local colors and in more prestigious profile views. Once again a painter of the imperial atelier deliberately manipulates opposing styles of representation to support by formal means the message of the painting, victory and defeat.[5] ❧ EK

1. The name is probably Dhola. (WT)
2. The others are the marginal figures on a Late Shah-Jahan Album page of about 1650 (Binney 1973, no.60) and a portrait of the same date identified, in the Emperor's hand, as that of Sayyid Hidayatullah Sadr and probably commissioned in honor of the grant of the title *sadr* in 1648 (Soudavar 1992, no.129 c). A further inscription on the latter work has occasionally been misread as an attribution to the distinctly different painter Dawlat (for example, in Verma 1994, p.129).
3. *Shah Jahan Nama*, pp.190–94; Sarkar 1972–4, I and II, pp.23–24; Sherwani & Joshi 1973–4, pp.275–77; Saksena 1976, p.147.
4. For a brief description of the actual fort see Navarane 1995, pp.217–18, 366 (illustration), figs 104, 105 (site) and fig.110 (plan). For the history of the fort, see Ahmad 1929–30.
5. For a general discussion of these principles see page 136, above.

41–42

Shah-Jahan visits the shrine of Khwaja Mu'inuddin Chishti at Ajmer

Ajmer, November 1654
Folios 205b–206a attributed to the 'Kashmiri Painter,' *circa* 1656
Image area 35.2 × 22.4 cm (folio 205b); 36.6 × 25.8 cm (folio 206a)

References: *'Amal-i Salih*, III, p.193; *Shah Jahan Nama*, p.502 and fig.22
Published: Leach 1987, pp.128–29 and fig.7

SEVERAL ALLEGORICAL IMAGES showing Shaykh [Khwaja] Mu'inuddin Chishti presenting the imperial crown to Jahangir or Shah-Jahan are known, for the Shaykh, who lived *circa* 1200, represented the spiritual source of Mughal imperial power. Robert Skelton has consequently identified the figure encountered by the Emperor in this scene as the Khwaja,[1] although the mystical Sufi saint Khizr (Figs 125, 126) is a more likely subject. Because he had drunk the waters of immortality, Khizr is associated with water, and here, as elsewhere, he is shown standing in a stream. Interpreted as an emblem of fertility, it is also thought that his appearance would cause nature to become green and lush.[2] Moreover, his name, *khizr*, is etymologically connected with the Arabic word for green (*akhzar*).[3] It is particularly appropriate, of course, that the aged Shah-Jahan should be shown encountering a symbol of immortality near the dynastic shrine of Ajmer, and the painting confirms that artists were not limited by strict historical reality when designing images for the emperors' historical chronicles.

While the distinctively rough style of these two pages indicates that they are by the same hand, differences of scale and size suggest that they may not have been intended to form a double-page. The date of the episode, 1654, also indicates that such a style was practiced during the final years of Shah-Jahan's reign. ❧ MCB

Fig.125 Unknown artist, *Khwaja Khizr*, circa 1640.

Fig.126 Khwaja Khizr (detail of **41**).

The importance accorded to the tombs of Sufis is a well known aspect of Indo-Muslim piety. It was believed that a measure of the *barakat* (spiritual charisma and power) of a saint persisted in his tomb, to which those who required his intercession would resort. The imperial Mughal ritual of pilgrimages to Chishti shrines (*dargahs*) reflected the specific relationship between Sufi shaykhs and rulers as exponents of spiritual and worldly power, which was already an object of continuous reflection and discussion during the Delhi Sultanate.[4]

Of the Mughals it was Akbar who first expressed special reverence for the Chishti saints as the most renowned Sufi order on the subcontinent. This was motivated not only by his personal religious convictions but also because it increased his prestige in the eyes of his subjects and secured broader support for his rule.[5] Particular emphasis was given to the emperor's pilgrimages to the *dargah* of Khwaja Mu'inuddin Chishti at Ajmer.[6] By visiting and supporting the tombs of Chishti shaykhs, Akbar's successors continued to seek their blessing and to associate themselves with their spiritual power.[7]

These two paintings were inserted into the Windsor manuscript to illustrate Shah-Jahan's visit to Ajmer in December 1636 when, on his return from Daulatabad to Agra (see **31**), he inaugurated a recently completed mosque, his main architectural addition to the *dargah*.[8] However, the ages of the Emperor and of Dara-Shikoh, who rides with him, clearly indicate that what is actually illustrated is the Emperor's last visit to Ajmer, namely that of November 1654.[9] This pilgrimage was – as the historians tell us – not entirely motivated by religious piety but was seen as an occasion to suppress the unauthorized activities of Rana Raj Singh of Udaipur, who had begun to rebuild his ancestral seat of Chittorgarh in violation of the imperial prohibition that had been in effect since Akbar's conquest of the fortress in 1568. The Rana was brought to bay, with Dara-Shikoh playing a prominent part in the negotiations. After his successful dealings with the Rana, Shah-Jahan went to Ajmer to visit the *dargah*, where he ordered the construction of a large gate with an inscription, proclaiming the victory of the 'sun of faith' over the darkness of infidelity.[10]

The imperial procession rides through the hilly region of Ajmer. Shah-Jahan is preceded on the left-hand page by part of his retinue, depicted on a smaller scale. The figures in the 'lesser' half of a double-page composition are often shown in this way (see **6–7**, **12–13**, and **27–28**) for reasons of hierarchy or perhaps, as Milo Beach suggests, because paintings not originally intended to appear together were combined. Footmen carry his ceremonial weapons, the *qur-i khassa*, wrapped up in cloth bags; they are followed by a line of horsemen, and behind them, by a line of elephants carrying other imperial insignia (see also **34**) as well as the *naqarakhana* or *nawbatkhana*, the orchestra consisting of kettle drums and other instruments.[11] On the right-hand page the Emperor passes the city of Ajmer; behind him rides his son, Dara-Shikoh. The mystical apparition of Khwaja Khizr directly in front of the Emperor clearly alarms the horses; their reaction is evident in the characteristic movement of their ears. The Emperor's head is surrounded by a particularly resplendent solar halo as if to underline his successful performance as 'the sun of faith.' The effect is heightened by the exquisite sunshade (*cf.* **29**)[12] decorated with a flowery plant and surrounded by orange and light-green flounces, which the Kashmiri Painter rendered with his customary masterful treatment of texture, a quality also apparent in the flowered robes of the imperial party and the saddle cloths of the elephants.

The painter's claims to topographical accuracy in his depiction of the city of Ajmer are not as great. All the buildings are entirely fictional; only the white color of the *dargah* complex in the center and the fact that the city is built at the foot of the Taragarh hill fort are true to the original. The landscape follows the scheme of a vertical plane structured as a succession of small mounds employed elsewhere in the Windsor manuscript (**34**).[13] The integration of rocky hills, intricately textured with various hatching patterns, is a more specific reference to the mountainous terrain of the region; the vivid green of the slopes on the right-hand page may indicate nature's response to the appearance of Khwaja Khizr.[14] As usual, the Kashmiri Painter populates his landscape with tiny, toy-like figures, amongst whom one notices two young men embracing across the backs of their horses, just behind the two towers of the city wall near the left margin. ❧ EK

1. Skelton 1988, p.184 and n.63; for a double-page image of Shaykh Mu'inuddin with Jahangir, see figs 3–4.
2. In this, Khwaja Khizr corresponds to the youthful Hindu ascetic Rishyasringa, whose appearance in early sections of the *Ramayana* is always accompanied by a gentle rain sent to nourish the earth; see, for example, Shastri 1962, I, p.28.
3. I am grateful to Wheeler Thackston for this observation.
4. Digby 1990.
5. Richards 1981, pp.256–58; Streusand 1989, pp.89–91, with further literature.
6. Currie 1989, with further literature.
7. Koch 1993, pp.13–14.
8. *Shah Jahan Nama*, pp.195–96. For the mosque, see Koch 1991a, pp.120–21; for mosque and *dargah*, see Asher 1992, pp.174–78.
9. *Shah Jahan Nama*, pp.501–502.
10. Tirmizi 1968, p.51; Asher 1992, p.178.
11. For the *naqara* as an imperial privilege, see page 166 and the literature quoted on page 195, note 5; *A'in-i Akbari*, I, pp.52–54; and *Tuzuk-i-Jahangiri*, I, p.205.
12. For the concept of shadow in the 'light rulership' of the Mughals, see page 160.
13. See also pages 139–40, above.
14. In his explanation of the color symbolism at Humayun's court, Khwandamir associates the color green with prophets and points out that Khizr 'obtained this appellation [green] because verdure grew round him wherever he sat. And it is said that on one occasion he sat on white leather, and through the grace of his feet this also became green'; *Qanun-Humayuni*, p.53.

43

Shah-Jahan honoring Prince Awrangzeb at Agra before his wedding

Agra, *Diwan-i 'Amm*, 27 April 1637
Folio 214b painted by Payag, *circa* 1640
Image area 34.1 × 23.4 cm

Inscription: [on platform below Emperor] *'amal-i Payâg* ('work of Payag')

References: *Padshahnama*, I/2, pp.255ff; *Shah Jahan Nama*, p.206 and fig.23
Published: S.C. Welch 1995, p.322 and fig.5; Dye 1991, p.124 and fig.4

3. (Afzal Khan) (*Diwan*)
9. (Asaf Khan) (*Wakil*)
12. (Prince Awrangzeb)
20. (Prince Dara-Shikoh)
28. (Hayat Khan)
37. (Ja'far Khan)
44. (Khan Dawran)
45. (Khan Jahan Barha)
52. (Makramat Khan) (*Mir Saman*)
60. (Prince Muradbakhsh)
64. (Musavi Khan) (*Sadr*)
65. (Mu'tamad Khan) (*Mir Bakhshi*)
68. (Shaykh Nazir)
80. (Sarafraz Khan)
85. (The Emperor Shah-Jahan)
87. (Prince Shah-Shuja')
88. (Shayista Khan)

IN ADDITION TO Prince Muradbakhsh, all the nobles mentioned in the *Padshahnama* account of this episode are included in this scene:[1] Khan Dawran, Sayyid Khan Jahan Barha, Shayista Khan, and Ja'far Khan. Other figures include the ministers shown in the centre: Asaf Khan, the *Wakil*, Afzal Khan, the *Diwan*, Mu'tamad Khan, the second *bakhshi*, Makramat Khan, the *Mir Saman*, and Musavi Khan, the *Sadr*. The only absentee is Mir Jumla, the *Mir Bakhshi*, who died the following August and was clearly represented here by Mu'tamad Khan.

In *The wedding procession of Prince Dara-Shikoh* (**23–24**), a comparably powerful group of nobles escorts Dara, further indicating that the marriages of the princes were among the most important state occasions. Nonetheless, it seems that the full court *darbar* was not assembled for this occasion, which accords with Jadunath Sarkar's comment that upon the Prince's arrival in Agra, 'Shah Jahan wrote him a most loving invitation in verse to come and see him quickly and without ceremony. Next day the prince had audience of his father.'[2]

This is the least distinguished picture by the artist Payag within the Windsor manuscript. It lacks his ebullient brushwork and ability to isolate and intensify idiosyncracies of appearance and character found, for example, in **39**. ❧ MCB

The reception of Awrangzeb in April 1637 on the occasion of his wedding at Agra is one of four illustrations in the Windsor manuscript dedicated to him.[3] Although Awrangzeb was the Emperor's less favored son, he was entrusted with important assignments at an early age, beginning with the nominal command of the Bundela campaign in 1635, when he was only seventeen years old.[4] In July 1636 he was appointed governor of the Deccan (see also **40**). During his governship he visited the court four times; two visits – that to Agra on the occasion of his wedding in 1637 (**43, 45**) and that to Lahore in 1640 (**44**) – are recorded by illustrations in the Windsor manuscript. Awrangzeb's governorship in the Deccan was temporarily ended in 1644, when he resigned from his post feeling that he was not sufficiently supported by the Emperor, who openly favored his older brother Dara-Shikoh.[5] Shah-Jahan's favoritism proved entirely counterproductive. The fact that he kept Dara at court denied the Prince the opportunity to assert himself politically, while Awrangzeb was able to carve out for himself a position similar to that of Shah-Jahan's own princehood (*cf.* **5**). Dara could appear in the limelight only in the context of his wedding, which was therefore given much attention in the *Padshahnama* illustrations (**21–22, 23–24, 25–26**), while the importance of Awrangzeb's – whose achievements and position demanded his depiction on other occasions – was minimized (**45**).

This difficult situation is reflected in the rendering of Awrangzeb's reception on arrival from the Deccan. Payag does not depict the moment when

Shah-Jahan embraced the Prince and 'showered numerous tokens of his love and tender regard on him'[6] but uses instead all the means at his disposal to slight the Prince in the pictorial language understood by those who mattered at court. The Prince of the royal blood is shown like any noble (**32**), or like the foreign envoy Muhammad-Ali Beg (**17**), making an obeisance outside the railing which demarcated the area for those of the highest rank. Although the viewer knows that Awrangzeb and the dignitaries who are with him are about to enter this innermost area, the Prince is nevertheless fixed in the memory in this humble position while his brothers, in particular his arch-rival Dara, tower high above him in the imperial sphere.

The imperial reserve shown towards the subject of this *darbar* scene seems also to have dampened Payag's artistic dedication. He arranges his figures conventionally and gives his best only in the treatment of texture. Here he excels, in the textiles – which include a carpet with the justice symbol of lion and ox at the center of a floral pattern[7] – as well as in the extraordinary velvets worn by the bearers of the insignia on the right and, with a stunning flowered gold brocade, covering the railing of the *jharoka*.

The *jharoka* itself is of the conventional type which had, by 1637, been replaced by a loggia in the Agra palace (Fig.104). Payag actualizes it with columns topped by oblong panels with flowery plants in *pietre dure* which he carefully shaped after the originals that featured conspicuously in a larger form elsewhere in the refashioned palaces, in the Shah Burj tower and in the hall in front of the Imperial Hammam (Fig.88). ❧ EK

1. See page 104.
2. Sarkar 1972–4, I, p.50.
3. The others are **29**, **44**, and **45**.
4. See **35**; and Sarkar 1972–4, I and II, p.II.
5. Sarkar 1972–4, I and II, pp.29, 41–43.
6. *Padshahnama*, I/2, pp.255ff and page 104 above; see also *Shah Jahan Nama*, p.206.
7. For these symbolic images see also pages 168 and 203, note 17, with further literature; see also pages 136–37, above. Payag uses this carpet pattern of flowers and acanthus leaves, with variations, in his other *darbar* scenes; *cf.* **39** and the *darbar* painting in the Bodleian Library, Oxford, MS.Ouseley, Add.173, no.13 (Fig.149; also reproduced in Koch 1988, pl.58; *Shah Jahan Nama*, pl.5).

44
The arrival of Prince Awrangzeb at the court at Lahore

Lahore, *Diwan-i 'Amm*, 9 January 1640
Folio 217b painted by Murar, *circa* 1645
Image area 34.7 × 23.8 cm

Inscription: [on platform, bottom center] *'amal-i banda-i dargâh Murâr* ('work of the slave of the court, Murar')

References: *Padshahnama*, II, p.174; *Shah Jahan Nama*, p.264 and fig.6

Fig.127 Railing of the *jharoka* (detail of **44**).

Fig.128 Unknown artist, grotesque pattern from the printshop of Macé Bonhomme, 1566.

4. (Alimardan Khan)
9. (Asaf Khan) (*Wakil*)
11. (Asalat Khan) (*Bakhshi*)
12. (Prince Awrangzeb)
20. (Prince Dara-Shikoh)
28. (Hayat Khan)
29. (Inayat Khan)
37. (Ja'far Khan)
41. (Jaswant Singh Rathor)
52. (Makramat Khan) (*Mir Saman*)
60. (Prince Muradbakhsh)
64. (Musavi Khan) (*Sadr*)
78. (Safshikan)
79. (Salabat Khan)
85. (The Emperor Shah-Jahan)
86. (Shahnawaz Khan)
87. (Prince Shah-Shuja')
88. (Shayista Khan)
90. (Wazir Khan)

WHILE THE HISTORICAL basis of this illustration is very specific, once again there is great flexibility with fact. At the *jharoka* level, Awrangzeb is embraced by his father, while his three brothers and their elderly grandfather Asaf Khan (who would die the following year) stand in attendance. The setting is the audience hall at Lahore, but the small number of nobles, and their specific

Fig.129 *Jharoka* of the *Diwan-i 'Amm* of the Lahore fort, sixteenth or early seventeenth century.

Fig.130 *Jharoka* of the *Diwan-i 'Amm* of the Lahore fort (detail of 44).

connection to either Awrangzeb or Lahore, suggests a relatively intimate event. The nobles standing in the most prestigious location, on the Emperor's right, for example, are not the chief government ministers, but Alimardan Khan, the Governor of Kashmir and, from 19 October 1639, Governor of the Punjab as well; Shayista Khan (Fig.131), who had served as Awrangzeb's deputy in Daulatabad when the Prince returned to Agra for his wedding; and Shahnawaz Khan Safavi, the Prince's new father-in-law. Only then do we encounter Musavi Khan, the *Sadr*, Ja'far Khan; and the *Mir Saman* Makramat Khan. (The absence of Islam Khan, the *Diwan*, is unexplained.) On the Emperor's left, the chief noble is Wazir Khan, the *subadar* (governor) of Lahore, whose mosque is one of the great monuments of that city; he is accompanied by Maharaja Jaswant Singh Rathor, a cousin of the Emperor; Safshikan Khan, son of the older brother of Shahnawaz Khan and now a cousin of Awrangzeb by marriage. The imperial enclosure is guarded by *Mir Bakhshi* Salabat Khan and Asalat Khan, the second *bakhshi*, who served together in these positions from August 1639 to 4 August 1644. Because Wazir Khan died in AH 1050 (AD 1640–41), and Asaf Khan the following year (on 21 November 1641), the episode depicted here almost certainly occurred between August 1639 and early 1641.

According to known chronicles, however, there was no moment when the three brothers were together for a major ceremony in Lahore. Thus, while this illustration probably refers to Awrangzeb's return to the city early in January 1640, Shah-Shuja' was, in fact, in Bengal at the time. As with the depiction of Khwaja Khizr (**41–42**) or the inclusion of Karan Singh of Mewar at the scene of Khurram's return from the Deccan (**8–9**), painters – and perhaps Murar especially – seem frequently to have been willing to include particularly important people in depictions of court events, even if their presence was in spirit rather than fact. ❧ MCB

The painting most likely records the visit of Awrangzeb, then governor of the Mughal territories in the Deccan, to the court, which was then residing at Lahore,[1] in January 1640. Still fearing reprisals from the Safavids after the Mughals had taken Qandahar from them,[2] Shah-Jahan had moved to the north of his empire in August 1638 and remained there, with his main headquarters at Lahore, until November 1642.[3]

Of Murar's four *darbar* scenes in the Windsor manuscript, Awrangzeb's reception at Lahore conforms least to the architectural formula laid down for such compositions.[4] In his concern to render what appear to be idiosyncratic features of the audience hall of Lahore, the artist shows more of its architecture than usual.

The most basic representation of the audience hall in Shah-Jahani *darbar* scenes involves framing the *jharoka* on both sides with two single pillars placed near the edge of the painting (see **19**). More often the two pillars appear as parallel pairs and closer to the *jharoka*.[5] This device serves to centralize an essentially flat composition where space is predominantly expressed by the 'layering' of figures and objects. Murar conforms to this convention in the frontal rendering of the two outer, green-and-red pillars of the hall which frame the composition, but he introduces a different perspective for the two inner pillars flanking the *jharoka*, showing them as three-dimensional and free-standing. These pillars are among several other strategically placed foreshortened features, such as the converging brackets of the *jharoka* and its flanking gallery, the canopy hung in front of the *jharoka*, and the bizarre stepped stool on which the bearers of the *chawri* (yak's-tail whisk) stand; all serve to guide the viewer's attention to the Emperor in the center.[6]

The artist seems to have included the pillars of the hall among the perspectivally rendered elements because he was clearly concerned to document their peculiar shape as fully as possible. The high red impost blocks inserted between the shafts and the brackets apparently reproduce an actual feature of the first wooden audience hall of Lahore, and are introduced here in appreciation of regional architecture: pillars of an identical shape supporting painted ceilings on beams (like that of the hall in the painting) are characteristic of the wooden architecture of Uchchh 'Sharif,' an ancient spiritual center with its own distinct building traditions in southern Punjab.[7]

To further characterize the hall as that in Lahore, Murar introduced other features. The birds painted on the eaves of the *jharoka* and the angels flanking it on both sides reflect typical elements of the Jahangiri program of mural decoration within the Lahore palace, and of the tilework decorating its outer wall.[8]

The *jharoka* conforms to the conventional type which now survives only at Lahore (Fig.129). Murar introduces some variation by decorating its dome – of which we see more than usual – with individual flowering plants, the favorite decorative motif of the period.[9] The *jharoka* gallery – an extravagant (and perhaps imaginary) construction whose yellow roof appears as if slid between the impost blocks and brackets of the pillars of the hall – has a balustrade decorated with the Europeanizing grotesque pattern (Fig.127) that Murar uses in two of his other *jharoka* scenes (**32, 38**). Here it makes a particularly odd contrast to the image of the two shaykhs below the *jharoka*.[10] ❧ EK

1. *Shah Jahan Nama*, p.264.
2. *Cf.* **17, 32**.
3. *Shah Jahan Nama*, pp.247, 298–99.
4. **8–9, 32**, and **38**.

Fig.131 Unknown artist, *Shayista Khan*, circa 1640.

5. *Cf.* **14**, **32**, and **43**. The fullest representation of Shah-Jahan's first wooden audience hall, showing the actual alignment of the four pillars of its shorter naves (perpendicular to the *jharoka*), appears in the double-page *darbar* scene of *Shah-Jahan honoring the religious orthodoxy* in the Freer Gallery of Art, Smithsonian Institution, Washington, DC, nos 42.17 and 18; illustrated in S.C. Welch 1978, pl.31–32; see also Fig.144. Abid made an unusual foreshortened rendering of the hall, now in the San Diego Museum of Art, Edwin Binney 3rd Collection; see Fig.146).
6. *Cf.* the discussion of this aspect of the composition of **10–11**, page 168, above.
7. The wooden architecture of Uchchh is still little studied; it seems to integrate certain Central Asian traditions. For illustrations of pillars see Khan 1980, pl.IV; Mufti 1991, p.142, fig.15.
8. *Cf.* **38**; Koch 1983. The Lahore hall also appears in the Bodleian *darbar* (MS.Ouseley, 173, no.13; for references see page 207, note 7), with angels as wall decoration, similarly placed to those in Murar's painting, though the upper red parts of the pillars are aligned with the green shafts to bring them closer to the more conventional shape of the Agra and Burhanpur pillars, which have red oblong cartouches in place of the impost blocks (*cf.* **14**, **17**, **32** and **43**).
9. For a discussion of the floral decoration, see in particular page 200, above.
10. *Cf.* **37**.

45
Shah-Jahan honoring Prince Awrangzeb at his wedding

Agra, *Diwan-i Khass*, 19 May 1637
Folio 218b attributed to Bhola, *circa* 1640
Image area 33.8 × 23.5 cm

References: *Padshahnama*, I/2, p.268; *Shah Jahan Nama*, pp.207–208 and fig.24

THE SETTING OF THIS moment during the wedding ceremonies of Awrangzeb and the grand-daughter of Mirza Rustam Safavi is identical to that of *Shah-Jahan honoring Dara-Shikoh at his wedding* (**25–26**), and in each the prince receives an ornamental head-piece of pearls, the *sehra*, which hangs before a bridegroom's face.[1] Both paintings also show elaborate fireworks along the river, depicted as if they were floral displays and illuminating garden structures on the opposite banks.

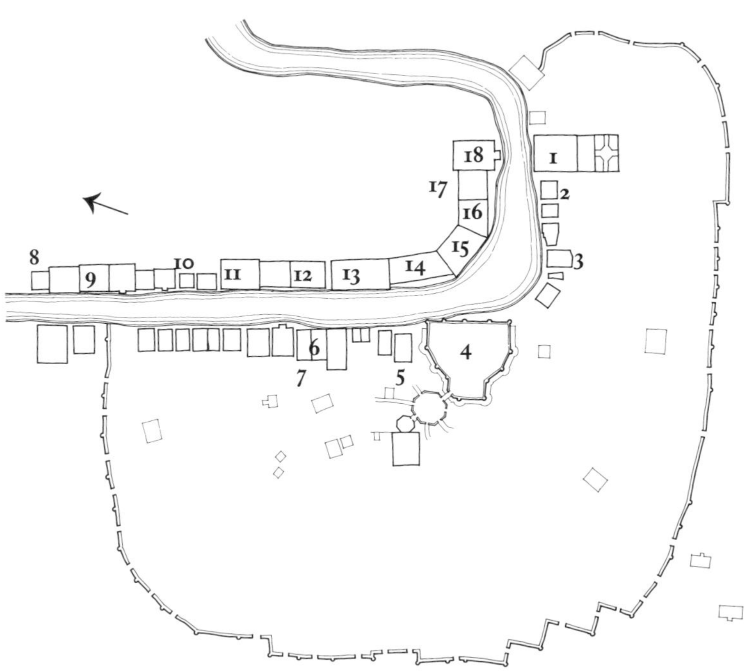

1. Taj Mahal
2. Bagh Khan-i 'Alam
3. Haveli of Azam Khan
4. Red Fort
5. Haveli of Dara-Shikoh
6,7. Havelis of Awrangzeb
8. Bagh-i Shahnawaz Khan
9. Ram Bagh [Bagh-i Nur Afshan]
10. Bagh-i Sultan Parvez
11. Tomb of I'timaduddawla
12. Bagh Padshahi
13. Moti Bagh Padshahi
14. [Bagh] Padshahi
15. Lal Bagh Padshahi
16. Second Chahar Bagh Padshahi
17. Chahar Bagh Padshahi
18. Bagh-i Mahtab Padshahi.

Fig.132 Line drawing after an 18th-century map of Agra. Maharaja Sawai Man Singh II Museum, Jaipur.

It would be hard to imagine that the painter of this illustration did not know the earlier work. Dara-Shikoh, however, as the elder prince and intended heir, is depicted in a double-page illustration – a format that creates a greater sense of spatial and ceremonial majesty.

Unlike the clear colors and crisp forms in Bulaqi's description of the earlier wedding, where neither the night setting nor the illumination by flares compromises the intensity of the white used to define the architecture, the unnamed painter here chooses to explore shadows and tones of grey. This is a trait seen also in *The wedding procession of Prince Shah-Shuja'* (**27–28**). In both works, while major figures are carefully detailed, the expressions of lesser attendants are unfocused and often blank. This is an unusual characteristic within this manuscript, but both scenes can be attributed to Bhola, whose *The weighing of Shah-Jahan on his forty-second lunar birthday* (**12–13**) exhibits an identical manner. MCB

On the occasion of his wedding in May 1637 Awrangzeb was ceremonially received by Shah-Jahan in the Agra fort and presented with a *sehra* in the same manner as his older brothers Dara-Shikoh and Shah-Shuja',[2] but the following *nikah* ceremony did not take place in the imperial palace. The official reason given was that the Emperor did not want to burden his daughter Jahanara – who had arranged the earlier festivities – with the organization of this one.[3] Consequently Shahnawaz Khan, the father of the bride, held the *nikah* in his house, which was situated on the opposite bank of the river at the northern end of the garden residences which lined the Jumna River (Fig.132).

Awrangzeb's veiling with the *sehra* took place on the same terrace of the Agra fort where his brothers' ceremonies had been held. Since the painter was obviously instructed not to devote a double page to it, he could devise a more symmetrical composition which, although stylistically later than Bulaqi's version showing Dara-Shikoh, is closer in concept to what seems to be the common prototype of both, a painting in the *Jahangirnama*, also attributed to Bulaqi, showing the same setting (Fig.15).[4] This painting, although badly worn, is one of those instances that reveal how strongly the *Padshahnama* illustrations depended on the innovations of Jahangir's period, in this case characterized by a detailed, realistic look at an identifiable palace setting, and the systematic use of architecture to structure the painting hierarchically, in that the Emperor and those closest to him are placed on the terrace above and the lesser participants of the event in the courtyard below. The atmospheric effects are a characteristic Shah-Jahani addition.

The scant attention given to the arrangement of Awrangzeb's wedding and the limited space accorded to its depiction suggest that the Prince was being deliberately marginalized (see page 206); the message is brought out clearly by the painters who make Shah-Jahan and Awrangzeb face each other during the ceremony with the greatest reserve, whereas they had emphasized the closeness of Shah-Jahan and his favorite Dara on the same occasion. ❧ EK

1. On the upper terrace are (from the left) Afzal Khan (*Diwan*), Asaf Khan (*Wakil*), Shah-Shuja', Awrangzeb, Shah-Jahan (seated), Muradbakhsh, Dara-Shikoh, Abdullah Khan *Firoz Jang*, Ja'far Khan, and Hayat Khan.
2. See **23** to **28**.
3. *Shah Jahan Nama*, p.207–208.
4. Raza Library, Rampur, MS.1, fol.7a, identified by Milo Beach (1992b, p.227) as the presentation of a monkey, a turkey and a pheasant to Jahangir. See also his discussion of **25–26** on page 184, above.

46
Shah-Jahan hunting lions at Burhanpur

Burhanpur, July 1630
Folio 220b attributed to Dawlat, *circa* 1635
Image area 35.3 × 22.9 cm

Reference: *Padshahnama*, I/I, pp.310–11
Published: S.C.Welch *et al.* 1987, p.291[1]

Fig.133 Dawlat, *Akbar fighting with Raja Man Singh* (detail of onlookers), circa 1598.

Fig.134 Detail of **46** showing attendants at a hunt.

WHILE DETAILS OF THIS HUNT seem to correspond to an event of July 1630,[2] the ages of the princes appear somewhat older than appropriate for that year. The style of this extraordinarily sensitive depiction is also puzzling. Individual Mughal painters often invented idiosyncratic character types which they used when specific portraits were not necessary; these lively figures with slightly heavy proportions, large noses, and expressive but awkwardly drawn hands are distinctively those of Dawlat, a painter whose career – like that of Balchand, Bishandas, or Payag – began under Shah-Jahan's grandfather Akbar late in the sixteenth century. His several illustrations in the imperial *Akbarnama* manuscript (for instance, Fig.133) show identical types of attendant figures, and their unique character appears more forcibly when Dawlat's courtiers are compared to those of Balchand (**5**) or Murar (**38**).

No signed historical or figure paintings by Dawlat after the early years of the seventeenth century were known until recently, although one would expect his career to have paralleled in length that of his artistic colleagues. A *Jahangirnama* page can be attributed to his hand,[3] however, and a superb portrait of Jahangir (Fig.139), inscribed to the artist and dated 1627, provides further confirmation that Dawlat remained active later than had previously been thought.[4] ❧ MCB

This painting was selected by those who assembled the Windsor manuscript to illustrate Lahawri's account of a lion hunt by Shah-Jahan at the imperial hunting ground of Bari in August 1637.[5] However, this is the first representation of a Mughal lion hunt to show the use of a net, a fact which points to an earlier hunt undertaken by the Emperor, in all likelihood that which took place in July 1630 near the Bagh-i Zaynabad at Burhanpur.[6] A few months earlier the Emperor had moved the court to the capital of the province of Khandesh in the north of the Deccan, from where he could supervise the campaign against the Nizamshahis of Ahmadnagar.[7] The successful lion hunt was certainly welcomed as a good omen for the outcome of the Deccan campaign.

The attention given to the net reinforces this identification.[8] All historians of Shah-Jahan's reign describe it as something new and unusual.[9] To explain the use of the device for hunting various kinds of feline 'like lions and tigers [*sher-u-babr*]' to their readers, they compare the netted enclosure – for which they employ the term *bâdar*[10] – to a *saraparda*, an enclosure of cloth used in the imperial camp, and similarly supported by wooden stakes (*sutun*) (*cf.* **6–7**). The texts also draw attention to the net's enormous dimensions, namely 10,000 *zira'padshahi* (approximately 8,200 metres or 5 miles) long and 6 *zira'* (almost 5 metres) high.[11]

By the time of Shah-Jahan the Mughals had established themselves firmly as rulers of their vast empire and this affected their attitude toward hunting. Hunting was still represented as an instrument of rule, but the concept that qualities of leadership could be ascertained through the display of personal bravery in the course of a hunt (see **30**) moved increasingly into the background while the interpretation of the hunt as a metaphor of good government received more emphasis (see **33**). The Mughals began to pay attention to the warnings of the *Mirrors for Princes* (books of counsel for rulers) that reckless and unguarded hunting presented an unnecessary risk to the life of the prince,[12] and increasingly favored less dangerous forms of the hunt. Hunting lions with nets was certainly a safer though less sporting technique than to approach them in a man-to-beast combat (as in **30**) or with a circle of beaters. Eventually the use of nets became the usual practice for Mughal lion hunts.[13]

Shah-Jahan's lion hunt takes place in a swampy landscape, depicted as tilted up in a steep vertical plane, a device which was unavoidable if the artist was to show the three elephants carrying the most prominent members of the imperial hunting party fully and in hierarchically correct profile, one above the other.[14] The contemporary viewer, familiar with the impact of such hierarchical impositions on the spatial logic of the composition, would without hesitation decode the arrangement of the main figures in the middle ground as the Emperor on his elephant facing a pack of lions, flanked on his right (that is, above) by the elephant carrying Prince Shah-Shuja' and on his left (below) by the elephants carrying the Princes Dara-Shikoh and Awrangzeb. The hierarchical arrangement also helps us to identify the princes. It is the youngest of the three, Awrangzeb, whose elephant is not shown fully but is almost entirely obscured by that of Dara-Shikoh, the heir apparent. The painter skillfully uses the net to separate the main scene from both the foreground, with its more freely grouped hunting assistants, and the background, where illusionistic interests could be openly expressed.

The net serves not only as a compositional device, therefore, but is also used by the artist to demonstrate that – while he had to consider certain formal conventions – he was entirely familiar with such basics of three-dimensional representations as the fact that the appearance of things changes with regard to their distance from the viewer. The stretch of net in the foreground is rendered painstakingly, including every detail of the wide-meshed fabric of strongly knotted white rope, which is looped on to wooden stakes reinforced with diagonal supports to make the net stand upright with no folding or sagging. That part of the net which marks the transition to the background is shown at reduced size and in perspective and undulates over the hills to embrace the middle-ground, before merging atmospherically with the light blue-green tonalities of the furthest landscape. That the painter's illusionistic talents could also assert themselves in the formally composed main zone of the hunt is evident in the meticulous treatment of the surface of the ground, dotted with bundles of wispy grass, a naturalistic reinterpretation of a time-honored device in Persianate landscapes (Fig.80), to create relief and texture, and in other natural elements, such as the small pond with sarus cranes,[15] and the tiny animals near the net in the background who are shown hiding or fleeing the hunters. The gnarled dry tree and the small bushy plant placed on top of the mound leading into the center of the picture – and thus giving some depth to the vertical terrain – are painted with the same eye-catching microscopic realism. Such detailed observation of nature is traditionally considered the undisputed domain of the painters of Northern Europe. This picture in particular demonstrates that the Europeans were being challenged successfully in their own field from such unlikely and distant quarters as Mughal India. ❧ EK

1. Welch attributes the page (without explanation) to Hashim.
2. See page 110 above.
3. Beach 1978, pp.113–16.
4. For further discussion of Dawlat, see page 216.
5. *Padshahnama*, 1/2, p.276.
6. The Bagh-i Zaynabad is situated opposite the Burhanpur fort on the other side of the Tapti River. It was in this garden that Shah-Jahan's favorite wife Mumtaz Mahal was temporarily buried after she had died in childbirth in June 1631, when the Emperor and his court were still residing in Burhanpur. See *Shah Jahan Nama*, p.70. Later in the year her body was brought to Agra to be buried in the Taj Mahal, the construction of which had just started.
7. See **15**, **17**, **18**, **31**, and **40**.
8. *Padshahnama*, 1/1, pp.310–11; *'Amal-i Salih*, 1, p.313; Qazwini, *Padshahnama*, fol.194a (re-foliated 195a). The texts omit to mention the princes in the context of this hunt. In 1630 Shah-Jahan was thirty-eight years old, Dara fifteen, Shah-Shuja' fourteen and Awrangzeb twelve. While the age of Shah-Jahan in the painting corresponds to his thirty-eight years, the princes are made to look somewhat older.
9. The method is not mentioned in the *A'in-i Akbari* among techniques of hunting lions (trans. 1, p.293).
10. This is not a Persian word. The term exists in Sanskrit where it means 'made of cotton,' which would describe the texture of the net.
11. The conversion of these figures is based on the assumption that the length of the *zira'* (or *gaz*) given for the net corresponds to the *zira'* used for architecture which, in Shah-Jahan's period, equalled approximately 32 inches or 0.8 to 0.82 metres. It is, however, possible that the measurements for the nets were given in a somewhat shorter *zira'*.
12. See Lahawri's comments on Jahangir's behaviour in his lion hunt (page 76), which are shaped on *adab* literature or Muslim books of counsel such as the *Qabusnama*, p.84. On the concept of *adab*, see Gabrieli 1979.
13. In later depictions of Mughal lion hunts nets feature less prominently in the backgrounds. See, for instance, the example in the Keir Collection, Richmond, Surrey, illustrated in Skelton 1976, cat. no.v.84, pls 130, 131. An even more guarded Mughal lion hunt is shown in the Chester Beatty Library, Dublin, where, in addition to the nets, a 'battle-array' of buffalos is used to keep the lions under control. For illustration see Leach 1995, 1, cat. no.4.12, color plate 76.
14. See pages 136–38 above for a discussion of the principles of Shah-Jahani representation.
15. *Grus antigonae*, studied by Jahangir, and one of the birds most favoured by Mughal painters to populate their landscapes. *Cf.* **30**.

THE ARTISTS OF THE PADSHAHNAMA

by Milo Cleveland Beach

There is little information about the lives of artists in Mughal India. From the time of Humayun, however, we know that certain favored painters were in the specific employ of the emperor, and that they were assigned workspace within the palace *karkhanas* (workshops). Organized under a studio head, they were also expected to travel when the emperor moved between capital cities or made extended journeys.[1] The official histories or memoirs of Humayun, Akbar, and Jahangir all mention the commissioning of specific illustrations, and often the painters to whom they were assigned.[2] The *A'in-i Akbari* additionally records that Akbar examined the work of his artists weekly, while images show Jahangir inspecting paintings, an activity mentioned also in the *Jahangirnama*.[3] When work was especially pleasing, as in the case of one illustration presented to Jahangir by the painter Bishandas, the artist might be rewarded with the gift of an elephant.[4] The emperors also collected (through gift, purchase, and the booty of battle) important illustrated manuscripts from other sources, and these were especially valued when they had belonged to one of Timur's descendants. In their turn, the Mughals gave manuscripts as diplomatic presents. This very modest amount of information has, however, to be pieced together from many sources: painting and painters are very seldom mentioned within the copious verbiage that documents Mughal rule.[5]

Paintings, of course, were themselves a further source of documentation. A major discovery of the recent research on the *Padshahnama* has been the realization that artists continually painted contemporary events with no idea of whether or how those illustrations might be used. Therefore, just as written accounts of court events were recorded daily by clerks and kept in the court archives, so paintings were created as ongoing documentation. When a decision was made that an official history was to be written, the assignment was given to a poet or other literary historian – a person distinct from the court recorder – and he was then given access to the archival materials as a source of information. The author would have worked with the head of the studios or the superintendent of that particular project to determine where to leave space for illustration within the texts – the procedure followed by Muhammad-Amin of Mashhad. Paintings would then have been drawn from among works on hand or commissioned afresh.

The inclusion of self-portraits in several court scenes suggests that, like the clerk, painters worked from direct observation. It would seem that this was most useful for the details of appearance or deportment, however, for artists worked within a standard overall compositional format whenever possible. Several drawings related to scenes in the Windsor *Padshahnama* are known (see Figs 82 and 83) and while these could have been preliminary studies, they would also have served as models for future compositions (and would usually have been available for this use to any of the painters in the imperial workshops). Nevertheless, the self-portraits and signatures indicate that artistic individuality was recognized and valued – individual painters were first lauded in the *A'in-i Akbari*.[6] It is only under Jahangir and Shah-Jahan, however, that inscriptions naming the painters are found as part of the artistic fabric of the image itself; under Akbar, by contrast, painters were more regularly named in marginal notations made for record-keeping by the clerks. (While few of the inscriptions in the *Padshahnama* may have been written by the painters they name, they were contemporaneous and carefully designed additions to the scenes and can be considered as reliable as literal signatures.)[7] It has been argued elsewhere that the paintings now found in the Windsor manuscript were not originally intended to illustrate that copy of the text (see pages 17–19). The absence on its pages of works signed or attributable to several major artists of the period (Govardhan, for example, or Hashim) therefore indicates only that appropriate works by those artists were not among the surviving images eventually placed within the Windsor volume.

While the fourteen artists that have been identified as having worked on the illustrations of the *Padshahnama* - and whose signed or attributable works are listed below – all worked within a carefully circumscribed stylistic range, recognition of even modest differences in artistic hands makes us acutely aware of the important human element behind a manner that may initially seem mechanically uniform. This was not an anonymous art form, but one in which artistic individuality was celebrated. The Emperor Jahangir had rejoiced in being able to recognize individual styles: 'I derive such enjoyment from painting and have such expertise in judging it that no work of past or present masters can be shown to me, without the name being mentioned, that I do not instantly recognize who did it. Even if it is a scene of several figures and each face is by a different master, I can tell who did which face. If in a single painting different persons have done the eye and eyebrows, I can distinguish who drew the face, and who made the eyes and eyebrows.'[8]

There is no indication that Shah-Jahan shared his father's intense personal involvement with the commissioning of painting, however, after the first year of his reign, so it may be that the most significant encouragement to painters during Shah-Jahan's rule was the energy generated by Jahangir's initial interest and support. For example, of the fourteen artists known to have worked on the *Padshahnama*, the majority had reached full maturity under Jahangir and none were unquestionably still active under Awrangzeb.[9] Thus, although Shah-Jahan clearly provided commissions and support for the painters that he had inherited with his father's workshops, no new, younger generation of master artists seems to have emerged under his patronage. This decline of imperial interest in painting, coupled with the increasingly self-referential nature of Mughal style – which under Shah-Jahan increasingly preferred to follow established formulas or copy earlier work rather than explore challenging alternatives – does suggest that the initial failure of later Mughal painting to uphold the standards of workmanship and perception brought to maturity under Jahangir should be traced to Shah-Jahan's patronage rather than that of Awrangzeb. (Ebba Koch has elsewhere mentioned Shah-Jahan's growing orthodoxy following the first decade of his rule, a development which would certainly have promoted this indifference to painting.)[10] Since Shah-Jahan's son and successor had little interest in any of the arts, the change of ruler offered no relief from this unprecedented situation.

1. *Tuzuk-i-Jahangiri*, II, p.20, describes the inspection of a painting while Jahangir was in Gujarat.
2. See *Tezkereh al Vakiat*, p.43, or *Tuzuk-i-Jahangiri*, II, p.108.
3. See note 2, above.
4. *Tuzuk-i-Jahangiri*, II, p.117.
5. For a discussion of Mughal historical manuscripts within the broad context of Islamic manuscript traditions, see Blair 1995, pp.90–108.
6. *A'in-i Akbari*, pp.113–15.
7. In the lists of key works given below, the term 'signed' indicates that an inscription, judged here to be reliable, gives the name of the painter. It may be either a statement by the emperor, a marginal notation by a clerk, or an actual signature. 'Dated' refers either to a date on the picture, or, for illustrations included within manuscripts, to the date when the text was copied.
8. Translated by Wheeler Thackston.
9. An important portrait of the young Awrangzeb enthroned has been convincingly attributed to Hashim, one of Shah-Jahan's greatest artists, so he evidently remained active in the first years of the new Emperor's reign; see S.C. Welch 1978, pl.37. This is the only known work attributable to Hashim during this period, however.
10. See pages 132 and 143, note 8, above.

ABID

16, 37

While Abid is known to have been the brother of Abu'l-Hasan *Nadiruzzaman*, who had begun his career at the end of Akbar's reign and became the favorite painter of the Emperor Jahangir,[1] no works can be assuredly attributed to him before the reign of Shah-Jahan.[2] Several paintings signed by Abid have inscriptions, of which the earliest is dated to the second year of the Emperor's reign (1629–30),[3] but these show such technical skill and artistic maturity that Abid was certainly active earlier.

Abid is responsible for *The death of Khan Jahan Lodi* (**16**), an image that can be described both in terms of its intensely powerful realism (note especially the actual beheading), and its mannered artificality (in the poses of the individual attendants or the even line-up of the distant army). The result is a visual tension and dynamic range found in no other painter of the time. In *Jahangir receives Prince Khurram* (**37**), the attendants, again depicted in an evenly spaced row along the bottom, include a wonderfully humorous laughing figure at the left and an uncannily three-dimensional pensive nobleman on the right (Figs 39 and 41). It is the realism of these figures as character types that interests Abid, not the uniqueness of their individual personalities, or the realism of their movements in or interactions across space. Compare the group of figures outside the gold railing here to those of Bichitr in his accession *darbar* (**10**), where jostling and movement is more expressively important. Abid's figures outside the railing are both more colorful and more precisely painted than the courtiers within; they are the men that seem most to have drawn his attention. The courtiers near the throne are depicted with an almost physical haziness that is quite the opposite of Bichitr's clarity and crispness.[4] Abid's interest in perfecting understanding of each man as a character type may be the legacy of his father, the Iranian emigrant artist Aqa-Riza. A master painter at the Mughal court, Aqa Riza could never free himself from his earlier training in Iran, where stress was given to the general rather than the unique attributes of human figures.[5]

OTHER KEY WORKS[6]

1629–30	*Shah-Jahan enthroned with Mahabat Khan and a shaykh*, signed and dated (Fig.142)
1635–36	*Shah-Jahan receives Prince Shah-Shuja' on his return from the Deccan*, signed and dated (Fig.145)
1639	*Shah-Jahan enthroned*, signed and dated (Fig.146)
circa 1640	*The Prophet and the Persian physician*, from a *Gulistan* of Sa'di MS., signed[7]

1. See Figs 13 and 122. For further references, see Beach 1978, pp.86–92; Beach 1980; Losty 1991; and Verma 1994, pp.47–55.
2. While there is no signature or corroborating factual evidence, stylistic and technical similarities suggest Abid was the painter of *The inscription of Jamshid*, from a *Bustan* of Sa'di manuscript dated 1605–6 (Soudavar 1992, no.137a).
3. See especially Figs 142, 145 and 146.
4. This quality constantly reappears in Abid's paintings, such as his illustration for a *Gulistan* of Sa'di manuscript of about 1640 (Soudavar 1992, no.136b), where he portrays literary archetypes rather than historically identifiable individuals.
5. For Aqa Riza, see also Beach 1978, pp.92–95; and Verma 1994, pp.62–69.
6. See also Beach 1978, p.85; and Verma 1994, pp.46–47.
7. Soudavar 1992, no.136b.

BALCHAND

5, 14, 15, 30

Balchand, too, was a member of a painter family, for inscriptions tell us that he was the brother of Payag.[1] But whereas Payag was flamboyant in his compositions, technique, and depiction of physical appearances, Balchand is comparatively reticent – and this comparison extends to the way each artist has depicted himself as a personality within a *darbar* scene in the *Padshahnama* (**5** and Fig.135, and **39** and Fig.141). Balchand's figures are full of movement, and there is wonderful contrast in his two audience scenes between the decorum of the courtiers assembled within the gold rail and the bustle and agitation of those on the outside (**5** and **14**). His figures also have character; they convince as specific portraits, and he does not idealize. The figure of Raja Bikramajit, who stands directly before Karan Singh in *Jahangir receives Prince Khurram on his return from the Mewar campaign* (**5**), for example, has a swarthy complexion seemingly marked by smallpox scars – a detail absent from other portraits such as that by La'lchand (Fig.78). There is no evidence that Balchand was an innovator, and his landscape scenes are almost Persian in the way the background remains a flat backdrop to the action rather than, as in the works of Payag, a dramatically expressive element of the scene. Nonetheless, few artists are as direct as Balchand, as unaffected, and as reliable in convincing us that (true or not) the scene once actually lay before his eyes.

Fig.135 Balchand, self portrait

OTHER KEY WORK[2]

circa 1596	*The building of Fatehpur-Sikri*, from an *Akbarnama* MS., signed[3]
1604–5	*The donkey's refusal*, from a *Nafahat al-uns* MS., signed and dated[4]
circa 1637	*The three younger sons of Shah-Jahan*, from the Minto Album, signed[5]
circa 1640	*A prince riding*, from a *Gulistan* of Sa'di, signed[6]

1. See the inscriptional information for **39** on page 201. A page from an early 17th-century *Shahnama* manuscript in the Nelson-Atkins Museum in Kansas City (54.81) is also inscribed as 'the work of Balchand and his brother.' The wording in both cases implies strongly that Payag was a younger sibling.
2. See also Beach 1978, pp.95–101; Verma 1994, pp.72–75; and Smart 1991.
3. Arnold & Wilkinson 1936, II, pl.24.
4. Beach 1992, fig.46.
5. Deneck 1967, fig.34.
6. Soudavar 1992, no.136d.

BICHITR

10

The accession portrait by Bichitr is the most famous painting in the Windsor Castle *Padshahnama*, and with good reason. It is technically brilliant, with superbly crafted, probing portraits of the major nobles and intensely precise descriptions of architectural decoration, textile patterns, and types of turban. Every part of the surface is visually alive; compare the rug at the bottom center, for example, with the simple, static, geometric rug design in one of Murar's audience scenes (**32**). Bichitr's knowledge of the effects of light is shown by the cast shadow beneath the footstool in the center of the picture; but whereas Abid and Payag both make their use of shading a dramatic and obvious addition to their scenes (**16** or **37**), Bichitr so subtly models his figures that we sense their weight and mass (and therefore their physical presence) before we notice that the artist is introducing a novel effect. These stylistic characteristics are found even in his first known works, dateable to the middle of Jahangir's reign (see below), so an earlier – although presently unknown – period of activity must certainly have existed.

Fig.136 Bichitr, self-portrait.

OTHER KEY WORKS[1]

circa 1615–18 *Jahangir prefers a Sufi shaykh to kings*, from the St Petersburg Album, signed[2]

circa 1623 *I'tibar Khan*, signed (Fig.73 and page 161 above)

1630–31 *Asaf Khan*, from the Minto Album, signed and dated[3]

1631 *Shah-Jahan*, from the Minto Album, signed and dated[4]

circa 1640 *Shah-Jahan with Asaf Khan*, from the Late Shah-Jahan Album, signed (Fig.19)

1. See also Verma 1994, pp.104–109; and Beach 1978, pp. 101–107.
2. Beach 1981, no.17a; Ettinghausen 1961.
3. Archer 1960, pl.27.
4. S.C. Welch 1963, no.43.

BISHANDAS

21, 22

While there is no signature of Bishandas (Fig.137) within the *Padshahnama* illustrations, a complicated series of interrelationships among other paintings suggests that he is responsible for the evocative, double-page illustration of musicians, bearers, and nobles carrying wedding presents to the home of Dara-Shikoh's future bride.[1] Bishandas' figure types are as distinctive as those of Payag or Bulaqi, but quite different; he is more deft than any of his contemporaries in his portrayals of women, for example. A fully mature artist by 1610 at the latest, Bishandas was active during a brief period under Jahangir when specific, individualized portraits of women were acceptable. In a culture where women were hidden in seclusion, he convincingly depicted court ladies with the same degree of individuality found in male portraits. His sensitivity to women, shown best in the great *Birth of Jahangir* (see below), is seen too in the *Padshahnamah* double page, and this interest immediately separates the artist from his younger contemporaries, for whom women have again become types (see **25**). It may also be Bishandas' training in the late sixteenth century – when painters focused their attention on movement, a clearly expressed narrative, figures turning in space, and lively facial expressions – that explains his skill in describing physical activity. The musicians accompanying the wedding procession are so vividly described that the painting effectively evokes the sounds and brouhaha of the ceremony.

Fig.137 Bishandas, self-portrait

OTHER KEY WORKS[2]

circa 1605 *The house of Shaykh Phul*, signed[3]

1608 *Raja Suraj Singh*, signed and dated (Fig.72 and page 161 above)

circa 1620 *The birth of Jahangir*, from a *Jahangirnama* MS., attributed[4]

circa 1625 *A procession*, probably from a *Jahangirnama* MS., attributed (Fig.102)

1. For this argument, see Beach 1978, pp.105–111.
2. See also Das 1971; Beach 1978, pp.107–11; and Verma 1994, pp.110–15.
3. Mehta 1926, pl.37.
4. Beach 1978, no.15; S.C. Welch 1978, fig.16.

BHOLA

12, 13, 27, 28, 45

Only one painting in the *Padshahnama* is inscribed with Bhola's name, but its inconsistencies help identify the artist's work elsewhere in the manuscript. In the left-half of *The weighing of Shah-Jahan on his forty-second lunar birthday* (**12** and **13**), for example, the closest members of the imperial circle have real physical presence, whereas the nobles at the bottom are a line of flat, silhouetted shapes; and while the central portraits are probing and individualistic, those of subsidiary figures are incoherent, with eyes that stare blankly into space, as if the artist were completely indifferent to the personalities involved. This is quite different from Abid, in whose works there is often greater interest in the lesser figures than those of the major courtiers (**37**); or Bichitr, who distributes his interest and skill evenly among his subjects (**10**). This highly personal manner of depicting minor figures, distinctive to Bhola, is also found in the additional scenes in the *Padshahnama* which are attributed here to his hand. No other works by the painter have been identified.[1]

1. See also Verma 1994, p.116 (where the name is listed as Bula). The attribution of a page from the British Museum *Shahnama* (MS.Add.5600) is not accepted here.

BULAQI

25, 26

More than any other artist, Bulaqi ('son of Hoshang') transmits to the viewer the grandeur of Mughal palace structures. In *Shah-Jahan honoring Prince Dara-Shikoh at his wedding*, the figures are smaller in scale in relation to the architecture than in the work of such artists as Murar (**38**), with whom Bulaqi has been confused, or Bhola, who depicts the identical palace space in **45**. (The lack of physical weight in Bulaqi's figures may be one reason for the confusion with Murar.) That Bulaqi's contemporaries recognized his success in depicting the surface power of palace architecture is evident from two additional illustrations by this painter that depict figures along the exterior wall of the fortress at Agra (Figs 157 and 158),[1] and a particularly majestic double-page scene, *A night celebration of the Prophet's birthday* (Fig.144), in which the architecture provides a strong framework that coordinates the many figures gathered in the *Diwan-i 'Amm* at Agra. Few other works by the artist have been identified, despite the first appearance of his signature in a manuscript dated 1598.

OTHER KEY WORKS[2]

1598 *A worshipper caught by a fisherman*, from a *Razmnama* MS., signed and dated[3]

circa 1625 *Jahangir receiving Muqarrab Khan with exotic animals from Goa*, probably from a *Jahangirnama* MS., signed (Fig.15)

circa 1635 *A night celebration of the Prophet's birthday*, attributed (Fig.144)

circa 1635 *Shah-Jahan watching an elephant fight*, signed (Fig.158)

1. For a prototype for the architectural scenes, see *The Emperor Jahangir giving audience*, by Abu'l-Hasan, from a *Jahangirnama* manuscript (Goswamy & Fischer 1987, pl.37).
2. See also Verma 1994, pp.116–17.
3. S.C. Welch 1959, fig.6.

DAWLAT

46

Whereas the unknown artist of *Shah-Jahan hunting* (**33**) limited his human figures to members of the imperial party, Dawlat (Fig.138), to whom *Shah-Jahan hunting lions at Burhanpur* is attributed here, forces us to pay attention to the attendants. He places them in the foreground, and then emphasizes their animated movements, their relaxed interactions, and their personal idiosyncrasies – two of the men wear European hats over their turbans, for example.[1] While the expansiveness of the space and the extraordinarily sensitive drawing of the vegetation shows Dawlat's awareness of developments in Mughal painting under Shah-Jahan, his figural style was established and already mature at the time of his contributions to the imperial *Akbarnama* of *circa* 1596 (Fig.133).[2]

A recently discovered portrait of Jahangir (Fig.139), signed by Dawlat and dated 1627,[3] expands considerably our knowledge of the artist; when seen together with the *Padshahnama* page, it is clear that Dawlat remained a major historical painter and portraitist until later than previously thought. There are also several superlative decorative margins in albums of the Shah-Jahan period that are signed by the artist and reveal a different aspect of his talent.[4]

Fig.138 Dawlat, self-portrait.

OTHER KEY WORKS[5]

circa 1596 *Akbar wrestling with Man Singh*, from an *Akbarnama* MS., signed (Fig.133)

1604–5 *Abu'l Adyan gives proof that fire submits to the will of God*, from a *Nafahat al-Uns* MS., signed and dated[6]

1605 *Sa'di and the dervish of Faryab*, from a *Bustan* of Sa'di, signed and dated[7]

circa 1610–15 *The salvation of one brother*, from a *Gulistan* of Sa'di MS., signed[8]

1627 *The Emperor Jahangir*, signed and dated (Fig.139)

1. An identical hat is worn by a figure in a signed page by Dawlat in a *Nafahat al-Uns* manuscript dated 1604–5; see Das 1991, fig.4. Dawlat is a painter particularly attentive to varieties of headgear.
2. See also Arnold & Wilkinson 1936, II, pls 27–28.
3. The inscription reads: *'amal-i kamtarin bandah-i dargah Daulat* ('The work of the most insig-nificant servant of the court, Dawlat, [Regnal] year 22'). In its composition, details such as the choice and arrangement of jewels, and the definition of surfaces through rather broadly defined planes rather than the accumulation of minute brushstrokes, the dated portrait is directly comparable to a large portrait of Jahangir on cloth that was recently sold in London (Sotheby's 1995, lot 85). The scale of the cloth painting, and therefore the character of its brushwork, is unprecedented for a known imperial work, but an attribution to Dawlat seems possible.
4. S.C. Welch *et al.* 1987, nos 17, 18, 29, 30, 33 and 34; but see also no.27. Other examples are in the Ardeshir Album (Sotheby's 1973, lot 41), and the Minto Album (Chester Beatty Library, Dublin, MS.7, no.10; and Victoria & Albert Museum, London, 24-1925).
5. See also Beach 1978, pp.113–16; Das 1991, pp.87–104; and Verma 1994, pp.126–30.
6. Wellesz 1952, fig.35.
7. Soudavar 1992, no.137j.
8. Beach 1978, no.16.

Fig.139 Dawlat, *Jahangir*, dated 1627.

DHOLA

40

The qualities of Dhola's *The surrender of the fort at Udgir to Khan Dawran* can best be described through comparison with three similar scenes in the Windsor manuscript by as many different painters: Balchand (**15**), Murar (**31**), and the otherwise unknown Tezdast (**35**). In Balchand's illustration, the fortress at Dharur is only one among many visual elements rather evenly distributed over the picture surface, while Murar's depiction of the Daulatabad fort overwhelms the landscape; Balchand presents a conservatively designed scene where every element is of interest, while Murar creates instead a vast open space around the towering fortified rock, and thereby gives sharp focus and a highly dramatic impact to his scene. The sense of a clear space through which the figures move continues in Tezdast's crystalline delineation of one of the forts in Orchha. And here, as at Daulatabad, distant armies are seen; quickly and lightly sketched, these figures animate the far distance. Dhola's depiction of Udgir closely follows the compositional type used by Tezdast, but the open, airy space created by his colleague has closed down, and there is far less attention to detail and precision in the individual figures. This movement away from technical bravado – accompanied by the return of an evenly colored background that flattens the space and a presentation of the sky as a decorative surface rather than open air – leads the way towards Mughal styles of the late seventeenth century. The only additional paintings that can presently be identified with Dhola (see below) indicate that he was an artist who remained active at the end of Shah-Jahan's reign.[1] His work is therefore important, marking the transition to the styles practiced by artists during Awrangzeb's rule.

OTHER KEY WORKS

circa 1650 *Sayyid Hidayatullah*, signed[2]

circa 1650 Marginal decorations from the Late Shah-Jahan Album, signed[3]

1. In Verma 1994, pp.126–30, works by Dhola are listed under Dawlat. For this confusion, see Beach 1978, p.114.
2. Colnaghi 1976, no.122; Brown 1975, pl.27.
3. Binney 1973, no.60.

THE 'KASHMIRI PAINTER'
7, 17, 34, 41, 42

That the mainstream imperial Mughal style did not come easily to this anonymous painter can be seen most effectively in *The submission of Rana Amar Singh of Mewar to Prince Khurram* (**6–7**), where his left-hand page is in direct artistic contrast to the main scene by La'lchand. Despite heavy use of shading on individual figures, this artist fails to impart a sense of physical mass to his subjects. When coupled with inconsistent proportions and his inability to endow the subsidiary figures with individuality, it would seem that the artist was struggling to meet the requirements of the imperial studio. Robert Skelton has suggested (in conversation) that this unnamed painter may have received his early training in Kashmir, where a group of artists were seeking to adapt Iranian and Central Asian styles to Mughal taste. Indeed, the Mughal style was so unusual within Islamic and Indian traditions that those who had received early training in other modes were often unable to direct their established skills to suit the demands of the Mughal court.

Shah-Jahan receives the Persian ambassador Muhammad-Ali Beg (**17**) can also be attributed to the Kashmiri Painter. As in his other works (**41–42**, for example), the important imperial figures are painted with skill and perception, whereas the minor subjects become generalized types, a trait he shares with Bhola (see **12–13**). Despite the rich portrayals of the imperial family, the highest nobility, and the ambassador, the relative proportions throughout are ungainly and physically inexplicable, the coloring rudimentary, and the clothing – even of the figures of Shah-Jahan and Dara-Shikoh, usually described with careful attention to pattern and physical texture – roughly executed. These features indicate that the scene is not the joint work of two painters with different levels of skill.

At present, no other works can be attributed to this artist.

LA'LCHAND
6

Neither as overwhelmingly precise as Murar, nor as technically brilliant as Bichitr, nor as dramatic and extrovert as Payag, the audience scene by La'lchand is an especially strong evocation of an actual historical event because so little of the artist's personality is inserted into it. That La'lchand modeled the scene closely on an earlier depiction of the same event presumably made for the *Jahangirnama* (Fig.17) is an indication of his acceptance of tradition; furthermore, his style seems little changed from the illustration of a different episode which he made himself for the earlier *Jahangirnama* manuscript (see below).[1]

La'lchand nonetheless experimented with new artistic ideas, as shown in his page from a *Gulistan* of Sa'di dated *circa* 1645 (see below). There he employs the chiaroscuro lighting effects more often seen in works by Payag, and the figures – as so often the case when texts are literary rather than historical – are types. Like other artists, La'lchand has a recognizably personal figural formula that he employs when specific portraits are not appropriate, or when the subjects are low in the social hierarchy.[2]

For an illustration of another work by La'lchand, see Fig.78.

OTHER KEY WORKS:[3]

circa 1620 *Jahangir giving books to shaykhs*, from a *Jahangirnama* MS., attributed[4]
circa 1640 *The Qazi of Hamadan in a drunken state*, from a *Gulistan* of Sa'di, signed[5]
circa 1640 *Qilich Khan Turani*, from the Kevorkian Album, signed[6]

1. *Jahangir giving books to shaykhs*, a page from the *Jahangirnama* placed in the St Petersburg Album, can now be attributed to La'lchand on the basis of its stylistic similarity to the *Padshahnama* page; see Beach 1981, no.17d.
2. For example, compare the standard bearers at the upper right of his *Padshahnama* audience scene to those in the *Gulistan* page (Soudavar 1992, no.136e).
3. For further information on La'lchand, see Verma 1994, pp.231–32. To the list of works given there, and the key works named above, the following should also be considered: three depictions of Mian Mir and other holy men in conversation, from the Ardeshir Album (Sotheby's 1973, lots 10–12); *Dara-Shikoh with Mian Mir and Mulla Shah* (Lowry 1988, no.72); and *Jahangir giving books to shaykhs* (Beach 1981, no.17d.
4. Beach 1981, no.17d.
5. Soudavar 1992, no.136e.
6. S.C. Welch *et al.* 1987, no.76.

MURAR
9, 23, 31, 32, 38, 44

Murar's *The arrival of Prince Awrangzeb at the court at Lahore* (**44**) depicts an event from 1640. It is the latest of the episodes illustrated by the artist within the Windsor manuscript, and there is so little difference in style in his four audience scenes that 1633–45 seems a reasonable range for the dates of his paintings in the book. (No earlier works by the artist have been identified.) Murar's *darbar* scenes carefully document each event (he even places himself in **9**, Fig.140): figures are precisely drawn and placed, and there is complete consistency down to the smallest detail and the lowliest courtier. In comparison with the portraits of the imperial family by Bulaqi (**25**), for example, Murar is far more observant of physical differences and character; but in contrast to Payag's work (**39**), the figures have neither mass nor life, and as personalities they seem free of the eccentricities beloved by Payag. When Payag's figures are not moving restlessly, the patterns of the fabrics that they wear move the viewers' eyes to produce a dynamic surface. Murar's *darbars*, by contrast, are still, even static. When free of well-established compositional formulas, however, and forced to be inventive, Murar's work is dazzlingly dramatic; none of his fellow artists so convincingly presented the power and impact of architectural masses achieved by Murar in *The siege of Daulatabad* (**31**).

Fig.140 Murar, self-portrait (detail of **9**).

OTHER KEY WORKS:[1]

circa 1640 *Hakim ul-Mulk*, signed[2]
circa 1640 *Sa'di's companions frightened by tribes from the hills*, from a *Gulistan* of Sa'di MS., signed[3]
circa 1640 *Khan Dawran Bahadur Nusrat Jang*, from the Kevorkian Album, signed[4]

1. See also Verma 1994, pp.274–76 (where the name is listed as Mirar).
2. This important single portrait is so titled in Goswamy & Fischer (1987, no.71). The figure can now be further identified as Hakim Abu'l-Qasim (or Abd al-Qasim), who received the title Hakim ul-Mulk at the accession *darbar* depicted in **10–11**; see *Shah Jahan Nama*, pp.21 and 213.
3. Soudavar 1992, no.136f.
4. S.C. Welch *et al.* 1987, no.71.

PAYAG

18, 36, 39, 43

Payag includes a self-portrait in *Jahangir presents Prince Khurram with a turban ornament* (**39** and Fig.141), and while the two courtiers standing in front of him wear a vest made apparently of European red velvet and a yellow coat with Mughal floral designs, Payag himself sports a jacket with a Turkish design popular in the reign of Sultan Murad IV (r. 1623–1640).[1] This may be a mark of his interest in fashion – he does not present himself as humbly as either Murar (**9** and Fig.140) or his own brother, Balchand (**5** and Fig.135). Indeed, there is nothing reticent about Payag's paintings. More than any other artist, he flaunts his interest in European style; throughout his works he uses spatial recession in a manner that places stress on events in the distance, his figures are dramatically modeled with light and shade, and his portraits accentuate individual physiognomies – sometimes to such a degree that they seem statements about personality types, rather than individual likenesses.

Paintings and prints introduced into India during Shah-Jahan's reign by missionaries and traders, and the works of amateurs traveling in Mughal territories, made Payag and others aware of the styles being developed in Europe by such disparate artists as Albrecht Altdorfer, Jacques Callot, or Adam Elsheimer.[2] Payag found these images an important extension to his innovative interest in powers beyond individual control. Stressing dramatic lighting effects, he created a sense both of the mystery of nature and of an uncontrollable power determining events. Crowds, for example – shown best by the military troops in the siege scene – were no longer individuals placed closely together; they were now part of an energetic mass to which specific personalities were subservient. The technique that achieved this is shown in the drawings of the English East India company merchant Peter Mundy, who recorded his own obervations in India with an amateur's understanding of current European style (Fig.117). The quickness of the sketch, and the ability to sense the movement by which pen or brush and pigment were applied to paper was also a novel contribution to Mughal illustration, which had heretofore stressed the painstakingly slow accumulation of images in successive layers and the obliteration of any sense of the texture of the pigment or of the artist's actions in painting.[3] The new techniques were explored by several artists, but the majority preferred to confine these quickly drawn sketches to distant background events, thereby exploiting their time-saving potential rather than their new expressive possibilities. Here, too, Payag was an exception. Even in his *darbar* scenes, the swirl and movement of pigment serve to emphasize his interest in the gestures and rustlings of the courtiers.

Another painting by Payag is illustrated below (Fig.151).

Fig.141 Payag, self-portrait (detail of **39**).

OTHER KEY WORKS:[4]

1598	page from a *Razmnama* MS., signedand dated[5]
circa 1615	*Raja Man Singh*, signed[6]
circa 1635	*Shah-Jahan standing on a globe*, signed[7]
circa 1640	*Holy men*, from the St Petersburg Album, signed[8]
circa 1640	*Galen watching a dispute*, from a *Gulistan* of Sa'di MS., signed[9]

1. Akurgal, Mango & Ettinghausen 1966, p.206.
2. See Beach 1992a, pp.133–150.
3. An important, hitherto unpublished *Jahangirnama* page in a private collection reveals the earliest phase of this development; see Fig.101.
4. Verma (1994, pp.326–29) unaccountably assigns the work of Payag to two separate artists, each with the same very unusual name; see also Beach 1978, pp.151–54; Dye 1991; and S.C. Welch 1995. While Payag began his career during Akbar's reign, it is an illustration made for the *Jahangirnama* that first shows his fully mature talents. *The Capture of Mirza Hasan* (Wilkinson 1948, pl.5), which is attributed here to Payag, illustrates an episode at the beginning of Jahangir's rule, although the painting must date from late in the reign. The work is not inscribed, but the smooth recession into space, the rejoicing in strongly characterized faces, and the movement within the dense mass of figures makes the scene immediately comparable to Payag's illustrations in the Windsor manuscript. That the building in the background is clearly recognizable as Akbar's tomb at Sikandara, outside Agra, indicates Payag's interest in physical accuracy. This is one of very few recognizable and accurate depictions of Mughal architecture by imperial artists of any period. Mirza Hasan was a follower of Jahangir's rebellious son Khusraw, and the episode is described in *Tuzuk-i Jahangiri*, I, p.54.
5. Wellesz 1952, pl.31.
6. Arnold & Wilkinson 1936, III, pl.76.
7. Beach 1992, fig.107.
8. Akimushkin 1996, pl.60.
9. Soudavar 1992, no.136 c.

RAMDAS

8, 11

It is probable that the Ramdas whose signature is on the page facing an audience scene by Murar (**9**) is the same artist that worked on a number of important imperial manuscripts during the late sixteenth century – several colleagues had similar careers. His style is technically solid, but he betrays little artistic personality. He worked certainly until about 1650, when a portrait of Shahnawaz Khan (see below) can be dated by the age of the subject.

OTHER KEY WORKS[1]

circa 1645	*Shahnawaz Khan*, from the Late Shah-Jahan Album, signed (see Binney 1973, no.62)

1. See also Verma 1994, pp.336–38, where the *Padshahnama* illustration is credited to a different and otherwise unknown painter also named Ramdas and dated to the late eighteenth century.

TEZDAST

35

Works by the artist who signed himself with the epithet Tezdast ('Swift of Hand') compare closely to those of Murar in the precision and weightlessness of the figures, and in their combination of careful physical description with an absence of interest in character or personality.

No additional works signed by or attributable to this painter are known.[1]

1. See also Verma 1994, p.365, where the artist's work is dated to the eighteenth century.

Fig.142 (see Appendix A).

Fig.143 (see Appendix B).

Fig.144 (see Appendix C).

Fig.145 (see Appendix D).

Fig.146 (see Appendix E).

MAJOR PAINTINGS RELATED TO THE WINDSOR PADSHAHNAMA MANUSCRIPT

by Milo Cleveland Beach

A.
Shah-Jahan enthroned with Mahabat Khan and a shaykh
Agra, *Diwan-i 'Amm*,
14 February 1628[1]
By Abid, dated 1629–30
Image area 30.2 × 19.9 cm

Inscriptions: [below throne] *raqam-i-'Âbid bar dar-i-Nâdir-al-Zamân Mashhad dar sana-i* II *bi-itmâm rasîd* ('by Abid, brother of Nadiruzzaman of Mashhad. Completed in the year II [Shahjahani = 1629–30]') [on shield to right] *Sayyid-Muzaffar Khan*

Mounted in the Late Shah-Jahan Album
Provenance: Vever, Demotte

Arthur M. Sackler Gallery, Smithsonian Institution, Washington, DC, acc.no.S86.406
Published: *Shah Jahan Nama,* pl.13; Lowry & Beach 1988, no.336; Marteau & Vever 1913, II, no.229, pl.CLXI

B.
Shah-jahan in darbar
Agra, *Diwan-i 'Amm*,
16 August 1628[2]
By an unknown artist, circa 1630
Image area 32 × 20.5 cm

Mounted in the St Petersburg Album

Institute of Oriental Studies, Academy of Sciences, St Petersburg, MS.E.14, p.34a
Published: Akimushkin *et al.* 1996, pl.125; Beach 1995a, fig.9; Akimushkin 1994, pp.103–104 and pl.49; Ivanova, Grek & Akimushkin 1962, pl.33

C.
A night celebration of the Prophet's birthday
Agra, *Diwan-i 'Amm*,
16 September 1633[3]
Attributed here to Bulaqi, son of Hoshang, circa 1635
Image area 30.5 × 23.1 cm

Detached from the St Petersburg Album

Freer Gallery of Art, Smithsonian Institution Washington, DC, acc.nos 42.17 and 42.18
Published: *Shah Jahan Nama*, pl.21; Begley 1987, pp.143–44 and figs 5–6; Beach 1981, nos 17e–f; S.C. Welch 1978, pls 31–32.[4]

D.
Shah-Jahan receives Prince Shah-Shuja' on his return from the Deccan
Lahore, *Diwan-i 'Amm*,
27 November 1634[5]
Signed by Abid, dated 1635–6[6]
Image area 30.3 × 21.4 cm

Bodleian Library, Oxford, MS.Ouseley Add. 173, fol.18[7]
Published: Begley 1987, pp.142–43 and fig.3;[8] Beach 1978, p.84; Binyon & Arnold 1921, pl.XXI

E.
Shah-Jahan enthroned
Agra, *Diwan-i 'Amm*,
26 January 1638[9]
By Abid, dated 1639
Image area 36.9 × 24.6 cm

Inscriptions: *raqam-i-walad-i-Aqa Riza Abid 23 shâbân* ('by 'Abid, son of Aqa Riza, in the regnal year of Shah-Jahan 13 [19 December 1639]')

Provenance: Ardeshir

San Diego Museum of Art, acc.no.1990: 0352
Published: *Shah Jahan Nama*, pl.25; Begley 1987, pp.144–46 and fig. 7; Das 1986, pl. XV; Desai 1985, no.12; Beach 1978, no.24; Binney 1973, no.58

Fig.147 (see Appendix F).

Fig.148 (see Appendix G).

Fig.149 (see Appendix H).

Fig.150 (see Appendix I).

Fig.151 (see Appendix J).

Fig.152 (see Appendix K).

F.
The surrender of a fortress
Badakhshan (perhaps at Bust), summer 1638[10]
By an unknown artist, circa 1640
Image area 34 × 24.2 cm

Provenance: Pozzi Collection

Musée Guimet, Paris, inv.no.MA3318
Published: Okada 1992, pl.51; Welch 1985, no.162b;[11] Blochet 1905–21, no.21 and pl.XXXIX

G.
Sa'id Khan defeats the armies of Siyawash
Near Qandahar (Afghanistan), April 1638[12]
By an unknown artist, circa 1640
Image area 33 × 23.6 cm

Mounted in the St Petersburg Album

Institute of Oriental Studies, Academy of Sciences, St Petersburg, MS.E.14, p.55a
Published: Akimushkin *et al.* 1996, pl.132;[13] Akimushkin 1994, p.108 and pl.62; Ivanova, Grek & Akimushkin 1962, pl. 44

H.
The Persian ambassador Yadgar Beg received at court
Lahore, *Diwan-i 'Amm*, 22 November 1638[14]
Attributed to Payag, circa 1640
Image area 34.5 × 23.8 cm

Bodleian Library, Oxford, MS.Ouseley Add.173, no.13[15]
Published: Topsfield 1994, no.14; *Shahjahannama*, pl.5; Begley 1987, pp.141–42 and fig.1; Beach 1978, p.84; Brown 1924, p. 87ff. and pl.XXIV; Binyon & Arnold 1921, pl.XXXVI

I.
Alimardan Khan arrives at court
Lahore, *Diwan-i 'Amm*, 22 November 1638[16]
By an unknown artist, circa 1640

Inscriptions:[17] [on robe of figure bowing, lower right] *Alimardan Khan*; [on figure in left center] *Ganj Ali Khan*

Provenance: Sita Ram Sahu Collection, Varanasi

Location unknown
Published: Topsfield 1994, p.34 and fig.6; Begley 1987, pp.141–42 and fig.2; Beach 1978, p. 84; Mehta 1926, pp.93–95 and pl.38

J.
Battle scene[18]
Attributed to Payag, circa 1640
Image area 34.3 × 23.9 cm

Provenance: Pozzi Collection

Private Collection
Published: S.C. Welch 1995, p.322 and pl.18; Beach 1992, pp.133–36 and fig.100; Begley 1987, pp.146–47 and fig.8; Beach 1978, no.25; S.C. Welch 1978, pl.33; S.C. Welch 1973, no. 66; Blochet 1930, pl.35

K.
Prince Muradbakhsh receiving Khusraw Sultan, son of Nazar Muhammad Khan
Encampment at Sirab (Balkh), June 1646[19]
By Fath Chand, circa 1648
Image area 34.9 × 24.1 cm

Inscription: [below throne] *amal-i Fath Chand* ('work of Fath Chand')[20]

Bharat Kala Bhavan, Varanasi, inv.no.5403
Published: Desai 1985, no.13; *Shah Jahan Nama*, pl.8; Krishnadasa 1955, pl.6

Fig.153 (see Appendix L).

Fig.154 (see Appendix M).

Fig.155 (see Appendix N).

Fig.156 (see Appendix O).

Fig.157 (see Appendix P).

Fig.158 (see Appendix Q).

L.
The battle of Shibarghan
Shibarghan (Balkh),
August 1646[21]
By Hunhar, circa 1648[22]
Image area 34.4 × 23.1 cm

Inscriptions:[23] [at top] 'The battle of Rustam Khan and Asalat Khan' [on horse, right margin] *Asalat Khân* [on robe lower right] *Rustam Khân* [lower right] *Amal-i Hunhar* ('work of Hunhar')

Metropolitan Museum of Art, New York, acc.no.1986.283
Published: Metropolitan Museum 1987, p.10

M.
Shah-Jahan receiving Prince Awrangzeb
Lahore, *Diwan-i 'Amm*,
22 March 1642[24]
By an unknown artist, circa 1645
Image area 30.5 × 22 cm

British Library, India Office Library and Records, MS.Add.Or.3853
Published: Falk & Archer 1981, no.80; Victoria and Albert Museum 1982, no.68

N.
Shah-Jahan receiving Prince Awrangzeb and Prince Muhammad Mu'azzam
Delhi, *Diwan-i 'Amm*,
12 January 1651[25]
By an unknown artist, circa 1660
Image area 33.5 × 25.5 cm

Mounted in the St Petersburg Album

Institute of Oriental Studies, Academy of Sciences, St Petersburg, MS. E.14, page 25a
Published: Akimushkin *et al.* 1996, pl.129; *Shah Jahan Nama*, pl.9; Ivanova, Grek & Akimushkin 1962, pl.34

O.
Darbar of Shah-Jahan
Image area 34.3 × 22.9 cm

Bharat Kala Bhavan, Varanasi
Published: Krishnadasa 1955, pl.8

P.
Shah-Jahan watching an elephant fight
Agra, possibly 15 October 1633[26]
Attributed to Bulaqi, son of Hoshang, ca. 1635
Image area 33 × 20.3 cm

Provenance: Latchford

Nasser D.Khalili Collection, London
Published: Pal *et al.* 1989, no.35; Spink & Son 1980, no.2

Q.
Shah-Jahan watching an elephant fight
Agra, Khass Mahal and riverbank, probably 22 August 1638[27]
By Bulaqi, son of Hoshang, dated 1639–40
Image area 34.8 1 × 21.9 cm

Inscriptions: [center] *Amal-i kamtarin-i khanaz dan Bulaqi walad-i Hoshang 1049* ('work of the most insignificant of the slaves, Bulaqi son of Hoshang 1049 [AD 1639–40]')

Provenance: Nasiruddin Shah; Petermann; Engel-Gros; Parish-Watson; Davis; Weaver[28]

Metropolitan Museum of Art, New York, acc.no.1989.135
Published: Metropolitan Museum 1990, pp.14–15; Sotheby's 1989, lot 63; Coomaraswamy 1924–25, I, pp.67–69, and II, pl.49

Fig.159 (see Appendix R).

Fig.160 (see Appendix S).

Fig.161 (see Appendix T).

Fig.162 (see Appendix U).

Fig.163 (see Appendix v).

Fig.164 (see Appendix w).

R.
Rustam Khan defeats the Qizilbash army
Near Qandahar (Afghanistan), June–July 1949[29]
By an unknown artist, circa 1650
Image area 33 × 23.5 cm

Mounted in the St Petersburg Album

Institute of Oriental Studies, Academy of Sciences, St Petersburg, MS. E14, p.54a
Published: Akimushkin *et al.* 1996, pl.133; Ivanova, Grek & Akimushkin 1962, pl.43

S.
Shah-Jahan in procession
Reportedly by Fazl, son of Abid,[30] circa 1655

Calico Museum, Ahmedabad

T.
Attendant figures
By an unknown artist, circa 1640

Raza Library, Rampur, Album 5, no.3

U.
Attendant figures
Possibly by Hunhar, circa 1645

Howard Hodgkin Collection, London

V.
Attendant figures
Possibly by Hunhar, circa 1645
Image area 24.7 × 17.5 cm

Museum Rietberg, Zurich
Published: Goswamy & Fischer 1987, no.101

W.
Attendant figures
circa 1650
Image area 37.5 × 22 cm

Art Institute of Chicago, acc.no.1975.555
Published: Beach 1978, no.26

X.
Attendant Figures
circa 1650

Indian Museum, Calcutta, inv.no.456
Published: Brown 1924, pls XLVI–XLVIII

Y.
Attendant Figures
By an unknown artist, circa 1650
Image area 33 × 23.5 cm

Mounted in the St Petersburg Album

Institute of Oriental Studies, Academy of Sciences, St Petersburg, MS.E14, p.13a
Published: Akimushkin *et al.* 1996, pl.132; Ivanova, Grek & Akimushkin 1962, pl.35

Fig.165 (see Appendix x).

Fig.166 (see Appendix y).

Fig.167 (see Appendix a).

MAJOR DRAWINGS RELATED TO THE WINDSOR PADSHAHNAMA MANUSCRIPT

a.
Shah-Jahan enthroned with Mahabat Khan and a shaykh
Agra, *Diwan-i 'Amm*,
8 March 1628[31]
By Anupchatar, circa 1630

British Museum,
acc.no.1920-9-17-013/27
Published: Martin 1971, II, pl.184; Binyon & Arnold 1921, pp.73–82 and pl.xx (where identifications are listed)

b.
Jahangir receives Prince Khurram on his return from the Deccan
Mandu, *Diwan-i 'Amm*,
12 October 1617[32]
By an unknown artist, circa 1640
Image area 29.6 × 20.8 cm

Published: Falk & Archer 1981, no.56;[33] Brown 1924, pl.LVIII
British Library, Indian Office Library Collection, Johnson Album 4, no.2
See above, Fig.82

c.
Jahangir receives Prince Khurram on his return from the Deccan
Mandu, *Diwan-i 'Amm*,
12 October 1617
By an unknown artist, circa 1640
Image area 24.45 × 16.8 cm

San Diego Museum of Art
Published: Binney 1973, no.55
See above, Fig.83

d.
Jahangir receives Prince Khurram on his return from the Mewar campaign
Ajmer, *Diwan-i 'Amm*,
19 February 1615[34]
Attributed to Murar, circa 1640
Image area 25.5 × 18 cm

Museum of Fine Arts, Boston,
acc.no.17.2696
Published: Coomaraswamy 1930, part VI, pl.XXXIII
See above, Fig.121

1. *Shah Jahan Nama*, p.19. The prominence of Mahabat Khan, who stands to the left of the Emperor, suggests that the episode is the restoration to him of the title *Khankhanan* on 14 February 1628, at the beginning of the accession ceremonies. This is confirmed by the unusual absence of the three older sons of Shah-Jahan. At the time they were traveling to Agra from Lahore with their grandfather, Asaf Khan.
2. *Shah Jahan Nama*, p.26. Begley (1987, p.151, n.16) identifies this episode as the second coronation ceremony of Shah-Jahan on 16 August 1628, but there is no distinctive feature of this rather routine painting to confirm this theory.
3. *Shahjahannama*, p.118 (see also Begley 1987, pp.143–44).
4. Welch attributes the scene to Murar, based on similarities perceived to **25–26** in the Windsor manuscript. That painting bears an inscription to the artist Bulaqi, son of Hoshang, however, and it is to him that this page is attributed here.
5. *Shahjahannama*, p.143
6. See Begley 1987, p.142.
7. The illustration has been crudely pasted onto pages of a nineteenth-century album that includes one other historical painting of Shah-Jahan's reign (see this Appendix H). The scene is surrounded by an original inner gold border remaining on the same sheet as the illustration; the original outer margin of the page, however, has been trimmed. The measurement of the original sheet including the fragmentary border is 31.5 × 22.8 cm.
8. Begley (1987, pp.142–43) believes this scene depicts Shah-Shuja' leaving Agra for the Deccan on 28 August 1633: however, the setting is not Agra but almost certainly Lahore (see also **44**).
9. *Shah Jahan Nama*, pp.219–20. Begley convincingly identifies this as the 46th solar birthday celebration on 25 January 1638 (Begley 1987, pp.144–46 and fig. 7). The figures in the throne area are (left to right) Abdullah Khan, Afzal Khan, Asaf Khan, Prince Muradbakhsh, Prince Dara-Shikoh, Shah-Jahan, Prince Shah-Shuja', an attendant, Ja'far Khan, Gaj Singh Rathor, and Jaswant Singh Rathor. At the bottom left, the saluting figure is Khan Dawran.
10. Since Qulij Khan is the most prominent equestrian figure, this probably represents his defeat of the fortress at Bust in Badakhshan – an identification proposed by Welch (S.C. Welch 1985, no. 162a). See *Shah Jahan Nama*, pp.230–31.
11. Welch assigns the work to Murar, an attribution not accepted here.
12. *Shahjahannama*, pp.226–27.
13. The main figure has been misidentified as Sayyid Khan Jahan Barha. It is actually Sa'id Khan, the Governor of Kabul (see *Ma'athir ul-Umara*, II, pp.674–78).
14. *Shah Jahan Nama*, pp.251–52. Topsfield identifies the scene as a different episode in the court reception of Alimardan Khan, who is identified by inscription in this Appendix I. The principals in the two illustrations are distinctly different men, however. Yadgar Beg and Alimardan Khan appeared at the imperial audience on the same day.
15. The illustration has been pasted into a crudely assembled album in European format which includes one other important historical scene from Shah-Jahan's reign (see this Appendix D).
16. *Shah Jahan Nama*, pp.251–52. Identification of the episode is confirmed by an inscription naming Alimardan Khan as the man making the gesture of homage. The close relationship of this composition and this Appendix H) suggests that the works may have been intended as a double page. This scene is clearly the right half of a double-page depiction, and both men arrived at court on 22 November 1638.
17. Given in Mehta 1926, p.93. Ganj Ali Khan was the father of Alimardan Khan.
18. The episode continues to elude convincing identification. For a summary of the proposals, see Begley 1978, pp.146–47; Beach 1978, pp.81–84.
19. *Shah Jahan Nama*, pp.342. The identification is by Krishnadasa (1955, pl.6), who also names the figure to the left of the platform as Alimardan Khan, and that to the right as Sa'adullah Khan.
20. For a list of known works related to Fath Chand, see Verma 1994, pp.155–56.
21. *Shah Jahan Nama*, pp.360–61. The episode is identified in Metropolitan Museum 1987, p.10. A closely related, but later drawing of the same episode is in the Fondation Custodia, Paris (Gahlin 1991, no.33).
22. On Hunhar, see Verma 1994, pp.176–79.
23. From Metropolitan Museum 1987, p.10.
24. Robert Skelton has suggested that this represents one of Awrangzeb's appearances at court in Lahore during 1649 (Victoria and Albert Museum 1982, no.68), although the *bakhshi* standing at the far left seems to be Salabat Khan, who died in 1644. Since one of the attendant courtiers (the man second from the left in the front row on the right) is Rao Chattarsal of Bundi, who accompanied Awrangzeb to court in March 1642, the scene may therefore represent Awrangzeb's arrival at court on 22 March 1642 (*Shah Jahan Nama*, p.292). For a biography of Rao Chattarsal, see *Ma'athir ul-Umara*, II, pp.722–24; and for a portrait, Grube, n.d., no. 249.
25. *Shah Jahan Nama*, p.452. Awrangzeb brought his two sons to court in 1651, but only one appears here, who, because of his apparent age, must be the younger son, Muhammad Mu'azzam (b. 1643). During the war of succession in which Awrangzeb imprisoned his father and defeated his brothers, the older Prince, Muhammad Sultan (b. 1639), sided with his uncle, Shah-Shuja', and on his father's succession was imprisoned for twelve years. His absence here suggests that the illustration was painted after he had fallen from favor – during his father's reign, not that of Shah-Jahan – and was intended for the *'Alamgirnama*. For a related drawing, see Binney 1973, no.64.
26. *Shah Jahan Nama*, pp.118–19. The identification is tentative. As in the almost identical scene in Appendix Q, the location is Agra, where Shah-Jahan lived between 19 June 1632 and early February 1634 and again from 16 January 1637 to 27 August 1638 (when the incident in the following illustration took place). The ages of the princes suggests the earlier residency; the dark-skinned figures at the lower right might then be Fath Khan and the young Husayn III, Nizamulmulk, on the occasion of the 43rd lunar birthday celebrations of the Emperor.
27. *Shah Jahan Nama*, pp.244–45. The third man from the left on the platform has long been identified as Alimardan Khan (Coomaraswamy 1924–5, pp.67–69), whereas he seems instead to be the Iranian ambassador Yadgar Beg (see also Appendix H). He arrived at court in Agra in early April 1638, and left on 22 August.
28. Sotheby's 1989, lot 63.
29. *Shah Jahan Nama*, pp.436–42. The commanding officer can be identified as Rustam Khan (see Arnold & Wilkinson 1936, III, pl.71).
30. I am most grateful to Dr B.N. Goswamy for informing me of this painting and its inscription.
31. This almost certainly represents the moment during the accession ceremonies, when Asaf Khan was appointed Wazir (*Shah Jahan Nama*, p.21).
32. Like Appendix b, it relates closely to **9**.
33. Falk and Archer note the inscription naming the painter Raja Manohar Singh, who is otherwise unknown. As the style does not relate to the Akbari painter Manohar, it is possible that the inscription needs careful examination.
34. This is probably a preliminary study for **38**.

Fig.168 The imperial family at the accession of Shah-Jahan in 1628 (detail of **10**).

Fig.169 The imperial family at Prince Dara-Shikoh's wedding in 1633 (detail of **25**).

Fig.170 The imperial family at Prince Awrangzeb's wedding in 1637 (detail of **43**).

Fig.171 The imperial family on Awrangzeb's arrival at court in 1640 (detail of **44**).

THE MUGHAL DYNASTY

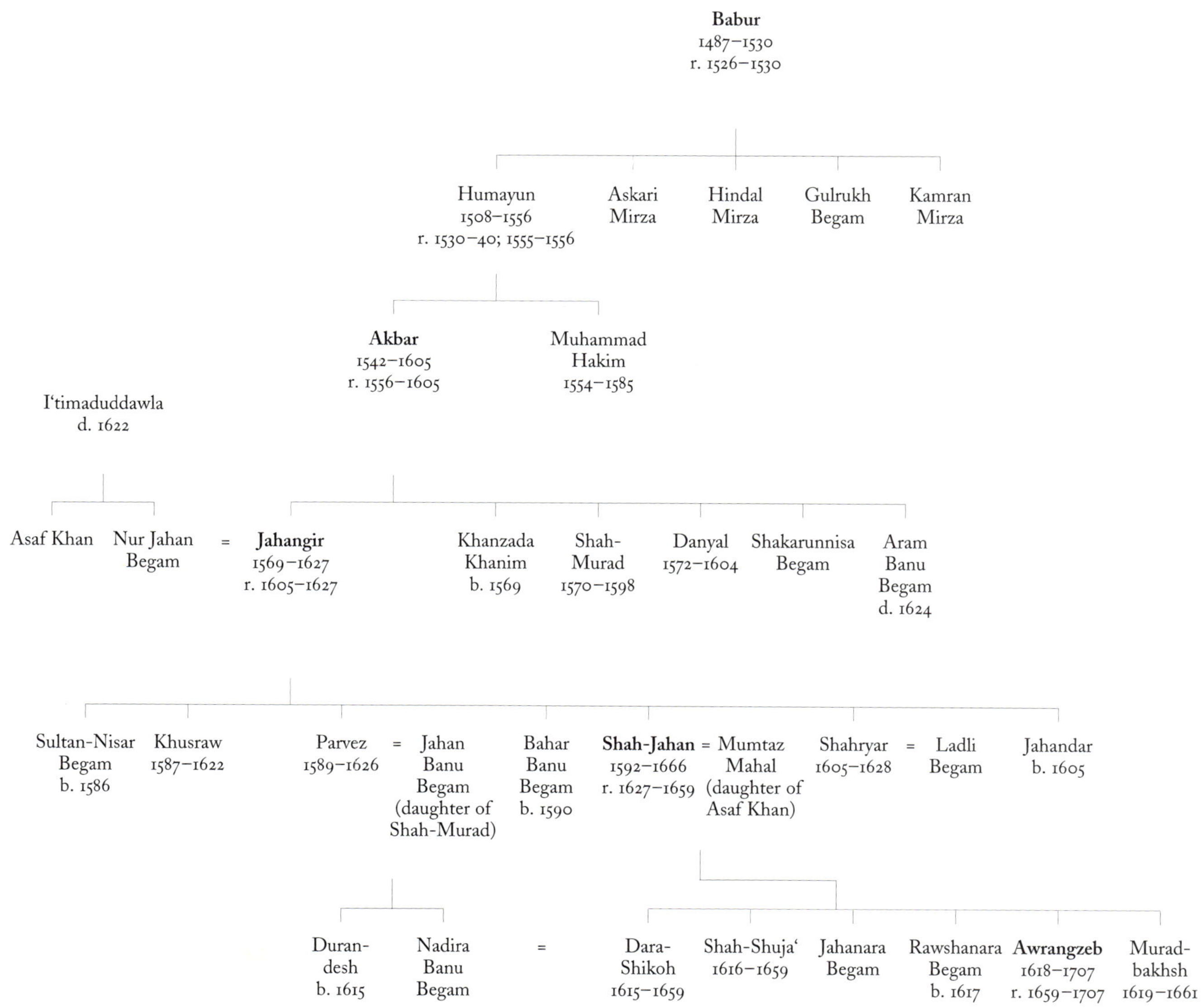

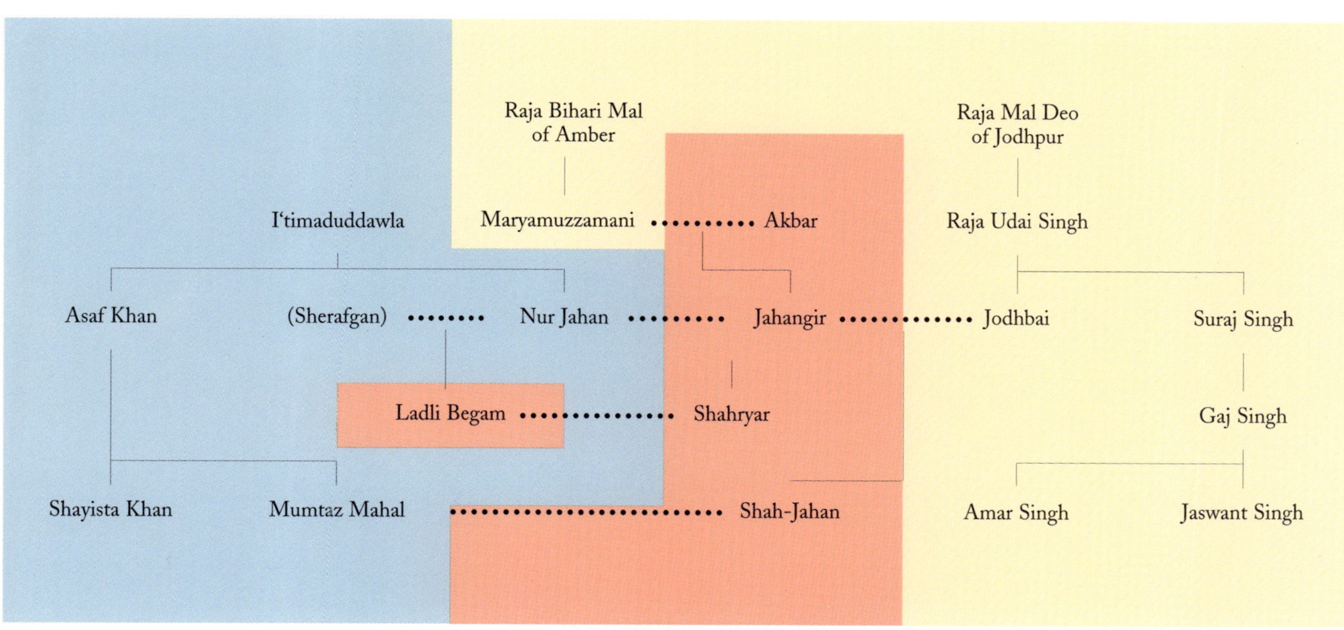

DYNASTIC MARRIAGES

Chart showing the intermarriages between the imperial Mughal family (pink), the family of I'timaduddawla (mauve), and the families of the Rajput rulers of Jodhpur and Amber (both yellow). The dotted lines represent marriages.

BIOGRAPHICAL LISTING

No.	Name[1]	Historical references See *Ma 'athir ul-Umara*	Padshahnama illustrations[2]	Comparative portraits[3]
I	**Abdullah Khan** *Firoz-Jang* (d. 1054/7 December 1644)	I, pp.97–105	5, (6), (9), (16), (19), (35), 38, (39), (45)	S.C. Welch *et al.* 1987, no.90 (Metropolitan Museum, New York acc.no.55.125.10.251)
2	**Abu'l-Hasan Turbati**, Khwaja (d.1042/1632–33) *Ruknussaltana*	I, pp.128–30	5, (9), (13), (17), (37), 38, (39)	Okada 1992, fig.178
3	**Afzal Khan Allami** (d.1048/7 January 1633) Mulla Shukrullah Shirazi *Mir Saman* 1628–1628/9 *Diwan-i Kul* 1628/9–1639	I, pp.149–153	5, (6), (9), 10, (13), (14), (17), (19), (21), (23), (25), (27), (32), (39), (43), (45)	Binyon & Arnold 192:, pl.xx (32)
4	**Alimardan Khan** (d.1067/1657–58)	I, pp.186–94	44	
5	**Allahverdi Khan** (d.1070/1660)	I, pp.668–72	10, (19), 38	Martin 1971, vol.II, pl.193
6	**Amar Singh Rathor,** Raja		(32)	
7	**Amar Singh Sisodia**, Rana		(6)	
8	**Ani Rai Singhdalan** Anup Singh Badjugar, Raja	I, pp.261–63	(30)	Binyon & Arnold 1921, pl.xx (35)
9	**Asaf Khan** (d.1051/1641–2)[4] Son of I'timaduddawla; brother of Nurjahan Begam (m.Jahangir); father of Mumtaz Mahal (m.Shah-Jahan), and Shayista Khan Abu'l-Hasan Asafjahi *Yaminuddawla* *Wakil* 1628–1641/2	I, pp.287–95	5, (9), (10), (13), (14),(17), (19), (21), (23), (25), (27), (32), (37), 38, (39), (43), (44), (45)	Okada 1992, fig.196; Martin 1971, II, pl.191; see also Figs 19, 33, and 120.
10	**Asalat Khan**[5]		10	
II	**Asalat Khan** (d.1057/1647–8) Mir Abdul-Hadi *Mir Bakhshi* 1645–1647/8	I, pp.295–97	(32),(33), 44	See Fig.112
12	**Awrangzeb**, Prince		(10), (13), (14), (17), (19), (25), (29), (43), (44), (45),	See Fig.18
13	**A'zam Khan** (d.1058/1648–49) Mir Muhammad Baqir Iradat Khan *Mir Saman* 1627/8–1628 (Erratum for *Mir Bakhshi* 1627/8–1628) *Diwan-i Kul* 1627–1628	I, pp.315–19	5, 10, (37)	Martin 1971, II, pl.189
14	**Bahadur Khan Rohilla** (d.1059/1649)	I, pp.340–48	10, (32), (37)	Kühnel & Goetz 1926, pl.35
15	**Balchand**		5	
16	**Bhao Singh**, Mirza		5	
17	**Bhim Singh Sisodia**, Maharaja	I, p.60	(9)	S.C. Welch *et al.* 1987, no.33; see also Fig.81
18	**Bikramajit**, Raja Sundardas Rai Raian	I, pp.412–19	5, (6), (9)	S.C. Welch *et al.* 1987, no.32; Leach 1986, no.24 and pl.VII; see also Fig.78
19	**Bithaldas Gaur**, Raja	I, pp.401–404	10	
20	**Dara-Shikoh**, Prince		(10), (13), (14),(17), (19), (23), (25), (27), (29), (32), (33), (41), (43), (44), (45), (46)	
21	**Dawlat Khan Mayi** Khawass Khan	I, pp.467–71	5	Binyon & Arnold 1921, pl.xx (19)

22	**Dilawar Khan Kakar**	I, pp.487–90	10	
23	**Firoz Khan** (d.1057/1647)	I, pp.564–65	5	
24	**Gaj Singh Rathor**, Raja (d.1048/1638) Son of Suraj Singh Rathor	I, pp.570–72	10, (19), (25), (32)	Beach 1978, no.34
25	**Gajpat Khan**		38	
26	**Gridhra La'l**		38	
27	**Habash Khan**		5	
28	**Hayat Khan** (d.1068/19 May 1658) Khidmatparast Khan	I, pp.623–24	(13), (14), (25), (30), (32), 38, (43), (44), (45)	Colnaghi 1976, no.113; see also Fig.107
29	**Inayat Khan** (d.1093/1682)	I, pp.678–80	44	
30	**Islam Khan Mashhadi** (d.1057/12 November 1647) Mir Abdus-Salam Ikhtisas Khan *Mir Bakhshi* 1634/35 *Diwan-i Kul* 1639/40–1645/6	I, pp.694–96	(9), 10, (14), 38	Okada 1992, fig.249; Falk & Archer 1981, p.412, no.108
31	**I'tibar Khan**	I, pp.704–705	5, (39)	Okada 1992, fig.197; see also Fig.73
32	**I'timad Khan** (?) Qizilbash Khan	II, pp.521–23	(13)	
33	**I'timad Rai**	I, p.99	(30)	
34	**I'timaduddawla** (d.1031/1622) Father of Asaf Khan, I'tiqad Khan, and Nur Jahan; grandfather of Shayista Khan and Mumtaz Mahal Ghiyas Beg Tehrani	II, pp.1072–79	5, (9), (37), 38, (39)	S.C. Welch *et al.* 1987, no.16
35	**I'tiqad Khan** (d.1060/1650) Son of I'timadaddawla; brother of Asaf Khan and Mumtaz Mahal Mirza Shapur	I, pp.714–15	38	
36	**Jadu Khan**		38	
37	**Ja'far Khan** (d.1081/1670) Son of Sadiq Khan; nephew and son-in-law of Asaf Khan *Mir Bakhshi* 1647/48–1649/50 *Diwan-i Kul* 1657–58	I, pp.722–23	10, (13), (23), (25), (32), (33), (43), (44), (45)	Martin 1971, II, pl.187; see also Figs 105 and 114
38	**Jahangir Padshah**		(5), (9), (30), (37), (38), (39)	See Fig.139
39	**Jai Singh Kachhwaha**, Mirza Raja (d.1077/1667)	I, pp.731–35	10, (29), (32)	
40	**Jannisar Khan**	I, pp.749–50	10	
41	**Jaswant Singh Rathor**, Maharaja (d.1089/1678) Grandson of Raja Surat Singh; son of Raja Gaj Singh	I, pp.754–56	(32), (44)	Leach 1986, no.88; Martin 1971, II, pl.96
42	**Karan Singh Sisodia** (d.1038/1628) Son of Rana Amar Singh; brother of Raja Bhim Singh	I, pp.761–64	5, (9), 38	Kühnel & Goetz 1926, pl.35
43	**Khan Alam**[6] Mirza Barkhurdar	I, pp.389–92	5, 10	Sotheby's 1996, lot 6; Okada 1992, no.192; Binyon & Arnold 1921, pl.xx (17)

No.	Name	Historical references See *Ma 'athir ul-Umara*	Padshahnama illustrations	Comparative portraits
44	**Khan Dawran Bahadur** *Nusrat-Jang* (d.1055/21 June 1645) Khwaja Sabir Nasiri Khan	I, pp.778–83	10, (18), (19), (31), (32), (35), (36), (40), (43)	S.C.Welch *et al.* 1987, no.71; Arnold and Wilkinson 1936, III, pl.72
45	**Khan Jahan Barha** (d.1055/1645) Sayyid Muzaffar Khan	I, pp.791–95	10, (13), 16, (19), (23), (25), (27), (29), (35), (43)	S.C.Welch *et al.* 1987, no.66
46	**Khan Jahan Lodi** (d.1040/1631) Pir Khan	I, pp.795–804	(16)	
47	**Khan Zaman** (d.1047/1637) Son of Mahabat Khan Amanullah	I, pp.212–19	31	
48	**Khidmat Khan**		5	
49	**Kishandas**		5	Marteau & Vever 1913, II, fig.254
50	**Madho Singh Hada** (d.1057/1647) Son of Rao Ratan Singh Hada	II, pp.1–3	(16)	Bautze 1986
51	**Mahabat Khan** (d.1044/1634) Father of Khan Zaman Zamana Beg	II, pp.9–28	5, (9), 10, (31), (37), 38	Kühnel & Goetz 1926, pl.35; see also Fig.142
52	**Makramat Khan** (d.1059/1649) *Mir Saman* 1636/7–1641/2	II, pp.264–76	(14), (32), (43), 44	Binyon & Arnold 1921, pl.XX (42)
53	**Mir Jumla** (d.1047/22 August 1637) Muhammad-Amin *Mir Saman* 1628/29–1634/5 *Mir Bakhshi* 1634/35–1637/8	II, pp.117–21	10, (13), (14), (19), (23)	Clarke 1922, pl.II, no.14
54	**Mirmiran** Son of a sister of Shah Tahmasp Safavi; m. daughter of Asaf Khan	I, pp.772–74	(6), (9), 38	
55	**Mu'azzam Khan** (d.1036/1626–27) Grandson of Shaykh Salim Chishti Shaykh Bayazid	II, pp.87–88	38	
56	**Muhammad-Ali Beg** (Persian Ambassador)		(13), (17)	S.C. Welch *et al.* 1987, no.87; see also Fig.87
57	**Muhammad Taqi** (d.1035/1626) Shahquli Khan	II, pp.212–14	(6)	
58	**Mukhlis Khan** (d.1047/1637–38)	II, pp.249–51	10, (13), (19)	Binyon & Arnold 1921, pl.XX (44)
59	**Muqarrab Khan** (d.1056/1646) Shaykh Hasan	I, pp.616–17	5, 10, (14), (37), (39)	Victoria and Albert Museum 1982, no.70; Arnold & Wilkinson, III, no.71
60	**Muradbakhsh**, Prince		(10), (13), (14), (17), (19), (23), (25), (27), (32), (43), (44), (45)	
61	**Murar**		9	
62	**Murshidquli Khan Turkman** (d.1047/1637–38) Murawwat Khan	II, pp.309–14	10, (13)	
63	**Murtaza Khan** (d.1025/1616–17) Shaykh Farid Bukhari	I, pp.521–27	5, (37)	Clarke 1922, pl.II, no.12
64	**Musavi Khan** (d.1054/16 April 1644) *Sadr* 1627/8–1644	II, pp.326–27	5, (9), 10, (14), (17), (23), (32), 38, (43), 44	Binyon & Arnold 1921, pl.XX (30)

65	**Mu'tamad Khan** (d.1049/1639) Muhammad Sharif *Mir Bakhshi* 1637/8–1639/40	II, pp.344–47	10, (32), (43)	
66	**Mu'taqid Khan** Mirza Maki	II, pp.347–50	10	
67	**Muzaffar Khan** Abdurrazzak Ma'muri	II, pp.357–59	10, (19)	
68	**Nazir, Shaykh**		(6), 10, (13), (26), (43)	
69	**Payag**		(39)	
70	**Qilich Khan Turani** (d.1064/1654)[7]	II, pp.541–44	(9), 10, 38	S.C. Welch *et al.* 1987, no.70
71	**Ramdas Kachhwaha**, Raja (d.1022/1613–14)	II, pp.587–89	(30)	
72	**Ratan Hada, Rao** (d.1040/1630–31) Father of Madho Singh Hada	II, pp.603–605	10	Bautze 1986
73	**Rai Khan**		5	
74	**Riza Bahadur** (d.1629–30) Khidmatparast Khan[8]	I, pp.811–13	5, (9), 10, (37)	
75	**Rustam Safavi**, Mirza (d.1051/1641–42) Daughters married Sultan Parvez, elder brother of Shah-Jahan, and Shah-Shuja'.	II, pp.631–37	5, (9), (14), 38	Okada 1992, fig.179
76	**Sadiq Khan** (d.1043/3 September 1633) Son of Aqa Tahir, brother of I'timaduddawla; m. daughter of I'timaduddawla; father of Ja'far Khan and Salabat Khan *Mir Bakhshi* 1628–1634	II, pp.656–58	5, (9), 10, (17), (19), (21), (37), (39)	Binyon & Arnold 1921, pls XX (27) and XXVII
77	**Safdar Khan** (d.1055/1645) Khwaja Qasim	II, pp.665–67	10	
78	**Safshikan** (d.1055/1645) Mirza Lashkari	II, pp.671–72	44	
79	**Salabat Khan** (d.1054/1644) Son of Sadiq Khan Raushan Damir *Mir Bakhshi* 1639/40–1644/45	II, pp.702–703	(33), 44	
80	**Sarafraz Khan**[9]	II, pp.714–16 (?)	10, (13), (14), (17), (22), (39), (43)	Leach 1986, no.27
81	**Sarbarah Khan**		5	
82	**Sardar Khan** (d.1063/1653)	II, pp.713–14	10	
83	**Sarla ?**		38	
84	**Sayf Khan** (d.1049/1639–40) m. daughter of Asaf Khan Mirza Safi	II, pp.689–92	10	
85	**Shah-Jahan Padshah**		4, (5), 6, (9), (10), (12), (14), (17), (19), (25), (29), (30), (32), (33), (37), (38), (39), (41), (43), (44), (45), (46)	See Figs 20, 105, 122

No.	Name	Historical references See *Ma'athir ul-Umara*	Padshahnama illustrations	Comparative portraits
86	**Shahnawaz Khan Safavi** (d.1069–1659) Son of Mirza Rustam Safavi; daughters m. Awrangzeb and Muradbakhsh	II, pp.767–71	10, (32), (44)	Leach 1986, fig.28ii; Binney 1973, no.62; Binyon & Arnold 1921, pl.XX (31)
87	**Shah-Shuja'**, Prince		(10), (13), (14), (17), (19), (23), (25), (27), (29), (32), (41), (43), (44), (45), (46)	Beach 1978, no.34
88	**Shayista Khan** (d.1105/1694) Son of Asaf Khan Mirza Abu-Talib	II, pp.825–36	10, (21), (23), (43), 44	Falk & Archer 1981, p. 412, no.109; Martin 1971, vol.II, pl.2C4; see also Fig.131
89	**Suraj Singh Rathor**, Raja (d.1028/1619) Sister m. Jahangir and was mother of Shah-Jahan	II, pp.914–17	5, (6), (9), (37), 38	S.C. Welch *et al.* 1987, no.28; see also Fig.72
90	**Wazir Khan** (d.1050/1640–41)[10] Hakim Alimuddin *Diwan-i Kul* 1627–8	II, pp.981–83	5, 10, (32), 38, 44	
91	**Zulfaqar Khan** Muhammad Beg		6	Skelton 1971, pl.LXXXVIII

1. See also *Ma'athir ul-Umara* and M. Athar Ali 1985. Those essential texts have provided the basis for the names (those commonly given are shown in bold type), alternate names and titles, and dates listed here. Titles for the chief ministers (those holding the positions of *Wazir*, *Diwan* or *Wakil*, *Mir Bakhshi*, and *Sadr* under the Emperor Shah-Jahan) are also listed. When known, the Muslim era date precedes that of the Christian calendar.
2. Illustration numbers without parentheses indicate that the identification is inscribed within the illustration; numbers in parentheses are for identifications made by comparison with separate inscribed portraits.
3. This is a selected list, with preference given to easily available publications when possible.
4. The two major ministerial positions at court – those of *Wazir* and *Diwan* – were frequently held within one family, that of I'timaduddawla (see the genealogy on page 229, lower chart). His brother, Aqa Tahir, was the father of Sadiq Khan, and several of his descendants served as *Mir Bakhshi* to the Emperor. I'timaduddawla and Aqa Tahir were the sons of Muhammad Sharif Herati.
5. The identity of this figure is uncertain. He may be a son of Khan Jahan Lodi (*Ma'athir ul-Umara*, I, p.804).
6. The inscribed depiction of Khan Alam in **5** is misleading, for he had been sent to Iran in 1613 on an embassy to Shah Abbas Safavi and returned only in 1620. The purpose of including his figure here is uncertain.
7. The figures identified by inscriptions on **10** and **38** as Qilich Khan appear to be different men, although Qilich Khan Turani held this title from the time he began service under Prince Khurram. Further information on the title *Qilich Khan* is therefore necessary.
8. The title *Khidmat Parast Khan* means 'the khan devoted to service', and following Riza Bahadur's death was held by Hayat Khan (*Ma'athir ul-Umara*, I, p.811).
9. The identity of the man identified by inscription as Sarafraz Khan is actually uncertain. The tentative reference given here is to Sarafraz Khan Chagta, but a second possibility is Abdullah, the son of Aziz Koka Muhammad Khan A zam, to whom Jahangir gave the titles Sardar Khan and Sarafraz Khan (M. Athar Ali 1985, p.354). No reference materials explain the frequency and prominence with which this man is depicted.
10. The man identified by inscription as Wazir Khan in **10** appears to be a different person than that so identified in the other illustrations listed.

GLOSSARY OF SELECTED TERMS

adab	politeness, good breeding
'alam	a standard, banner
amir	a military commander, also a title of rank
ashrafi	a coin
aspak	a type of curved tent
bâdar	a large net used as a circular barricade in hunting
bagh	a garden, usually a formal one
bakhshi	a political or military official
begam	'Lady,' a title of respect
charkhi	a pinwheel type of firework mounted on sticks and used to control elephants
charqab	a type of garment or textile awarded by the emperor, perhaps an embroidered cloud collar
chatr	a parasol
chhajja	projecting eaves supported by brackets
Chihil Sutun	literally a 'forty-pillared' pavilion or hall of Persian design which became the favored constructional type for the *Diwan-i 'Amm* (q.v.) during Shah-Jahan's reign
chinikhana	'china room'; also a decorative motif employing wall niches, which originally held bottles, vases and other artifacts
darbar, durbar	an official audience given by the emperor
dargah	a shrine
darshan	'sight,' or 'viewing' (see *jharoka*)
Diwan	the Finance Minister (see also *Wazir*)
Diwan-i Khass	the Hall of Private Audiences
Diwan-i 'Amm	the Hall of Public Audiences
farman	a decree
firangi	'Frank,' a European
ghari	a measurement of time equivalent to 24 minutes
haveli	a mansion
jahaz	wedding gifts
jali	a pierced stone screen
jamdhar	a type of dagger
jharoka	the raised seat or viewing window used by the Mughal emperor for official appearances
jharoka-i darshan	a viewing window on the outer wall of the palace
jharoka-i khass-u-'amm	the raised seat of the emperor in the *Diwan-i 'Amm* (q.v.), projecting from a raised gallery and sheltered by a cupola
jigha	a turban ornament
julus	the accession of the emperor
kangura	battlements, crenelations
khan	a title of military or political rank
khanazad	'house born,' used of someone born to imperial service by virtue of having a parent in service
khass	personal, belonging or pertaining personally to the ruler
khawass	the court elite
khutba	part of the Friday congregational prayer in which the name of the ruler is announced
khwaja	a title of learning, generally accorded to Sufi masters
körünüsh	(*kornish*) the formal 'kowtow' or prostration made before the emperor
lakh or lac	100,000
mahout	the driver of a tamed elephant
mansab	rank in the Mughal meritocracy defined by two numbers, indicating personal rank (*zat*) and the number of horsemen the holder was expected to field in battle, e.g. 2500/2000
mansabdar	a holder of office and rank (see above)
maund	(from the Persian *man*) a measurement of weight equivalent to a little over 82 pounds or 38.25 kilograms
mihrab	the niche in the *qibla* (q.v.) wall of a mosque that indicates the direction of Mecca
mir	an abbreviated form of amir (q.v.); the title *mir* also precedes the given names of *sayyids* (q.v.)
Mir Saman	the Imperial Steward
Mir Bakhshi	the chief *bakhshi* and fiscal administrator of the empire
mithcal	(from the Persian *mithqal*) a measurement of weight, equivalent to 0.161 ounces, or 4.6 grams
mohur	(from the Persian *muhr*) a coin
muraqqa'	an album of paintings and/or calligraphic specimens
naqara	an Indian kettle drum
naqarakhana	'drum-house,' the imperial orchestra that accompanied the ceremonial proceedings of the court; also the structure in which the orchestra sat
nawbat	the sounding of drums and clarions at a nobleman's or emperor's gate; also used loosely for the *nawbatkhana* (q.v.)
nawbatkhana	comparable to *naqarakhana* (q.v.)
Nawroz	the Persian New Year, the vernal equinox
nazr	an amount of money or equivalent given in fulfillment of a vow; also a gift presented to the emperor
nikah	the wedding ceremony
nisar	gifts strewn over the recipient
padshah	emperor
pahr	'watch,' a three-hour division of the day
parchin kari	stone intarsia
pargana	an administrative district or estate
peshkash	a present
phul-katara	an Indian thrusting dagger
pietre dure	a highly specialized form of stone intarsia (q.v.)
qarina	'counter image,' a compositional scheme of bilateral symmetry highly favored during Shah-Jahan's reign
qibla	the direction, towards the Kaaba in Mecca, which all Muslims face when praying
rai	a variant derivative of *raja* (q.v.)
raja	(Sanskrit) a ruler in Rajputana
Rajput	Hindu warrior nobles of Western India
rana	a variant derivative of *raja* (q.v.)
rao	a variant derivative of *raja* (q.v.)
rupee	a coin
sachiq	gifts sent from a bridegroom's house to the bride's prior to the wedding
Sadr	the Supervisor of Religious Endowments
sahib	'Sir,' a title of respect
sangandaz	embrasures
saraparda	a wall of canvas
sayyid	a lineal descendant of the Prophet Muhammad
sehra	a ceremonial veil worn by a bridegroom
shah	king, ruler, prince
shamsa	a sunburst design
shaykh	a title given to a figure of religious prominence
sher hajji	an outer fortification wall
suba	a province
subadar	a governor of a *suba* (q.v.)
surkh	a measurement of the weight of gems, equivalent to 0.875 of a carat
sutun	a column or stake
tanga	a coin
tânk	a measurement of the weight of gems, equivalent to 24 *surkhs* (q.v.) or 21 carats
tola	a measurement of weight, equivalent to slightly more than two mithcals (q.v.)
tümän tugh	the banner of a commander of 10,000 troops
Wakil	the Prime Minister
Wazir	the Finance Minister (see also *Diwan*)
zamindar	a land holder

CONCORDANCES

CONCORDANCE
BY PLATE/CATALOGUE NUMBER

1	FOLIO 1B	RL OMS 1603
2	FOLIO 2A	RL OMS 1604
3	FOLIO 2B	RL OMS 1605
4	FOLIO 3A	RL OMS 1606
5	FOLIO 43B	RL OMS 1607
6	FOLIO 46B	RL OMS 1608
7	FOLIO 47A	RL OMS 1609
8	FOLIO 48B	RL OMS 1610
9	FOLIO 49A	RL OMS 1611
10	FOLIO 50B	RL OMS 1612
11	FOLIO 51A	RL OMS 1613
12	FOLIO 70B	RL OMS 1614
13	FOLIO 71A	RL OMS 1615
14	FOLIO 72B	RL OMS 1616
15	FOLIO 92B	RL OMS 1617
16	FOLIO 94B	RL OMS 1618
17	FOLIO 98B	RL OMS 1619
18	FOLIO 102B	RL OMS 1620
19	FOLIO 116B	RL OMS 1621
20	FOLIO 117A	RL OMS 1622
21	FOLIO 120B	RL OMS 1623
22	FOLIO 121A	RL OMS 1624
23	FOLIO 122B	RL OMS 1625
24	FOLIO 123A	RL OMS 1626
25	FOLIO 124B	RL OMS 1627
26	FOLIO 125A	RL OMS 1628
27	FOLIO 126B	RL OMS 1629
28	FOLIO 127A	RL OMS 1630
29	FOLIO 134A	RL OMS 1631
30	FOLIO 135B	RL OMS 1632
31	FOLIO 144A	RL OMS 1633
32	FOLIO 147B	RL OMS 1634
33	FOLIO 165A	RL OMS 1635
34	FOLIO 166B	RL OMS 1636
35	FOLIO 174A	RL OMS 1637
36	FOLIO 176B	RL OMS 1638
37	FOLIO 192B	RL OMS 1639
38	FOLIO 194B	RL OMS 1640
39	FOLIO 195A	RL OMS 1641
40	FOLIO 204B	RL OMS 1642
41	FOLIO 205B	RL OMS 1643
42	FOLIO 206A	RL OMS 1644
43	FOLIO 214B	RL OMS 1645
44	FOLIO 217B	RL OMS 1646
45	FOLIO 218B	RL OMS 1647
46	FOLIO 220B	RL OMS 1648

CONCORDANCE
BY PADSHAHNAMA FOLIO NUMBER

FOLIO 1B	1	RL OMS 1603
FOLIO 2A	2	RL OMS 1604
FOLIO 2B	3	RL OMS 1605
FOLIO 3A	4	RL OMS 1606
FOLIO 43B	5	RL OMS 1607
FOLIO 46B	6	RL OMS 1608
FOLIO 47A	7	RL OMS 1609
FOLIO 48B	8	RL OMS 1610
FOLIO 49A	9	RL OMS 1611
FOLIO 50B	10	RL OMS 1612
FOLIO 51A	11	RL OMS 1613
FOLIO 70B	12	RL OMS 1614
FOLIO 71A	13	RL OMS 1615
FOLIO 72B	14	RL OMS 1616
FOLIO 92B	15	RL OMS 1617
FOLIO 94B	16	RL OMS 1618
FOLIO 98B	17	RL OMS 1619
FOLIO 102B	18	RL OMS 1620
FOLIO 116B	19	RL OMS 1621
FOLIO 117A	20	RL OMS 1622
FOLIO 120B	21	RL OMS 1623
FOLIO 121A	22	RL OMS 1624
FOLIO 122B	23	RL OMS 1625
FOLIO 123A	24	RL OMS 1626
FOLIO 124B	25	RL OMS 1627
FOLIO 125A	26	RL OMS 1628
FOLIO 126B	27	RL OMS 1629
FOLIO 127A	28	RL OMS 1630
FOLIO 134A	29	RL OMS 1631
FOLIO 135B	30	RL OMS 1632
FOLIO 144A	31	RL OMS 1633
FOLIO 147B	32	RL OMS 1634
FOLIO 165A	33	RL OMS 1635
FOLIO 166B	34	RL OMS 1636
FOLIO 174A	35	RL OMS 1637
FOLIO 176B	36	RL OMS 1638
FOLIO 192B	37	RL OMS 1639
FOLIO 194B	38	RL OMS 1640
FOLIO 195A	39	RL OMS 1641
FOLIO 204B	40	RL OMS 1642
FOLIO 205B	41	RL OMS 1643
FOLIO 206A	42	RL OMS 1644
FOLIO 214B	43	RL OMS 1645
FOLIO 217B	44	RL OMS 1646
FOLIO 218B	45	RL OMS 1647
FOLIO 220B	46	RL OMS 1648

CONCORDANCE
BY ROYAL LIBRARY ACCESSION NUMBER

RL OMS 1603	1	FOLIO 1B
RL OMS 1604	2	FOLIO 2A
RL OMS 1605	3	FOLIO 2B
RL OMS 1606	4	FOLIO 3A
RL OMS 1607	5	FOLIO 43B
RL OMS 1608	6	FOLIO 46B
RL OMS 1609	7	FOLIO 47A
RL OMS 1610	8	FOLIO 48B
RL OMS 1611	9	FOLIO 49A
RL OMS 1612	10	FOLIO 50B
RL OMS 1613	11	FOLIO 51A
RL OMS 1614	12	FOLIO 70B
RL OMS 1615	13	FOLIO 71A
RL OMS 1616	14	FOLIO 72B
RL OMS 1617	15	FOLIO 92B
RL OMS 1618	16	FOLIO 94B
RL OMS 1619	17	FOLIO 98B
RL OMS 1620	18	FOLIO 102B
RL OMS 1621	19	FOLIO 116B
RL OMS 1622	20	FOLIO 117A
RL OMS 1623	21	FOLIO 120B
RL OMS 1624	22	FOLIO 121A
RL OMS 1625	23	FOLIO 122B
RL OMS 1626	24	FOLIO 123A
RL OMS 1627	25	FOLIO 124B
RL OMS 1628	26	FOLIO 125A
RL OMS 1629	27	FOLIO 126B
RL OMS 1630	28	FOLIO 127A
RL OMS 1631	29	FOLIO 134A
RL OMS 1632	30	FOLIO 135B
RL OMS 1633	31	FOLIO 144A
RL OMS 1634	32	FOLIO 147B
RL OMS 1635	33	FOLIO 165A
RL OMS 1636	34	FOLIO 166B
RL OMS 1637	35	FOLIO 174A
RL OMS 1638	36	FOLIO 176B
RL OMS 1639	37	FOLIO 192B
RL OMS 1640	38	FOLIO 194B
RL OMS 1641	39	FOLIO 195A
RL OMS 1642	40	FOLIO 204B
RL OMS 1643	41	FOLIO 205B
RL OMS 1644	42	FOLIO 206A
RL OMS 1645	43	FOLIO 214B
RL OMS 1646	44	FOLIO 217B
RL OMS 1647	45	FOLIO 218B
RL OMS 1648	46	FOLIO 220B

LIST OF FIGURES

Unless otherwise stated, all images are from Mughal India, and all are in opaque watercolor on paper. Images from the Windsor *Padshahnama* itself are cross-referenced to the full-page plates and the technical information presented in the sections entitled *The Manuscript* (pages 22–111) and *Catalogue* (pages 157–211).

Frontispiece Jahangir and Prince Khurram (Shah-Jahan) in the *jharoka* (detail of **5**).

Fig.1 The death of Khan Jahan Lodi (detail of **16**).

Fig.2 Folios 1b–2a of the Windsor *Padshahnama* manuscript, showing the *shamsa*s and 18th-century margins (see **1–2**).

Fig.3 Folios 2b–3a of the manuscript, showing portraits of Timur and Shah-Jahan with illuminated frames and 18th-century margins (see **3–4**).

Fig.4 Folio 96b of the Windsor *Padshahnama* with eleven lines of script.

Fig.5 Folio 49b of the Windsor *Padshahnama* with three lines of script.

Fig.6 Detail of the left-hand edge of **13**, showing figures extending over the marginal ruling.

Fig.7 Courtiers attend the weighing of Shah-Jahan (detail of **13**).

Fig.8 Unknown artist, *Yusuf is presented with a mirror by the traveler from Canaan*, from a manuscript of the *Tuhfat al-ahrar* ('Gift of the Free') of Jami. Bukhara, Uzbekistan, text dated 1515–16, illustrations added in the mid-16th century; folio area 24.3 × 15.9 cm. Arthur M. Sackler Gallery, Smithsonian Institution, Washington, DC (acc.no. S86.0052, fol.58b).

Fig.9 Unknown artist, *Audience scene*, from a manuscript of the *Dewal Devi Khizr Khan*. Text dated 1567–68; folio area 32 × 21.3 cm. National Museum of India, New Delhi.

Fig.10 Unknown artist, *The throne of prosperity*, from a manuscript of the *Nujum al-Ulum* ('The Star of the Sciences'). Bijapur, Indian Sultanate, dated 1570–71; folio area 25.8 × 16 cm. Chester Beatty Library, Dublin (MS.2, fol.191r).

Fig.11 Unknown artist, *The court of the queen of Sheba*, from a manuscript of the *Sulaymannama* ('The Story of Solomon') of Sharafuddin Musa. Ottoman Turkey, before 1512; folio area 44.3 × 31 cm. Chester Beatty Library, Dublin (MS.406, fol.2a).

Fig.12 Husayn Naqqash and Kesu, *Akbar receiving prisoners-of-war from Gujarat* (right half), from an *Akbarnama* manuscript. Circa 1585; Victoria and Albert Museum, London (acc.no.2-1896 I.S.113/117).

Fig.13 Abu'l-Hasan, *Ceremonies at the accession of Jahangir*, from a *Jahangirnama* manuscript. Dated 1605 (left half) and circa 1618 (right half); image area 37.8 × 22 cm. From the St Petersburg Album, assembled mid-18th century. Institute for the Peoples of Asia, Academy of Sciences, St Petersburg (MS.E.14, pages 21–22).

Fig.14 Attributed to Manohar, *Prince Khusraw brought to Jahangir as a captive*, from a *Jahangirnama* manuscript. Circa 1610; image area 35 × 23 cm. Raza Library, Rampur (Album 1, fol.6a).

Fig.15 Bulaqi, 'son of Hoshang,' *Jahangir receiving Muqarrab Khan with exotic animals from Goa*, from a *Jahangirnama* manuscript. Circa 1625; image area 32.8 × 19.9 cm. Raza Library, Rampur (Album 1, fol.7a).

Fig.16 La'lchand, *The submission of Rana Amar Singh of Mewar to Prince Khurram*, from the Windsor *Padshahnama* (see **6**).

Fig.17 Nanha, *The submission of Rana Amar Singh of Mewar to Prince Khurram*, from a *Jahangirnama* manuscript. Circa 1625; image area 31 × 19.8 cm. Victoria and Albert Museum, London (acc.no. I.S.185-1984).

Fig.18 Attributed to Hunhar, *Awrangzeb in a shaft of light*. Circa 1660; image area 29 × 17 cm. From the St Petersburg Album, assembled mid-18th century. Freer Gallery of Art, Smithsonian Institution, Washington, DC (acc.no.F1996.1b).

Fig.19 Bichitr, *Shah-Jahan with Asaf Khan*. Circa 1640; image area 26.3 × 18.7 cm. From the Late Shah-Jahan Album, assembled circa 1650. Arthur M. Sackler Gallery, Smithsonian Institution, Washington, DC (acc.no.S86.0403).

Fig.20 Unknown artist, *Shah-Jahan with Asaf Khan*. Early 18th century; image area 27.5 × 19.5 cm. From the St Petersburg Album, assembled mid-18th century. Institute for the Peoples of Asia, Academy of Sciences, St Petersburg (MS.E.14, page 27).

Fig.21 Unknown artist, *Bahadur Shah enthroned*. Circa 1710; image area 42.2. × 30.5 cm. Collection of Sven Gahlin, London.

Fig.22 Unknown artist, *Prince Azimushshan enthroned with Khwaja Khizr*. Circa 1713; image area 34.7 × 22.3 cm. Bibliothèque Nationale, Paris (MSS Or.Smith-Lesoeuf 249 pièce 6557).

Fig.23 Unknown artist, *The wedding procession of Dara-Shikoh*. Lucknow, mid-18th century; image area 40 × 28 cm. Collection of Hashem Khosrovani.

Fig.24 Unknown artist, *Krishna celebrates Holi festival*. Punjab Hills at Mandi, circa 1740. Private collection.

Fig.25 Attributed to Payag, *Officers and wise men*. Circa 1640; folio area 55.5 × 35 cm. From the Ardeshir Album, assembled circa 1725. San Diego Museum of Art, acc.no.1990:0350. Edwin Binney 3rd Collection.

Fig.26 The fort at Daulatabad (detail of **31**).

Fig.27 Kesu Kalan and Madhu Kalan, *Akbar receiving the submission of the rebel brothers Ali Quli Khan Zaman and Bahadur Khan in 1561*, from a manuscript of the *Akbarnama*. Circa 1590. Victoria & Albert Museum, London (acc.no.2-1896 I.S.20/117).

Fig.28 Manohar, *Akbar receiving Mirza Aziz Koka*. Circa 1602–4; image area 18.4 × 12.1 cm. Cincinnati Art Museum, Gift of John J. Emery (acc.no.1950.289).

Fig.29 Manohar, *Darbar of Jahangir*. 1620s; image area 34.5 × 19.5 cm. Museum of Fine Arts, Boston, Francis Bartlett Donation of 1912 and Picture Fund (acc.no.14.657).

Fig.30 Pieter van der Borcht, *Pietatis Concordiae* or *The peace among the animals according to Isaiah*, first title page of volume 1 of the Royal Polyglot Bible, edited by B.A. Montanus, published in Antwerp by C. Plantin, 1569; engraving by Pieter van der Heyden. British Library, London, 6.h.4.

Fig.31 Grisaille wall-painting of lion and ox (detail of **39**).

Fig.32 Manohar, *Majnun mourns his father's death*, from a manuscript of the *Khamsa* of Nizami. Circa 1595; image area 13.6 × 9 cm. British Library, London (MS.Or.12208, fol.132a).

Fig.33 Bichitr, *Asaf Khan*. Late 1620s; image area 27.5 × 19.6 cm. From the Minto Album. Victoria and Albert Museum, London (acc.no.IM 26-1925).

Fig.34 Govardhan, *Music at an encampment*. 1620s; image area 23 × 16.7 cm. From the Minto Album. Chester Beatty Library, Dublin (MS.7A.11).

Fig.35 Sebastian Münster, Map of Ostia, from the *Cosmographia*, published in Basle, 1628. Woodcut; image area approximately 15 × 12 cm. Österreichische Nationalbibliothek, Vienna (acc.no.393.218 – DK).

Fig.36 Hilly landscape (detail of **34**).

Fig.37 Sebastian Münster, Map of Istria (detail), from the *Cosmographia*, published in Basle, 1550; approximate size 15.5 × 14.5 cm. Österreischische Nationalbibliothek, Vienna (acc.no.393.212 – DK).

Fig.38 Raja Bikramajit and Afzal Khan offering tribute (detail of **9**).

Fig.39 A laughing shaykh (detail of **37**).

Fig.40 Religious men awaiting alms after the weighing of Shah-Jahan (detail of **13**).

Fig.41 Courtiers at a *darbar* (detail of **37**).

Fig.42 Musicians at the wedding of Dara-Shikoh (detail of **26**).

Fig.43 Female musicians celebrate the wedding of Awrangzeb (detail of **45**).

Fig.44 Musician with a kettledrum (detail of **24**).

Fig.45 The *nawbat* ensemble (detail of **12**).

Fig.46 Musicians in a wedding procession (detail of **22**).

Fig.47 The imperial encampment outside Hoogly (detail of **20**).

Fig.48 The fortress at Orchha (detail of **35**).

Fig.49 Europeans escaping from a mined building at Hoogly (detail of **20**).

Fig.50 An explosion at the siege of Qandahar (detail of **18**).

Fig.51 The death of the rebellious Bundelas (detail of **36**).

Fig.52 Imperial soldiers scale the walls of Fort Dharur (detail of **15**).

Fig.53 The elephant Surat Sundar controlled by men with *charkhi*s (detail of **29**).

Fig.54 Black buck antelope by a stream (detail of **33**).

Fig.55 A pair of sarus cranes (detail of **30**).

Fig.56 Ducks in a marshy creek (detail of **30**).

Fig.57 Lions defending their cubs (detail of **46**).

Fig.58 Rabbits by a pool on the battlefield (detail of **18**).

Fig.59 The countryside beyond Fort Dharur (detail of **15**).

Fig.60 Fireworks above Agra at the wedding of Awrangzeb (detail of **45**).

Fig.61 The riverfront at Agra (detail of **29**).

Fig.62 Farmers fetching water (detail of **33**).

Fig.63 The Jumna River at Agra (detail of **21**).

Fig.64 Dancers celebrate the weighing of Shah-Jahan (detail of **12**).

Fig.65 Manuscript of the Windsor *Padshahnama* of Abdul-Hamid Lahawri, written in *nastaliq* script by Muhammad-Amin of Mashhad in AH 1067 (AD 1656–7), 58.5 × 37 cm (full page), 34.6 × 19 cm (text panel).

Fig.66 Lacquer binding of the Windsor *Padshahnama*.

Fig.67 Folio 50a, detail showing the seal of Asafuddawla.

Fig.68 Husayn Naqqash, *Shamsa*, from a manuscript of the *Diwan* of Amir Khusraw Dihlavi. Text dated 1597–8; folio area 28.4 × 19 cm. Walters Art Gallery, Baltimore (MS.W.624, fol.42).

Fig.69 Govardhan, *Timur presenting the imperial crown to Babur*. Circa 1635; image area 29.3 × 20.2 cm. From the Minto Album. Victoria and Albert Museum, London (acc.no.I.M.8-1925).

Fig.70 Timur enthroned (detail of **3**).

Fig.71 Timur enthroned (detail of Fig.69).

Fig.72 Bishandas, *Suraj Singh Rathor* (detail). Dated 1608. From the Jahangir Album. Berlin, Staatsbibliothek zu Berlin – Preussischer Kulturbesitz Orientabteilung (page 22b).

Fig.73 Bichitr, *I'tibar Khan* (detail). Circa 1623; image area 19.8 × 11.8 cm. Musées Nationales de France – Guimet, Paris (acc.no.7162).

Fig.74 Flowering plants in marble relief, dado of the Shah Burj Pavilion. Agra fort, 1630s; height of dado approximately 120 cm.

Fig.75 Floral wall decoration (detail of **5**).

Fig.76 *Jali* (perforated stone) screens in the courtyard of the Shah Burj Pavilion. Agra fort, 1630s; approximate height 205 cm.

Fig.77 *Jali* screen (detail of **5**).

Fig.78 La'lchand, *Raja Bikramajit*. Circa 1630; image area 17.4 × 10 cm. Cleveland Museum of Art, Purchase from the J.H. Wade Fund (acc.no.1945.170).

Fig.79 Distant rock formation (detail of **7**).

Fig.80 Sultan Muhammad, *Bahram hunting the lion*, from a manuscript of the *Khamsa* of Nizami dated 1539–43; image area 28.5 × 18 cm. British Library, London (MS.Or.2265, fol.202b).

Fig.81 Hunhar, *Maharaja Bhim of Udaipur*. Circa 1650; image area 37.9 × 24.7 cm. From the Late Shah-Jahan Album. Museum of Fine Arts, Boston, Gift of John Goelet (acc.no.60.172).

Fig.82 Unknown artist, *Jahangir receives Prince Khurram on his return from the Deccan*. Circa 1640; drawing with gouache and gold in places, image area 29.6 × 20.8 cm. British Library, London (India Office Library, Johnson Album 4, no.2).

Fig.83 Unknown artist, *Jahangir receives Prince Khurram on his return from the Deccan*. Circa 1640; drawing, image area 24.4 × 16.8 cm. San Diego Museum of Art , acc.no.1990:0351.

Fig.84 Wall-painting (detail of **8**).

Fig.85 Painting below *jharoka* (detail of **10**).

Fig.86 Diagram of the central perspective elements of the composition in **10**.

Fig.87 Hashim, *Muhammad-Ali Beg*. Circa 1632; image area 22.8 × 14.6. From the Minto Album. Victoria and Albert Museum, London (acc.no.I.M.25-1925).

Fig.88 Colonnade from the hall in front of the Imperial Hammam. Agra fort, 1630s. Victoria and Albert Museum, London (reserve collection, no.167).

Fig.89 Façade of the *Diwan-i Khass*. Agra fort, 1630s.

Fig.90 View into the corner of the *Diwan-i Khass* and south wing of the *Machchhi Bhawan*. Agra fort, 1630s.

Fig.91 Elevation drawing of the riverfront terrace of the Agra fort, with the *Diwan-i Khass* and Imperial Hammam, 1630s.

Fig.92 Raiders scaling the walls of the fort at Dharur (detail of **15**).

Fig.93 The capture of three enemy elephants (detail of **15**).

Fig.94 Etienne de Laune, *Ad Locutio*. France, dated 1570; engraving, image area 6.6 × 8 cm. Private collection.

Fig.95 Detail of **16**.

Fig.96 Severed heads (detail of **16**).

Fig.97 Figures arranged behind landscape 'segments' (detail of 18).

Fig.98 Unknown artist, *Timur conquers Baghdad* (detail), from a manuscript of the *Zafarnama* by Sharafuddin Ali Yazdi. Dated 1436; image area 28.8 × 20.3 cm. Metropolitan Museum of Art, New York (acc.no.55.121.17, fol.317b).

Fig.99 A European carrying a lacquer box (detail of **19**).

Fig.100 A distant encampment (detail of **20**).

Fig.101 Unknown artist, *Battle scene* from a *Jahangirnama* manuscript. Circa 1620; image area 32.8 × 20.8 cm. Collection of Terence McInerney, New York.

Fig.102 Attributed to Bishandas, *A procession*, probably from a *Jahangirnama* manuscript. Circa 1625; image area 32.9 × 22 cm. Raza Library, Rampur (Album 5, fol.4a).

Fig.103 The riverfront at Agra (detail of **21**).

Fig.104 Elevation drawing of the façade of the *Diwan-i 'Amm* and adjoining wings. Agra fort, 1630s.

Fig.105 Unknown artists, *Ja'far Khan* (left) and *Shah-Jahan* (right). Circa 1635; image area 3.6 × 3 cm (left and right). From the St Petersburg Album, assembled mid-18th century. Institute of Oriental Studies, Academy of Sciences, St Petersburg (MS.E.14, page 63).

Fig.106 Hayat Khan (detail of **30**).

Fig.107 Murar, *Hayat Khan*. Circa 1630; image area 15 × 7.5 cm. Private collection.

Fig.108 Unknown artist, *Khurram slays the lion attacking Anup Rai*. Circa 1610; brush and ink with color on paper; image area 17.9 × 18.5 cm. Private collection.

Fig.109 The fortifications of Daulatabad (detail of **31**).

Fig.110 Georg Braun and Frans Hogenberg, Map of Würzburg, from the *Civitates orbis terrarum*, published in Cologne, 1572–1618. Woodcut, 19.5 × 23.7 cm. Österreischische Nationalbibliothek, Vienna (acc.no.393.706 –D).

Fig.111 Throne-*jharoka* of Shah-Jahan in the *Diwan-i 'Amm*. Red Fort, Delhi, completed 1648.

Fig.112 Unknown artist, *Asalat Khan* (detail). Circa 1645; image area 16.1 × 9.2 cm. Freer Gallery of Art, Smithsonian Institution, Washington, DC (acc.no.F29.81).

Fig.113 Asalat Khan (detail of **33**).

Fig.114 Chitarman, *Ja'far Khan* (detail). Circa 1645; image area 17.7 × 10.3 cm. British Museum, London (acc.no.1920-9-17-013 (36).

Fig.115 Ja'far Khan (detail of **33**).

Fig.116 The imperial army storming through the jungle of Bundelkhand (detail of **35**).

Fig.117 Peter Mundy, *A royal progression*. English (in India), dated 1632; ink; 39.0 × 21.5 cm. Bodleian Library, Oxford (Rawlinson MS.A.315, no.13).

Fig.118 The capture of the rebel women's elephants (detail of **36**).

Fig.119 Asaf Khan (detail of **37**).

Fig.120 Unknown artist, *Asaf Khan*. Circa 1616. From the Jahangir Album. Berlin, Staatsbibliothek zu Berlin, Preussischer Kulturbesitz Orientabteilung (page 20b).

Fig.121 Attributed to Murar, *Jahangir receives Prince Khurram on his return from the Mewar campaign*. Circa 1640; ink drawing; image area 25.5 × 18 cm. Museum of Fine Arts, Boston (acc.no.17.2696).

Fig.122 Abu'l-Hasan, *Shah-Jahan in his twenty-fifth year*. Circa 1618; image area 20.6 × 11.5. From the Minto Album. Victoria and Albert Museum, London (acc.no.I.M.14-1925).

Fig.123 Jahangir wearing a miniature portrait medallion (detail of **39**).

Fig.124 Isaac Oliver, *An unknown lady.* English, circa 1615; image area 5.4 × 4.5 cm. Victoria and Albert Museum, London (acc.no.P.24–1932).

Fig.125 Unknown artist, *Khwaja Khizr.* Circa 1640; image area 16.2 × 9.8 cm . Chronos Collection. Metropolitan Museum of Art, New York.

Fig.126 Khwaja Khizr (detail of **41**).

Fig.127 Railing of the *jharoka* (detail of **44**).

Fig.128 Unknown artist, grotesque pattern forming the letter T, from an alphabet from the print shop of Macé Bonhomme, Lyons, 1566; image area 3 × 3 cm. After A.F. Butsch, *Handbook of Renaissance Ornament*, reprinted New York, 1969, pl.149.

Fig.129 *Jharoka* of the *Diwan-i 'Amm* of the Lahore fort. Second half of the 16th or first quarter of the 17th century, with later alterations.

Fig.130 *Jharoka* of the *Diwan-i 'Amm* of the Lahore fort (detail of **44**).

Fig.131 Unknown artist, *Shayista Khan.* Circa 1640; image area 30.4 × 21.5 cm. Bibliothèque Nationale, Paris (acc no.Od. 44).

Fig.132 Line drawing after an 18th-century map of Agra. Maharaja Sawai Man Singh II Museum, Jaipur.

Fig.133 Dawlat, *Akbar fighting with Raja Man Singh* (detail of onlookers) from an *Akbarnama* manuscript. Circa 1598; image area 43 × 26 cm. Chester Beatty Library, Dublin (MS2, fol.168v).

Fig.134 Attendants at a hunt (detail of **46**).

Fig.135 Balchand, self-portrait (detail of **5**).

Fig.136 Bichitr, self-portrait (detail of **10**).

Fig.137 Dawlat, *Bishandas*, from the margins of the Jahangir Album. Gulistan Library, Tehran (detail of page 44).

Fig.138 Dawlat, self-portrait, from the margins of the Jahangir Album. Gulistan Library, Tehran (detail of page 44).

Fig.139 Dawlat, *Jahangir.* Dated 1627; image area 7.6 × 5.8 cm. Collection of Sven Gahlin, London.

Fig.140 Murar, self-portrait (detail of **9**).

Fig.141 Payag, self-portrait (detail of **39**).

Fig.142 (Appendix A) Abid, *Shah-Jahan enthroned with Mahabat Khan and a shaykh.* Dated 1629–30; image area 30.2 × 19.9 cm. From the Late Shah-Jahan Album. Arthur M. Sackler Gallery, Smithsonian Institution, Washington, DC (acc.no.S86.406).

Fig.143 (Appendix B) Unknown artist, *Shah-Jahan in darbar.* Circa 1630; image area 32 × 20.5 cm. From the St Petersburg Album. Institute of Oriental Studies, Academy of Sciences, St Petersburg (MS.E14, p.34a).

Fig.144 (Appendix C) Bulaqi, *A night celebration of the Prophet's birthday.* Circa 1635; image area 30.5 × 23.1 cm. Detached from the St Petersburg Album. Freer Gallery of Art, Smithsonian Institution, Washington, DC (acc. nos 42.17 and 42.18).

Fig.145 (Appendix D) Abid, *Shah-Jahan receives Prince Shah-Shuja' on his return from the Deccan.* Dated 1635–6; image area 30.3 × 21.4 cm. Bodleian Library, Oxford (MS.Ouseley Add.173, fol.18).

Fig.146 (Appendix E) Abid, *Shah-Jahan enthroned.* Dated 1639; image area 36.9 × 24.6 cm. San Diego Museum of Art, acc.no.1990:0352.

Fig.147 (Appendix F) Unknown artist, *The surrender of a fortress.* Circa 1640; image area 34 × 24.2 cm. Musée Guimet, Paris (inv. no.MA 3318).

Fig.148 (Appendix G) Unknown artist, *Sa'id Khan defeats the armies of Siyawash.* Circa 1640; image area 33 × 23.6 cm. From the St Petersburg Album. Institute of Oriental Studies, Academy of Sciences, St Petersburg (MS.E.14, p.55a).

Fig.149 (Appendix H) Payag, *The Persian ambassador Yadgar Beg received at court.* Circa 1640; image area 34.5 × 23.8 cm. Bodleian Library, Oxford (MS.Ouseley Add.173, no.13).

Fig.150 (Appendix I) Unknown artist, *Alimardan Khan arrives at court.* Circa 1640. Formerly Sita Ram Sahu Collection, Varanasi.

Fig.151 (Appendix J) Payag, *Battle scene.* Circa 1640; image area 34.3 × 23.9 cm. Private Collection.

Fig.152 (Appendix K) Fath Chand, *Prince Muradbakhsh receiving Khusraw Sultan, son of Nazar Muhammad Khan.* Circa 1648; image area 34.9 × 24.1 cm. Bharat Kala Bhavan, Varanasi (inv. no.5403).

Fig.153 (Appendix L) Hunhar, *The battle of Shibarghan.* Circa 1648; image area 34.4 × 23.1 cm. Metropolitan Museum of Art, New York, Purchase, Bernard and Audrey Aronson Charitable Trust Gift, in memory of her beloved husband, Bernard Aronson, 1986 (acc.no.1986.283).

Fig.154 (Appendix M) Unknown artist, *Shah-Jahan receiving Prince Awrangzeb.* Circa 1645; image area 30.5 × 22 cm. British Library, London, India Office Library (MS.Add. Or.3853).

Fig.155 (Appendix N) Unknown artist, *Shah-Jahan receiving Prince Awrangzeb and Prince Muhammad Mu'azzam.* Circa 1660; image area 33.5 × 25.5 cm. From the St Petersburg Album. Institute of Oriental Studies, Academy of Sciences, St Petersburg (MS.E.14, page 25a).

Fig.156 (Appendix O) Unknown artist, *Darbar of Shah-Jahan.* Image area 34.3 × 22.9 cm. Bharat Kala Bhavan, Varanasi.

Fig.157 (Appendix P) Bulaqi, *Shah-Jahan watching an elephant fight.* Circa 1635; image area 33 × 20.3 cm. Nasser D.Khalili Collection, London.

Fig.158 (Appendix Q) Bulaqi, *Shah-Jahan watching an elephant fight.* Dated 1639–40; image area 34.8 × 21.9 cm. Metropolitan Museum of Art, New York, Harris Brisbane Dick, Louis V. Bell, Pfeiffer, and Dodge Funds (acc.no.1989.135).

Fig.159 (Appendix R) Unknown artist, *Rustam Khan defeats the Qizilbash army.* Circa 1650; image area 33 × 23.5 cm. From the St Petersburg Album. Institute of Oriental Studies, Academy of Sciences, St Petersburg (MS.E.14, page 54a).

Fig.160 (Appendix S) Reportedly by Fazl, son of Abid, *Shah-Jahan in procession.* Circa 1655. Calico Museum, Ahmedabad.

Fig.161 (Appendix T) Unknown artist, *Attendant figures.* Circa 1640. Raza Library, Rampur (Album 5, no.3).

Fig.162 (Appendix U) Possibly by Hunhar, *Attendant figures.* Circa 1645. Howard Hodgkin Collection.

Fig.163 (Appendix V) Possibly by Hunhar, *Attendant figures.* Circa 1645; image area 24.7 × 17.5 cm. Museum Reitberg, Zurich.

Fig.164 (Appendix W) Unknown artist, *Attendant figures.* Circa 1650; image area 37.5 × 22 cm. Art Institute of Chicago, Kate S. Buckingham Collection (acc.no.1975.555).

Fig.165 (Appendix X) Unknown artist, *Attendant figures.* Circa 1650. Indian Museum, Calcutta (inv. no.456).

Fig.166 (Appendix Y) Unknown artist, *Attendant figures.* Circa 1650; image area 33 × 23.5 cm. From the St Petersburg Album. Institute of Oriental Studies, Academy of Sciences, St Petersburg (MS.E.14, page 13a).

Fig.167 (Appendix a) Anupchatar, *Shah-Jahan enthroned with Mahabat Khan and a shaykh.* Circa 1630. British Museum (acc.no.1920-9-17-013/27).

Fig.168 The imperial family at the accession of Shah-Jahan in 1628 (detail of **10**).

Fig.169 The imperial family at Prince Dara-Shikoh's wedding in 1633 (detail of **25**).

Fig.170 The imperial family at Prince Awrangzeb's wedding in 1637 (detail of **43**).

Fig.171 The imperial family on Awrangzeb's arrival at court in 1640 (detail of **44**).

CONTEMPORARY HISTORIES OF SHAH-JAHAN

Padshahnama of Qazwini
Mirza Muhammad-Amin of Qazwin was appointed Imperial Historiographer in the eighth year of Shah-Jahan's reign (1635) and prepared a history of its first decade, the *Padshahnama*.

Padshahnama of Lahawri
Abdul-Hamid of Lahore (d. 1654) wrote a history of the first two decades of the Emperor's reign which was revised by the vizier Sa'dullah Khan. It was continued to cover the third decade by Abdul-Hamid's pupil, Muhammad Waris (d. 1680) and revised by Fazil Khan Ala'ulmulk Tuni.

The *Shahjahannama* of Sadiq Khan
Sadiq Khan, Shah-Jahan's bakhshi and *waqayi' nawis* (reporter of significant events) in Agra, wrote a *Shahjahannama* that covered events up to 1658.

The *Mulakhkhas* and *Shahjahannama* of Inayat Khan
Muhammad-Tahir 'Ashna' Inayat Khan wrote an abridgement of Abdul-Hamid Lahawri and Muhammad Waris' *Padshahnama* covering the first three decades of the reign, entitled *Mulakhkhas* and *Shahjahannama*.

The *'Amal-i-Salih* of Kanbo
Muhammad Salih Kanbo of Lahore's *'Amal-i-Salih* was a detailed history of the reign of Shah-Jahan completed in 1659.

The *Chahar Chaman* of Rai Chandarbhan 'Barahman'
While not a history per se, the *Chahar Chaman* by Rai Chandarbhan 'Brahman,' written soon after 1647, provides descriptions of court festivals and daily occupations.

In addition to these works there are three versified histories: the *Zafarnama-i-Shah-Jahani* of Muhammad-Jan Qudsi Mashhadi (d. 1646), which was never completed; the *Padshahnama* of Yahya Kashi (d. 1653); and an uncompleted *Shahjahannama* by Abu-Talib Kalim of Hamadan and Kashan (d. 1652).

BIBLIOGRAPHY

PRIMARY HISTORICAL TEXTS

A'in-i Akbari
Abu'l Fazl 'Allami, *The A'in-i Akbari*, Persian text edited by H. Blochmann, 2 volumes, Calcutta, 1867–77; translated in 3 volumes by (I) H. Blochmann, revised and edited by D.C. Phillot, 1927, (II, III) by H.S. Jarrett, edited by Jadunath Sarkar, 1948–9; reprinted New Delhi, 1977–8.

Akbarnama
Abu-l-Fazl, *Akbarnama*, translated by H. Beveridge, 3 volumes, Calcutta, 1897–1939; reprinted Delhi, 1972–3; second reprint 1979.

'Alamgirnama
Muhammad Kazim, *'Alamgirnama*, Persian text edited by Khadim Husain and 'Abd al-Hayy, Calcutta, 1865–8.

'Amal-i Salih
Muhammad-Salih Kanbo, *'Amal-i Salih or Shahjahannama*, revised Persian text edited by Wahid Quraishi, based on the Calcutta edition of 1912–46 by Ghulam Yazdani; second edition, 3 volumes, Lahore, 1967–72.

Baburnama
The Baburnama – Memoirs of Babur, Prince and Emperor, edited and annotated by Wheeler M. Thackston, Washington, New York and Oxford, 1995.

Chahar Chaman
Chandarbhan Brahman, *Chahar Chaman*, British Library, Oriental and India Office Collections, Persian MS.Add.16863.

Diwan-i Kalim
Abu-Talib Kalim, *Diwan*, Persian text edited by Partaw Baiza'i, Tehran, 1336/1957.

Humayunnama
Gulbadan Begam, *The History of Humayun*, translated by Annette S. Beveridge, London, 1902; reprinted Delhi, 1972.

Jahangirnama
Jahangirnama: Tuzuk-i Jahangiri, Persian text edited by Muhammad Hashim, Tehran, 1359/1940–41.

Kautiliya Arthasastra
Kautiliya Arthasastra, edited, translated, and with commentary by R.P. Kangle, 3 volumes, Bombay, 1965–72.

Ma'asir-i-'Alamgiri
Saqi Must'ad Khan, *A History of The Emperor Aurangzib-'Alamgir (Reign 1658–1707 AD)*, translated and annotated by Jadunath Sarkar, Calcutta, 1947; reprinted New Delhi, 1986.

Ma'athir ul-Umara
Samsam al-Daula Shah-Nawaz Khan, *Ma'athir ul-Umara*, translated by H. Beveridge and Baini Prashad, 2 volumes, Calcutta, 1911–41; reprinted Patna, 1979.

Manu-smriti
[Manu], *The Laws of Manu*, translated by G. Bühler, Oxford, 1886.

Mir'at al-'Alam
Muhammad Bakhtawar Khan, *Mir'at al-'Alam, History of Awrangzeb: 1658–1668*, Persian text edited by Sajida S. Alvi, 2 volumes, Lahore, 1979.

Muntakhab al-Lubab
Muhammad Hashim Khafi Khan, *Muntakhab al-Lubab*, Persian text edited by Kabir al-Din Ahmad, Calcutta, 1869–1925. The portions relating to Awrangzeb have been translated as *History of 'Alamgir* by S. Moinul Haq, Karachi, 1975.

Muntakhab-ut-Tawarikh
'Abd al-Qadir Bada'oni, *Muntakhab ut-Tawarikh*, translated in 3 volumes by (I) G.S.A. Ranking, 1898, (II) W.H. Lowe, second edition, 1924, (III) T.W. Haig, 1925; reprinted Delhi, 1973.

Nasihat al-Muluk
al-Ghazali, *Ghazali's Book of Counsel for Kings (Nasihat al-Muluk)*, translated by F.R.C. Bagley, Durham, 1964; reprinted London, 1971.

Padshahnama
'Abd al-Hamid Lahawri, *The Badshah Namah*, Persian text edited by M. Kabir al-Din Ahmad and M. 'Abd al-Rahim, 2 volumes, Calcutta, 1867–72.

Padshahnama (Kalim)
Abu-Talib Kalim, *Padshahnama*, British Library, Oriental and India Office Collections, Persian MS.Ethé 1570.

Padshahnama (Qazwini)
Muhammad Amin Qazwini or Amina-i Qazwini, *Padshahnama*, British Library, Oriental and India Office Collections, Persian MS.Or.173.

Padshahnama (Waris)
Muhammad Waris, *Padshahnama*, British Library, Oriental and India Office Collections, Persian MS.Add.6556.

Qabusnama
Kai Ka'us ibn Iskandar, *A Mirror for Princes: Qabusnama*, translated by R. Levy, London, 1951.

Qanun-i Humayuni
Ghiyas al-Din Khwandamir, *Qanun-i Humayuni*, translated by Baini Prashad, Calcutta, 1940.

Rajatarangini
Kalhana, *Kalhana's Rajatarangini or Chronicles of the Kings of Kashmir*, translated by M.A. Stein, 2 volumes, 1900; reprinted Lahore, 1991.

Ruqa'at-i 'Alamgiri
[Awrangzeb], *Ruka'at-i 'Alamgiri or Letters of Aurangzebe*, translated by J.H. Bilimoria, Bombay, 1908; reprinted Delhi 1972.

Shah Jahan Nama
'Inayat Khan, *The Shah Jahan Nama of 'Inayat Khan: An Abridged History of the Mughal Emperor Shah Jahan, Compiled by his Royal Librarian*, translated by A.R. Fuller, revised and edited by Wayne E. Begley and Z.A. Desai, New Delhi, Oxford and New York, 1990.

Ta'rikh-i Akbari
Muhammad 'Arif Qandahari, *Ta'rikh-i Akbari*, Persian text edited by S. Mu'in ud-Din Nadwi, Azhar 'Ali Dihlawi and Imtiyaz 'Ali 'Arshi, Rampur, 1382/1962.

Tawarikh-i Shahjahani
Muhammad Sadiq Khan, *Tawarikh-i Shahjahani*, British Library, Oriental and India Office Collections, Persian MS.Or.174.

Tezkereh al Vakiat
Jouher, *Tezkereh al Vakiat or Private Memoirs of the Moghul Emperor Humayun*, translated by Charles Stewart, New York, 1969; reprinted Delhi, 1970.

Tuzuk-i-Jahangiri
Tuzuk-i-Jahangiri, or The Memoirs of Jahangir, translated by Alexander Rogers, edited by Henry Beveridge, 2 volumes, London, 1909–14.

TRAVEL ACCOUNTS AND GEOGRAPHICAL WORKS

Bernier 1891
François Bernier, *Travels in the Mogul Empire, AD 1656–1668*, translated by Irving Brock and annotated by Archibald Constable, London, 1891; third reprint New Delhi, 1972.

Braun & Hogenberg 1572–1618
Georg Braun and Frans Hogenberg, *Civitates orbis terrarum*, 6 volumes, Cologne, 1572–1618.

Downton 1939
Nicholas Downton, *The Voyage of Nicholas Downton to the East Indies 1614–15*, edited by William Foster, London, 1939.

Finch 1985
'William Finch 1608–11' in *Early Travels in India: 1583–1619*, edited by W. Foster, 1921; reprinted New Delhi, 1985, pp.122–187.

Ibn Batuta 1993
Muhammad Ibn Batuta, *The Travels of Ibn Batuta AD 1325–1354*, translated by H.A.R. Gibb, 3 volumes, Cambridge, 1958–71; reprinted New Delhi, 1993.

Manrique 1927
Sebastian Manrique, *Travels of Sebastian Manrique 1629–1643*, translated by C. Ekford Luard assisted by Father H. Hosten, S.J., 2 volumes, Oxford, 1927.

Manucci 1889
Niccolao Manucci, *Storia do Mogor*, translated by William Irvine, 2 volumes, London, 1889.

Monserrate 1992
Father Anthony Monserrate, S.J., *The Commentary of Father Monserrate, S.J., on His Journey to the Court of Akbar*, translated by J.S. Hoyland and annotated by S.N. Banerjee, London and Calcutta, 1922; reprinted New Delhi, 1992.

Mundy 1907–24
Peter Mundy, *The Travels of Peter Mundy in Europe and Asia 1608–1667*, 6 volumes, London, 1907–24.

Münster 1550
Sebastian Münster, *Cosmographia*, Basle, 1550.

Pelsaert 1972
Francisco Pelsaert, *Jahangir's India or The Remonstrantie of Francisco Pelsaert*, translated by W.H. Moreland and P. Geyl, 1925; reprinted Delhi, 1972.

Roe 1967/1990
Sir Thomas Roe, *The Embassy of Sir Thomas Roe to the Court of the Great Mogul 1615–1619*, edited by William Foster, London 1899; reprinted Nendeln, 1967; second edition Oxford, 1926; reprinted New Delhi, 1990.

Tavernier 1925
Jean-Baptiste Tavernier, *Travels in India*, translated by V. Ball, second edition edited by W. Crooke, 2 volumes, London and New York, 1925.

SECONDARY SOURCES

Ackerman 1964–5
Phyllis Ackerman, 'Standards, Banners, and Badges' in *A Survey of Persian Art from Prehistoric Times to the Present*, edited by Arthur U. Pope and Phyllis Ackerman, London and New York, 1938–9; second edition, 13 volumes, London and New York, 1964–5, VI, pp.2766–82.

Ahmad 1929–30
Khwaja Muhammad Ahmad, 'Inscriptions from Udgir, Bidar District', *Epigraphia Indo-Moslemica*, 1929–30, pp.18–31.

Akimushkin 1994
Oleg F. Akimushkin, *The St Petersburg Muraqqa*, Lugano, 1994.

Akimushkin *et al.* 1996
Oleg F. Akimushkin, Gauvin Bailey, Milo Beach, Navina Najat Haidar, Anatoly A. Ivanov, Terence McInerney, Marie Lukens Swietochowski, Stuart Cary Welch, *The St Petersburg Muraqqa' – Album of*

Indian and Persian miniatures from the 16th through the 18th century and specimens of Persian calligraphy by 'Imad al-Hasani, Milan, 1996.
Akurgal, Mango & Ettinghausen 1966
Ekrem Akurgal, Cyril Mango, and Richard Ettinghausen, *Treasures of Turkey*, Geneva, 1966.
Alvi & Rahman 1968
M.A. Alvi and A. Rahman, *Jahangir – The Naturalist*, New Delhi, 1968.
Anawati 1978
G.C. Anawati, ''Isa', *Encyclopaedia of Islam*, second edition, IV, Leiden, 1978, pp.81–86.
Andrews 1987
Peter Alford Andrews, 'The Generous Heart or the Mass of Clouds: The Court Tents of Shah Jahan', *Muqarnas*, 4, 1987, pp.149–65.
Andrews, forthcoming
—, *Felt Tents and Pavilions: The Nomadic Tradition and its Interaction with Princely Tentage*, forthcoming.
Ansari 1974
Muhammad Azhar Ansari, *Social Life of the Mughal Emperors, 1526–1707*, Allahabad, 1974.
Archer 1957
William George Archer, *Indische Miniaturen*, Bern, 1957.
Archer 1960
—, *Indian Miniatures*, New York, 1960.
Arnold & Wilkinson 1936
Thomas Arnold and J.V.S. Wilkinson, *The Library of A. Chester Beatty: A Catalogue of the Indian Miniatures*, 3 volumes, London, 1936.
Asad 1988
Majda Asad, *Indian Muslim Festivals and Customs*, translated by M.R. Sharma, New Delhi, 1988.
Asher 1992
Catherine B. Asher, *Architecture of Mughal India*, Cambridge, New York and Oakleigh, 1992.
Asher 1993
—, 'Sub-Imperial Palaces: Power and Authority in Mughal India', *Premodern Islamic Palaces*, edited by G. Necipoğlu, *Ars Orientalis*, 23, 1993, pp.281–302.
Ashton 1950
Sir Leigh Ashton, *The Art of India and Pakistan: A Commemorative Catalogue of the Exhibition Held at the Royal Academy of Arts, London, 1947–48*, London, 1950.
M. Athar Ali 1985
M. Athar Ali, *The Apparatus of Empire: Awards of Ranks, Offices and Titles to the Mughal Nobility, 1574–1658*, Delhi, 1985.
Auboyer 1975
Jeannine Auboyer, *Rareties of the Musée Guimet*, New York, 1975.
Aziz 1947
Abdul Aziz, *Arms and Jewellery of the Indian Mughuls*, Lahore, 1947.
Aziz n.d.
—, *Thrones, Tents and Their Furniture Used by the Indian Mughuls*, Lahore, no date.
Balfour 1967
Edward Balfour, *The Cyclopaedia of India and of Eastern and Southern Asia*, 3 volumes, London, 1885; reprinted Graz, 1967.
Barrett & Gray 1963
Douglas E. Barrett and Basil Gray, *Painting of India*, Geneva and Cleveland, OH, 1963.
Bautze 1986
Joachim Bautze, 'Portraits of Rao Ratan and Madho Singh Hara', *Berliner Indologische Studien*, 2, 1986, pp.87–106.
Beach 1975
Milo Cleveland Beach, 'The Context of Rajput Painting', *Ars Orientalis*, 10, 1975, pp.11–18.
Beach 1978
—, *The Grand Mogul: Imperial Painting in India 1600–1660*, Williamstown, MA, 1978.
Beach 1980
—, 'The Mughal Painter Abu'l Hasan and Some English Sources for His Style', *The Journal of the Walters Art Gallery*, 38, 1980, pp.7–33.
Beach 1981
—, *The Imperial Image: Paintings for the Mughal Court*, Washington, DC, 1981.
Beach 1987a
—, *Early Mughal Painting*, Cambridge, MA, 1987.
Beach 1987b
—, 'The Study of Fatehpur-Sikri' in *Fatehpur-Sikri*, edited by Michael Brand and Glenn D. Lowry, Bombay, 1987, pp.13–24.
Beach 1992a
—, *Mughal and Rajput Painting*, Cambridge and New York, 1992.
Beach 1992b
—, 'Jahangir's Jahangir-Nama' in *The Powers of Art: Patronage in Indian Culture*, edited by Barbara Stoler Miller, Delhi and New York, 1992, pp.224–34.
Beach 1995a
—, 'Characteristics of the St Petersburg Album', *Orientations*, January 1995, pp.66–79.
Beach 1995b
—, 'All the Great Men', *Country Life*, 24, 1995, pp.54–57.
Begley 1987
Wayne E. Begley, 'Illustrated Histories of Shah Jahan: New Identifications of Some Dispersed Paintings and the Problem of the Windsor Castle *Padshahnama*' in Skelton *et al.* 1987, pp.139–52.
Bence-Jones 1982
Mark Bence-Jones, *The Viceroys of India*, New York, 1982.
Bibliothèque Nationale 1986
Bibliothèque Nationale, *A la Cour du Grand Moghol*, Paris, 1986.
Binney 1973
Edwin Binney, 3rd, *Indian Miniature Paintings from the Collection of Edwin Binney, 3rd*, Portland, OR, 1973.
Binstock 1995
Benjamin Binstock, 'Alois Riegl in the Presence of *The Nightwatch*', *October*, 74 (Fall 1995), pp.36–44.
Binyon & Arnold 1921
Laurence Binyon and T.W. Arnold, *The Court Painters of the Grand Moguls*, London and New York, 1921.
Blair 1995
Sheila S. Blair, *A Compendium of Chronicles: Rashid al-Din's Illustrated History of the World*, The Nasser D. Khalili Collection of Islamic Art, XXVII, London, 1995.
Blochet 1905–12
Edgar Blochet, *Catalogue des Manuscrits Persans de la Bibliothèque Nationale, Paris*, 2 volumes, Paris, 1905–12.
Blochet 1928
—, *Les peintures orientales de la Collection Pozzi*, Paris, 1928.
Blochet 1930
—, *Collection Jean Pozzi: Miniatures Persanes et Indo-Persanes*, Paris, 1930.
Bosworth 1986
C.E. Bosworth, 'Lakab', *Encyclopaedia of Islam*, second edition, V, Leiden, 1986, pp.618–31.
Brand 1991
Michael Brand, 'Mughal Ritual in Pre-Mughal Cities: The Case of Jahangir in Mandu' in *Environmental Design*, 1–2: *Mughal Architecture: Pomp and Ceremonies*, 1991, pp.8–17.
Brand & Lowry 1985
Michael Brand and Glenn D. Lowry, editors, *Fatehpur Sikri: A Sourcebook*, Cambridge, MA, 1985.
Brend 1980
Barbara Brend, 'Rocks in Persian Miniature Painting' in Watson 1980, pp.111–37.
Brend 1982–3
—, 'The Weighing of Khurram Mirza', *Oriental Art*, XXVIII/4, 1982–3, pp.352–55.
Brend 1995
—, *The Emperor Akbar's Khamsa of Nizami*, London, 1995.
British Museum 1976
British Museum, *Paintings from the Muslim Courts of India*, London, 1976.
Brown 1975
Percy Brown, *Indian Painting under the Mughals, AD 1550 to AD 1750*, Oxford and New York, 1924; reprinted New York, 1975.
Burton-Page 1983
John Burton-Page, 'Dawlatabad: Monuments', *Encyclopaedia of Islam*, second edition, II, Leiden, 1983, p.180.
Burton-Page 1986
—, 'Daulatabad' in *Islamic Heritage of the Deccan*, edited by George Michell, Bombay, 1986, pp.16–25.
Burton-Page 1991
—, 'Maratib', *Encyclopaedia of Islam*, second edition, VI, Leiden, 1991, pp.536–37.
Busse 1968
Heribert Busse, 'Fürstenspiegel und Fürstenethik im Islam', *Bustan*, 1, 1968, pp.12–19.
Carroll 1972
David Carroll, *The Taj Mahal – India under the Moguls*, New York, 1972.
Chandra 1976
Pramod Chandra, *The Tuti-nama of the Cleveland Museum of Art and the Origins of Mughal Painting*, Graz, 1976.
Chopra 1976
P.N. Chopra, *Life and Letters Under the Mughals*, New Delhi, 1976.
Clark 1973
Kenneth Clark, *Landscape into Art*, London, 1949; 5th printing, London, 1973.
Clarke 1922
C. Stanley Clarke, *The Wantage Bequest*, London, 1922.
Colnaghi 1976
P. & D. Colnaghi & Co. Ltd., London, *Persian and Mughal Art*, London, 1976.
Colnaghi 1978
—, *Indian Painting – Mughal and Rajput and a Sultanate Manuscript*, London, 1978.
Coomaraswamy 1924–5
Ananda K. Coomaraswamy, 'Two Mughal Paintings with Portraits of 'Ali Mardan Khan' in *Yearbook of Oriental Art and Culture*, edited by Arthur Waley, 2 volumes, London, 1924–5.
Coomaraswamy 1930
—, *Catalogue of the Indian Collections in the Museum of Fine Arts, Boston, Part VI, Mughal Painting*, Cambridge, MA, 1930.
Cooper 1993
Ilay Cooper, 'Sikhs, Saints, and Shadows of

Angels', *South Asian Studies*, 9, 1993, pp.11–28.
Currie 1989
P.M. Currie, *The Shrine and Cult of Mu'in al-Din Chishti of Ajmer*, Delhi, 1989.
Curtis & Reade 1995
J.E. Curtis and J.E. Reade, *Art and Empire: Treasures from Assyria in the British Museum*, New York, 1995.
Czuma 1975
Stanislaw Czuma, *Indian Art from the George P. Bickford Collection*, Cleveland, OH, 1975.
Das 1971
Asok Kumar Das, 'Bishan Das', *Chhavi – Golden Jubilee Volume*, edited by Anand Krishna, Banaras, 1971, pp.183–91.
Das 1986
—, *Splendour of Mughal Painting*, Bombay, 1986.
Das 1991
—, 'Daulat' in Pal 1991, pp.87–104.
Delhi 1911
'Standards and Insignia', *Delhi: Coronation Durbar, Catalogue of Loan Exhibition of Antiquities*, Delhi, 1911.
Deneck 1967
Marguerite-Marie Deneck, *Indian Art*, London, 1967.
Desai 1985
Vishakha N. Desai, *Life at Court: Art for India's Rulers, 16th–19th Centuries*, Boston, 1985.
Digby 1990
Simon Digby, 'The Sufi Shaykh and the Sultan: A Conflict of Claims to Authority in Medieval India', *Iran*, 28, 1990, pp.71–81.
Dye 1991
Joseph M. Dye, 3rd, 'Payag' in Pal 1991, pp.119–34.
Eaton 1993
Richard Eaton, *The Rise of Islam and the Bengal Frontier 1204–1760*, Berkeley, Los Angeles and London, 1993.
Edwards 1960
Michael Edwards, *The Orchid House: Splendors and Miseries of the Kingdom of Oudh, 1827–1857*, London, 1960.
Enderlein & Hickman 1979
Volkmar Enderlein and Regina Hickman, *Indische Albumblätter – Miniaturen und Kalligraphien aus der Zeit der Moghul-Kaiser*, Leipzig and Weimar, 1979.
Ettinghausen 1961
Richard Ettinghausen, 'The Emperor's Choice', *De Artibus Opuscula XL: Essays in Honor of Erwin Panofsky*, edited by Millard Meiss, New York, 1961, pp.99–120.
Falk & Archer 1981
Toby Falk and Mildred Archer, *Indian Miniatures in the India Office Library*, London, 1981.
Findly 1993
Ellison Banks Findly, *Nur Jahan: Empress of Mughal India*, New York and Oxford, 1993.
Gabrieli 1979
F. Gabrieli, 'Adab', *Encyclopaedia of Islam*, second edition, I, Leiden, 1979, pp.175–76.
Gahlin 1991
Sven Gahlin, *The Courts of India: Indian Miniatures from the Collection of the Fondation Custodia, Paris*, Zvolle and Paris, 1991.
Gamerman 1994
Amy Gamerman, 'The Gallery: Curator Decodes a Mughal Hunt Painting', *The Wall Street Journal*, Thursday, 23 June 1994.
Gibson 1989
Walter S. Gibson, *'Mirror of the Earth': The World Landscape in Sixteenth-Century Flemish Painting*, Princeton, NJ, 1989.
Gole 1989
Susan Gole, *Indian Maps and Plans: From Earliest Times to the Advent of European Surveys*, New Delhi, 1989.
Gombrich 1976
Ernst Gombrich, *The Heritage of Apelles: Studies in the Art of the Renaissance*, Oxford, 1976.
Goswamy 1975
B.N. Goswamy, *Pahari Paintings of the Nala-Damayanti Theme*, New Delhi, 1975.
Goswamy & Fischer 1987
B.N. Goswamy and Eberhard Fischer, *Wonders of a Golden Age: Painting at the Court of the Great Mughals: Indian Art of the 16th and 17th Centuries from Collections in Switzerland*, Zurich, 1987.
Grube n.d.
Ernst J. Grube, *Islamic Paintings from the 11th to the 18th Centuries in the Collection of Hans P. Kraus*, New York, no date.
Guy & Swallow 1990
John Guy and Deborah Swallow, editors, *Arts of India, 1550–1900*, London, 1990.
Hanaway 1971
William L. Hanaway, Jr., 'The Concept of the Hunt in Persian Literature', *Boston Museum Bulletin*, 69, 1971, pp.21–81.
Hardy 1985
Peter Hardy, 'Abu'l Fazl's Portrait of the Perfect Padshah: A Political Philosophy for Mughal India or a Personal Puff for a Pal?' in *Religion and Religious Education*, edited by Christian W. Troll, *Islam in India: Studies and Commentaries*, II, 1985, pp.114–37.
Harvey 1980
P.D.A. Harvey, *The History of Topographical Maps: Symbols, Pictures, and Surveys*, London, 1980.
Herklots & Sharif 1972
G.A. Herklots and Ja'far Sharif, *Islam in India or Qanun-i-Islam: The Customs of the Musalmans of India*, Oxford, 1921; reprinted New Delhi, 1972.
Huart 1993
Cl. Huart, 'Tugh', *Encyclopaedia of Islam*, first edition, VIII, Leiden, 1927; reprinted 1993, p.820.
Husain 1937
Muhammad Ashraf Husain, *An Historical Guide to the Agra Fort Based on Contemporary Records*, Delhi, 1937.
Irvine 1962
William Irvine, *The Army of the Indian Moghuls: Its Organization and Administration*, London, 1903; reprinted New Delhi, 1962.
Irvine 1971
—, *Later Mughals*, 2 volumes, edited and augmented by Jadunath Sarkar, New Delhi, 1971.
Islam 1970
Riazul Islam, *Indo-Persian Relations: A Study of the Political and Diplomatic Relations Between the Mughul Empire and Iran*, Tehran, 1970.
Islam 1993
—, 'Mughals. External Relations', *Encyclopaedia of Islam*, second edition, VII, Leiden, 1993, pp.316–20.
Ivanov, Grek & Akimushkin 1962
A.A. Ivanova, T.V. Grek, O.F. Akimushkin, *Al'bom Indiiskikh i Persidskikh Miniatyur XVI–XVIII vv.*, edited by L.T. Gyuzal'yan, Moscow, 1962.
Iverson 1993
Margaret Iverson, *Alois Riegl: Art History and Theory*, Cambridge, MA, and London, 1993.
Jahn 1975
Johannes Jahn, *Antike Tradition in der Landschaftsdarstellung bis zum fünfzehnten Jahrhundert*, Berlin, 1975.
Khan 1980
Ahmad Nabi Khan, *Uchchh: History and Architecture*, Islamabad, 1980.
Khandalavala 1953–5
Karl Khandalavala, 'A Moghul Miniature of Prince Khurram Slaying a Lion', *Bulletin of the Baroda Museum and Picture Gallery*, 10–11, 1953–5, pp.1–5.
Khosla 1976
R.P. Khosla, *Mughal Kingship and Nobility*, Delhi, 1934; reprinted 1976.
Koch 1982a
Ebba Koch, 'The Influence of the Jesuit Missions on Symbolic Representations of the Mughal Emperors' in *The Akbar Mission and Miscellaneous Studies*, edited by Christian W. Troll, *Islam in India: Studies and Commentaries*, I, New Delhi, 1982, pp.14–29.
Koch 1982b
—, 'The Baluster Column: A European Motif in Mughal Architecture and Its Meaning', *Journal of the Warburg and Courtauld Institutes*, 45, 1982, pp.251–62.
Koch 1982 c
—, 'The Lost Colonnade of Shah Jahan's Bath in the Red Fort of Agra', *Burlington Magazine*, 124/951, 1982, pp.331–39.
Koch 1983
—, 'Jahangir and the Angels: Recently Discovered Wall Paintings Under European Influence in the Fort of Lahore' in *India and the West: Proceedings of a Seminar Dedicated to the Memory of Hermann Goetz*, edited by Joachim Deppert, New Delhi 1983, pp.173–95.
Koch 1987
—, 'Notes on the Painted and Sculptured Decoration of Nur Jahan's Pavilions in the Ram Bagh (Bagh-i Nur Afshan) at Agra' in Skelton *et al.* 1987, pp.51–65.
Koch 1988
—, *Shah Jahan and Orpheus: The Pietre Dure Decoration and the Programme of the Throne in The Hall of Public Audiences at the Red Fort of Delhi*, Graz, 1988.
Koch 1991a
—, *Mughal Architecture: An Outline of Its History and Development: 1526–1858*, Munich, 1991.
Koch 1991b
—, 'The Copies of the Qutb Minar', *Iran*, 29, 1991, pp.95–108.
Koch 1993
—, 'The Delhi of the Mughals Prior to Shahjahanabad as Reflected in the Patterns of Imperial Visits' in *Felicitation Volume in Honour of Professor S. Nurul Hasan*, edited by A.J. Qaisar and S.P. Verma, Jaipur, 1993.
Koch 1994
—, 'Diwan-i 'Amm and Chihil Sutun: The Audience Halls of Shah Jahan', *Muqarnas*, 11, 1994, pp.143–65.
Koch, forthcoming (a)
—, 'Mughal Palace Gardens from Babur to Shah Jahan (1526–1648)', *Muqarnas*, 14 (1997), forthcoming.
Koch, forthcoming (b)
—, 'Dara Shikoh Shoots Nilgais: Hunt and Landscape in Mughal Painting', *Freer Occasional Paper*, forthcoming.

Koch, forthcoming (c)
—, 'Renaissance Calendar Illustrations and the Representation of the Mughal Hunt', *Islam and the Italian Renaissance*, papers of a colloquium held at the Warburg Institute and at the Victoria and Albert Museum, London, 15–16 March 1996, edited by Anna Contadini and Nicholas Mann, forthcoming.

Koch, forthcoming (d)
—, 'The Mughal Waterfront Garden' in *Theory and Design of Gardens During the Times of Great Muslim Empires*, edited by A. Petruccioli, Leiden, forthcoming.

Komala 1984
Wilma Komala, 'The Windsor Castle *Badshah Nama* and Its Place in the Development of Historical Painting During the Reign of Shah Jahan, 1628–1658', doctoral dissertation, Ann Arbor, MI, 1984.

Koreny 1985
Fritz Koreny, *Albrecht Dürer und die Tier-und Pflanzenstudien der Renaissance*, Munich, 1985.

Krishnadasa 1955
Rai Krishnadasa, *Mughal Miniatures*, New Delhi, 1955.

Kühnel & Goetz 1926
Ernst Kühnel and Hermann Goetz, *Indian Book Painting from Jahangir's Album in the State Library in Berlin*, London, 1926.

Leach 1981
Linda York Leach, 'Later Mughal Painting', in *Arts of India*, edited by B. Gray, Oxford, 1981.

Leach 1986
—, *Indian Miniature Paintings and Drawings – The Cleveland Museum of Art Catalogue of Oriental Art: Part One*, Cleveland, OH, 1986.

Leach 1987
—, 'Painting in Kashmir from 1600 to 1650' in Skelton *et al.* 1987, pp.124–31.

Leach 1995
—, *Mughal and Other Indian Paintings from the Chester Beatty Library*, 2 volumes, London, 1995.

Lentz 1993
Thomas W. Lentz, 'Dynastic Imagery in Early Timurid Wall Painting', *Muqarnas*, 10, 1993, pp.253–65.

Lentz & Lowry 1989
Thomas W. Lentz and Glenn D. Lowry, *Timur and the Princely Vision: Persian Art and Culture in the Fifteenth Century*, Los Angeles and Washington, DC, 1989.

Losty 1982
Jeremiah P. Losty, *The Art of the Book in India*, London, 1982.

Losty 1990
—, 'From Three-Quarter to Full Profile in Indian Painting: Revolutions in Art and Taste', *Das Bildnis in der Kunst des Orients*, edited by M. Kraatz, J. Meyer zur Capellen and D. Seckel, Stuttgart, 1990, pp.153–66.

Losty 1991
—, 'Abu'l Hasan' in Pal 1991, pp.69–86.

Lowry 1988
Glenn D. Lowry with Susan Nemazee, *A Jeweler's Eye: Islamic Arts of the Book from the Vever Collection*, Washington, DC, and Seattle, WA, 1988.

Lowry & Beach 1988
Glenn D. Lowry and Milo Cleveland Beach with Roya Marefat and Wheeler M. Thackston, *An Annotated and Illustrated Checklist of the Vever Collection*, Washington, DC, Seattle, WA, and London, 1988.

Luard 1989
C.E. Luard, *Imperial Gazetteer of India: Central India*, 1908; reprinted New Delhi, 1989.

McInerney 1991
Terence McInerney, 'Manohar' in Pal 1991, pp.53–68.

Maclagan 1990
Sir Edward Douglas Maclagan, *The Jesuits and the Great Mogul*, London, 1932; reprinted Gurgaon, 1990.

Marteau & Vever 1913
Georges Marteau and Henri Vever, *Miniatures Persanes Tirées des Collections de mm. Henry d'Allemagne, Claude Anet ... et Exposées au Musée des Arts Decoratifs juin–octobre 1912*, 2 volumes, Paris, 1913.

Martin 1971
F.R. Martin, *The Miniature Painting and Painters of Persia, India and Turkey from the 8th to the 18th Century*, 2 volumes, London, 1912; reprinted 1971.

Mehta 1926
Nanalal Chamanlal Mehta, *Studies in Indian Painting: A Survey of Some New Material Ranging from the Commencement of the VIIth Century to Circa 1870 AD*, Bombay, 1926.

Metropolitan Museum 1987
The Metropolitan Museum of Art, *Recent Acquisitions: A Selection 1986–1987*, New York, 1987.

Metropolitan Museum 1990
—, *Recent Acquisitions: A Selection 1989–1990*, New York, 1990.

Mubarak Ali 1986
Mubarak Ali, *The Court of the Great Mughals Based on Persian Sources*, Lahore, 1986.

Mufti 1991
Kamran Mufti, 'Concept of Tomb Architecture at Uchch' in *Sultanate Period Architecture in Pakistan*, edited by Siddiq-a-Akbar, Abdul Rehman and Muhammad Ali Tirmizi, Lahore, 1991, pp.134–43.

Navarane 1995
Mukund S. Navarane, *Forts of Maharashtra*, New Delhi, 1995.

Necipoğlu 1993
Gülru Necipoğlu, 'Framing the Gaze in Ottoman, Safavid, and Mughal Palaces' in *Pre-Modern Islamic Palaces*, edited by G. Necipoğlu, *Ars Orientalis*, 23, 1993, pp.303–42.

Nicolle & McBride 1993
David Nicolle and Angus McBride, *Mughul India: 1504–1761, Men-At-Arms Series*, 263, London, 1993.

Nur Bakhsh 1903–4
Nur Bakhsh, 'The Agra Fort and Its Buildings', *Archaeological Survey of India, Annual Report*, 1903–4, pp.164–93.

Okada 1992
Amina Okada, *Indian Miniatures of the Mughal Court*, translated by Deke Dusinberre, New York, 1992.

Pächt 1950
Otto Pächt, 'Early Italian Nature Studies and the Early Calendar Landscape', *Journal of the Warburg and Courtauld Institutes*, 13, 1950, pp.13–47.

Pal 1991
Pratapaditya Pal, editor, *Master Artists of the Imperial Mughal Court*, Bombay, 1991.

Pal *et al.* 1989
Pratapaditya Pal, Janice Leoshko, Stephen Markel and Joseph M. Dye, 3rd, *Romance of the Taj Mahal*, London and Los Angeles, 1989.

Pinder-Wilson 1957
R.H. Pinder-Wilson, 'Three Illustrated Manuscripts of the Mughal Period', *Ars Orientalis*, 2, 1957, pp.413–22.

Plato 1987
Plato, *The Republic*, translated with an introduction by Desmond Lee, Harmondsworth, 1955; reprinted with additional revisions, 1987.

Pope-Hennessy 1989
John Pope-Hennessy, *The Portrait in the Renaissance*, Washington, DC, 1966; reprinted Princeton, NJ, 1989.

Prasad 1922
Beni Prasad, *History of Jahangir*, London, Bombay, Calcutta and Madras, 1922.

Ranjitsinh 1989
M.K. Ranjitsinh, *The Indian Blackbuck*, New Delhi, 1989.

Richards 1981
J.F. Richards, 'The Formulation of Imperial Authority Under Akbar and Jahangir' in *Kingship and Authority in South Asia*, edited by J.F. Richards, *South Asian Studies*, 3, Madison, WI, 1978; second edition, 1981, pp.252–85.

Riegl 1902
Alois Riegl, 'Das holländische Gruppenporträt', *Jahrbuch der kunsthistorischen Sammlungen des Allerhöchsten Kaiserhauses*, 23, 1902, pp.71–278.

Robinson 1972
B.W. Robinson, 'Shah Abbas and the Mughal Ambassador Khan Alam: The Pictorial Record', *Burlington Magazine*, 114, 1972, pp.58–63.

Saksena 1976
Banarsi Prasad Saksena, *History of Shahjahan of Dihli*, Allahabad, 1932; reprinted 1976.

Sarkar 1972–4
Jadunath Sarkar, *History of Aurangzib*, second edition, 5 volumes, Calcutta, 1919–25; reprinted Bombay, 1972–4.

Schäfer 1986
Heinrich Schäfer, *Principles of Egyptian Art*, 1919, translated into English and edited with an introduction by John Baines, Oxford, 1974; reprinted with revisions, 1986.

Schimmel 1992
Annemarie Schimmel, *A Two-Colored Brocade: The Imagery of Persian Poetry*, Chapel Hill, NC, and London, 1992.

Sen 1984
Geeti Sen, *Paintings from the Akbar Nama: A Visual Chronicle of Mughal India*, Varanasi, 1984.

Seyller 1990
John Seyller, 'Codicological Aspects of the Victoria and Albert Museum *Akbarnama* and Their Historical Implications', *Art Journal*, XLIX/4, 1990, pp.379–87.

Seyller 1991
—, 'Hashim' in Pal 1991, pp.105–118.

G.N. Sharma 1954
G.N. Sharma, *Mewar & the Mughal Emperors (1526–1707 AD)*, Agra, 1954.

R. Sharma 1962
Ram Sharma, *The Religious Policy of the Mughal Emperors*, Bombay, 1962.

Shastri 1962
Hari Prasad Shastri, translator, *Ramayana of Valmiki*, 3 volumes, London, 1962.

Sherwani 1983
H.K. Sherwani, 'Dawlatabad', *Encyclopaedia of Islam*, second edition, II, Leiden, 1983, pp.179–80.

Sherwani & Joshi 1973–4
H.K. Sherwani and P.M. Joshi, editors, *History of Medieval Deccan (1295–1724)*, 2 volumes, Hyderabad, 1973–4.
'Shir u-Kurshid', forthcoming
'Shir u-Kurshid', *Encyclopaedia of Islam*, second edition, forthcoming.
Sims 1992
Eleanor Sims, 'Ibrahim-Sultan's Illustrated *Zafarnama* of 1436 and its Impact on the Muslim East' in *Timurid Art and Culture: Iran and Central Asia in the Fifteenth Century*, edited by Lisa Golombek and Maria Subtelny, Leiden, New York and Köln, 1992, pp.132–43.
Skelton 1972
Robert Skelton, 'A Decorative Motif in Mughal Art' in *Aspects of Indian Art: Papers Presented in a Symposium at the Los Angeles County Museum of Art, October 1970*, edited by Pratapaditya Pal, Leiden, 1972, pp.147–52.
Skelton 1976
—, 'Indian Painting of the Mughal Period' in *Islamic Painting and the Arts of the Book*, edited by B.W. Robinson, London, 1976.
Skelton 1980
—, 'Landscape in Indian Painting' in Watson 1980, pp.150–71.
Skelton 1988
—, 'Imperial Symbolism in Mughal Painting' in *Content and Context of Visual Arts in the Islamic World*, edited by Priscilla P. Soucek, University Park, PA, and London, 1988, pp.177–87.
Skelton *et al.* 1987
Robert Skelton, Andrew Topsfield, Susan Stronge, Rosemary Crill and Graham Parlett, editors, *Facets of Indian Art: A Symposium Held at the Victoria and Albert Museum on 26, 27, 28 April and 1 May 1982*, London, 1986; reprinted New Delhi, 1987.
Smart 1973
Ellen S. Smart, 'Four Illustrated Mughal Baburnama Manuscripts', *Art and Archaeology Research Papers*, 3, 1973, pp.54–8.
Smart 1979
—, 'A Recently Discovered Mughal Hunting Picture by Payag', *Art History*, 2/4, 1979, pp.396–400.
Smart 1991
—, 'Balchand' in Pal 1991, pp.135–48.
Somani 1976
Ram Vallabh Somani, *History of Mewar*, Jaipur, 1976.
Sotheby & Co. 1973
Sotheby & Co., London, *Important Mughal Miniatures* (catalogue of the sale of 26 March 1973), London, 1973.
Sotheby's 1973
Sotheby's, London, *Important Oriental Miniatures & A Mughal Manuscript* (catalogue for the sale of 10 July 1973), London, 1973.
Sotheby's 1986
Sotheby's, London, *Fine Oriental Manuscripts & Miniatures* (catalogue for the sale of 22 May 1986), London, 1986.
Sotheby's 1988
Sotheby's, London, *Oriental Manuscripts and Miniatures* (catalogue for the sale of 11 April 1988), London, 1988.
Sotheby's 1989
Sotheby's, New York, *Indian, Himalayan & Southeast Asian Art* (catalogue of the sale of 22 March 1989), New York, 1989.
Sotheby's 1995
Sotheby's, London, *Oriental Manuscripts & Miniatures* (catalogue of the sale of 18 October 1995), London, 1995.
Sotheby's 1996
Sotheby's, *London, Persian and Indian Manuscripts and Miniatures from the Collection formed by the British Rail Pension Fund (catalogue of the sale of 23 April 1996)*, London, 1996.
Soucek 1980
Priscilla Soucek, 'The Role of Landscape in Iranian Painting to the 15th Century' in Watson 1980, pp.86–110.
Soudavar 1992
Abolala Soudavar with a contribution by Milo Cleveland Beach, *Art of the Persian Courts: Selections from the Art and History Trust Collection*, New York, 1992.
Soudavar, forthcoming
Abolala Soudavar, *The Saga of Abu-Sa'id Bahādor Khān: The Abu-Sa'idnāmé* in *The Court of the Ilkhans 1290–1340*, edited by Julian Raby and Teresa Fitzherbert, Oxford Studies in Islamic Art, XII, Oxford, forthcoming.
Spink 1980
Spink & Son Ltd, London, *Islamic Art from India*, London, 1980.
Streusand 1989
Douglas E. Streusand, *The Formation of the Mughal Empire*, Delhi and New York, 1989.
Strzygowski 1922
Josef Strzygowski, *Die Landschaft in der nordischen Kunst*, Leipzig, 1922.
Strzygowski & Glück 1923
Josef Strzygowski and Heinrich Glück, *Die indischen Miniaturen im Schlosse Schönbrunn*, Leipzig and Vienna, 1923.
Subhan 1960
John A. Subhan, *Sufism: Its Saints and Shrines*, Lucknow, 1960.
Suleiman 1969
Hamid Suleiman, *Miniatyury k Baburname*, Tashkent, 1969.
Tillotson 1987
G.H.R. Tillotson, *The Rajput Palaces: The Development of an Architectural Style 1450–1750*, New Haven and London, 1987.
Tirmizi 1968
S.A.I. Tirmizi, *Ajmer Through Inscriptions*, New Delhi, 1968.
Topsfield 1994
Andrew Topsfield, *Indian Paintings: from Oxford Collections*, Oxford, 1994.
Topsfield & Beach 1991
Andrew Topsfield and Milo Cleveland Beach, *Indian Paintings and Drawings from the Collection of Howard Hodgkin*, New York and London, 1991.
Verma 1976
Som Prakash Verma, 'Ensigns of Royalty at the Mughal Court (in the Sixteenth Century)', *Islamic Culture*, 50, 1976, pp.41–45.
Verma 1978
—, *Art and Material Culture in the Paintings of Akbar's Court*, New Delhi, 1978.
Verma 1994
—, *Mughal Painters and Their Work: A Biographical Survey and Comprehensive Catalogue*, Delhi, 1994.
Victoria and Albert Museum 1982
Victoria and Albert Museum, *The Indian Heritage: Court Art and Life Under Mughal Rule*, London, 1982.
Wade 1995
Bonnie C. Wade, 'Music Making in Mughal Painting, *Asian Art and Culture*, Fall 1995, pp.68–97.
Waley 1994
Muhammad Isa Waley, *Manuscripts of the Middle East*, 6, 1992, Leiden, 1994.
Warnke 1993
Martin Warnke, *Hofkünstler: Zur Vorgeschichte des modernen Künstlers*, Cologne, 1985; translated into English by David McLintock as *The Court Artist: On the Ancestry of the Modern Artist*, Cambridge, New York, Oakleigh and Paris, 1993.
Watson 1980
William Watson, editor, *Landscape Style in Asia, Colloquies on Art & Archaeology in Asia*, IX, London, 1980.
William Watson 1980
William Watson, 'Landscape Elements in the Early Buddhist Art of China' in Watson 1980, pp.1–17.
A. Welch 1995
Anthony Welch, 'A Shamsa for Shah Jahan', *Indian Painting: Essays in Honour of Karl J. Khandalavala*, edited by B.N. Goswamy in association with Usha Bhatia, New Delhi, 1995, pp.450–63.
S.C. Welch 1959
Stuart Cary Welch, 'Early Mughal Miniature Paintings from Two Private Collections Shown at the Fogg Art Museum', *Ars Orientalis*, 3, 1959, pp.133–46.
S.C. Welch 1963
—, *The Art of Mughal India: Painting & Precious Objects*, New York, 1963.
S.C. Welch 1973
—, *A Flower from Every Meadow: Indian Paintings from American Collections*, New York, 1973.
S.C. Welch 1978
—, *Imperial Mughal Painting*, New York and London, 1978.
S.C. Welch 1985
—, *India: Art and Culture 1300–1900*, New York, 1985.
S.C. Welch 1995
—, 'The Two Worlds of Payag: Further Evidence on a Mughal Artist' in *Indian Art and Connoisseurship: Essays in Honour of Douglas Barrett*, edited by John Guy, London, 1995, pp.320–41.
S.C. Welch *et al.* 1987
Stuart Cary Welch, Annemarie Schimmel, Marie L. Swietochowski and Wheeler M. Thackston, *The Emperors' Album: Images of Mughal India*, New York, 1987.
Wellesz 1952
Emmy Wellesz, *Akbar's Religious Thought as Reflected in Mogul Painting*, London, 1952.
Wensinck & Johnstone 1991
A.J. Wensinck and Penelope Johnstone, 'Maryam', *Encyclopaedia of Islam*, second edition, VI, Leiden, 1991, pp.628–32.
Wescoat 1991
James L. Wescoat, 'Gardens of Invention and Exile: The Precarious Context of Mughal Garden Design During the Reign of Humayun (1530–1556)', *Journal of Garden History*, X/2, 1991, pp.106–117.
Wilkinson 1948
J.V.S. Wilkinson, *Mughal Painting*, London, 1948.

INDEX

Numbers in italics refer to pages on which figures appear; bold numbers refer to the main colour illustrations in the section *The Manuscript*.

PHOTOGRAPHIC ACKNOWLEDGEMENTS

Azimuth Editions, the picture researcher and the authors wish to thank the institutions and indivuals who have kindly provided photographic material or artwork for use in this book. Museums and private owners are given in the *List of figures* on pages 237–39. Additional sources, photographic and copyright acknowledgements are given below.

Figs 8, 19, 142 Smithsonian Institution, Washington, DC, Courtesy of the Arthur M. Sackler Gallery; Fig.9 copyright National Museum, New Delhi; Figs 10, 11, 34, 133 Reproduced by kind permission of the Trustees of the Chester Beatty Library, Dublin; Figs 12, 17, 27, 33, 69, 87, 122, 124 Victoria & Albert Museum, London; Figs 13, 20, 105, 143, 148, 155, 159, 166 Institute of Oriental Studies, Russian Academy of Sciences, St Petersburg; Figs 14, 15, 102, 161 By kind permission of the Chairman of the Rampur Raza Library Board and the Ministry of Human Resource Development, Department of Culture, Government of India; Figs 18, 112, 144 Smithsonian Institution, Washington, DC, Courtesy of the Freer Gallery of Art; Figs 22, 131 Bibliothèque Nationale, Paris; Fig.25 San Diego Museum of Art (Edwin Binney 3rd Collection); Fig.28 Cincinatti Art Museum; Figs 29, 81 Courtesy, Museum of Fine Arts, Boston © 1996 Museum of Fine Arts, Boston. All rights reserved; Figs 30, 32, 80, 82, 154 British Library, London; Fig.68 Walters Art Gallery, Baltimore; Figs 72, 120 Photographs © Bildarchiv Preussischer Kulturbesitz, Berlin; Figs 73, 147 Musées Nationales de France – Guimet, Paris, © Photo RMN; Figs 74, 76, 89, 90, 111, 129 photographs by Ebba Koch; Fig.78 © The Cleveland Museum of Art , 1996; Figs 91, 104, 132 drawings by Richard A. Barraud; Figs 98, 153, 158 All rights reserved, The Metropolitan Museum of Art, New York; Fig.108 The Harvard University Art Museums, Cambridge, Massachusetts; Figs 114, 167 British Museum, London. Photographs © copyright British Museum; Figs 117, 145, 149 Bodleian Libary, University of Oxford; Fig.162 Photography by Prudence Cuming Associates Limited; Fig.163 Rietberg Museum, Zürich. Photograph © Wettstein & Kauf; Fig.164 Art Institute of Chicago. Photograph © 1996 The Art Institute of Chicago. All Rights Reserved.